# The Biosynthesis of
# Secondary Metabolites

T0205455

# The Biosynthesis of
# Secondary Metabolites

**R. B. HERBERT**

Senior Lecturer
School of Chemistry
University of Leeds

SECOND EDITION

**CHAPMAN & HALL**

London · Glasgow · Weinheim · New York · Tokyo · Melbourne · Madras

'All we did was to do some decent experiments and have the luck to hit upon a substance with astonishing properties.'

Howard Florey, on penicillin

'I am, in point of fact, a particularly haughty and exclusive person, of pre-Adamite ancestral descent . . . I can trace my ancestry back to a protoplasmal primordial atomic globule. Consequently my family pride is something inconceivable.'

Pooh-Bah, in *The Mikado*
by W.S. Gilbert and A. Sullivan

To Margaret, Julie and Andrew

**Published by Chapman & Hall, 2-6 Boundary Row, London SE1 8HN, UK**

Chapman & Hall, 2-6 Boundary Row, London SE1 8HN, UK

Blackie Academic & Professional, Wester Cleddens Road, Bishopbriggs, Glasgow G64 2NZ, UK

Chapman & Hall GmbH, Pappelallee 3, 69469 Weinheim, Germany

Chapman & Hall Inc., One Penn Plaza, 41st Floor, New York, NY10119, USA

Chapman & Hall Japan, Thomson Publishing Japan, Hirakawacho Nemoto Building, 6F, 1-7-11 Hirakawa-cho, Chiyoda-ku, Tokyo 102, Japan

Chapman & Hall Australia, Thomas Nelson Australia, 102 Dodds Street, South Melbourne, Victoria 3205, Australia

Chapman & Hall India, R. Seshadri, 32 Second Main Road, CIT East, Madras 600 035, India

First edition 1981
Second edition 1989
Reprinted 1994

© 1981,1989 R.B. Herbert

Typeset in 10pt English Times by Colset Private Limited , Singapore
Printed in Great Britain by St Edmundsbury Press Ltd, Bury St Edmunds, Suffolk

ISBN  0  412  27720  4

A Catalogue record for this book is available from the British Library

Library of Congress Cataloging-in-Publication Data
Herbert, R.B. (Richard B.)
The biosynthesis of secondary metabolites.

Includes bibliographies and index.
1. Metabolism, Secondary.     2. Biological products—
Synthesis. 3. Biosynthesis.     I. Title.
[DNLM: 1. Biochemistry.     2. Metabolism.     QU 120 H537b]
    QH521.H47     1989     574.19'29     88-29961
ISBN 0-412-27720-4

# Contents

# Preface

The chief purpose of such new editions is to bring the text up to date. This I have tried to do. The danger with such endeavours is that the book becomes corpulent. This I have tried to avoid. In so doing I am evermore conscious of how much material has had to be omitted. As before the references (ca. 770 of them) which are cited are intended in part to provide the reader with access to material that has been omitted.

I am deeply grateful to Mrs Marjorie Romanowicz for preparing the typed manuscript for this edition with customary accuracy and efficiency, to my wife Margaret for checking the references, and to research students Karl Cable, Lucy Hyatt, Mashupye Kgaphola and Andrew Knaggs for helping with the checking of the manuscript.

Richard Herbert
*April 1988*

# Preface to the first edition

This is a book about experiments and results of experiments. The results described are the fruit of thirty years' labour in the field of secondary metabolism.

Secondary metabolism, more than any other part of the chemistry of life, has been the special preserve of organic chemists. Investigation of secondary metabolism began with curiosity about the structures of compounds isolated from natural sources, i.e. secondary metabolites. Coeval with structure determination there has been a curiosity about the origins and mechanism of formation of secondary metabolites (or natural products as they have been called). It is the experimental outcome of this curiosity that is described here.

This account is primarily intended to be an introduction to the biosynthesis of secondary metabolites. I have also endeavoured, however, to make the book as comprehensive as possible. This has meant that some of the material has had to be presented in abbreviated form. The abbreviated material is largely confined to particular sections of the book. The paragraphs marked with a vertical rule can be omitted by the reader wishing to acquire a general introduction to the subject.

A blend of the most significant and the most recent references is cited to provide the reader with ready access to the primary literature. This is clearly most necessary for the material presented in abbreviated form. Relevant reviews are also cited.

In compiling this account I am indebted to the following people: Mrs M. Romanowicz for typing the manuscript with accuracy and speed, to my wife, Margaret, for checking the references, to final year PhD students, Stuart H. Hedges and William J.W. Watson for critically reading and checking the typed manuscript, and John E. Cragg for proofreading.

Richard Herbert
*June 1980*

# 1 *Introduction*

## 1.1 PRIMARY AND SECONDARY METABOLISM

### 1.1.1 Introduction

Since prehistoric times man has used plant extracts to heal and to kill. Folklore abounds in references to the use of plant extracts in the healing of a variety of illnesses; examples of applications as agents of death range from that of calabar beans and hemlock as judicial poisons to that of the South American curare arrow poisons [1]. In modern times organic compounds isolated from cultures of micro-organisms, as well as from plants, have been used for the cure of disease (e.g. penicillin and tetracycline antibiotics). These organic compounds from natural sources form a large group known as natural products, or secondary metabolites.

Study of the metabolism, fundamental and vital to living things, has led to a detailed understanding of the processes involved. A complex web of enzyme-catalysed reactions is now apparent, which begins with carbon dioxide and photosynthesis and leads to, and beyond, diverse compounds called primary metabolites, e.g. amino acids, acetyl-coenzyme A, mevalonic acid, sugars, and nucleotides [2, 3]. Critical to the overall energetics involved in metabolism is the coenzyme, adenosine triphosphate (ATP), which serves as a common energy relay and co-operates, like other coenzymes, with particular enzymes in the reactions they catalyse.

This intricate web of vital biochemical reactions is referred to as primary metabolism. It is often displayed usefully in chart form [4], and to the eye appears very much like an advanced model railway layout, not least because of the way primary metabolism proceeds in cycles (e.g. the citric acid cycle). The organic compounds of primary metabolism are the stations on the main lines of this railway, the compounds of secondary metabolism the termini of branch lines. Secondary metabolites are distinguished more precisely from primary metabolites by the following criteria: they have a restricted distribution being found mostly in plants and micro-organisms, and are often characteristic of individual genera, species, or strains; they are formed along specialized pathways from primary metabolites. Primary metabolites, by contrast, have a broad distribution in all living things and are intimately involved in essential life processes (for further discussion of parts of primary metabolism see sections 1.1.2 and 5.1). It follows that secondary metabolites are non-essential to life although they are important to the organism that produces them. What this importance is, however, remains, very largely, obscure.

It is interesting to note that secondary metabolites are biosynthesized essentially from a handful of primary metabolites: $\alpha$-amino acids, acetyl-coenzyme A, mevalonic acid, and intermediates of the shikimic acid pathway. It is these starting points for the elaboration of secondary metabolites which allow their classification, and also their discussion as discrete groups (Chapters 3 to 7). In the remainder of this chapter various aspects of biosynthesis of general importance to the discussion in Chapter 3 and succeeding chapters is reviewed. The first examples of primary and secondary metabolite biosynthesis will be found in sections 1.1.2 and 1.1.3. Chapter 2 is devoted to a brief discourse on the various techniques used in studying the biosynthesis of secondary metabolites.

### 1.1.2 Fatty acid biosynthesis [2, 3, 7]

Fatty acids, e.g. stearic acid (*1.1*) and oleic acid (*1.2*), are straight chain carboxylic acids found predominantly as lipid constituents. They are primary

$$CH_3(CH_2)_{16}CO_2H \qquad CH_3(CH_2)_7 CH = CH(CH_2)_7 CO_2H$$

*(1.1)* Stearic acid       *(1.2)* Oleic acid

metabolites formed under enzyme catalysis by linear combination of acetate units. In this they are similar to the polyketides which are secondary metabolites (Chapter 3). Like many primary metabolites, their biosynthesis is understood in intricate detail. Much less detail is generally available on secondary metabolite biosynthesis.

Fatty acids are synthesized in a multienzyme complex from a crucially important primary metabolite, acetyl-coenzyme A (*1.8*). The principal source of acetyl-CoA (*1.8*) is pyruvic acid (*1.5*) and the conversion of (*1.5*) into (*1.8*) involves the coenzymes, thiamine pyrophosphate (*1.3*)* and lipoic acid (*1.6*) (Scheme 1.1). The key to the action of thiamine is the ready formation of the zwitterion (*1.4*) at the beginning and end of the reaction cycle. The lipoic acid (*1.6*) is enzyme linked via the side chain of a lysine residue (*1.7*). The disulphide functionality is thus at the end of a long (14 Å) arm. It has been suggested that this arm allows the lipoate to swing from one site to another within the multienzyme complex and transfer (and oxidize) the

---

*$\textcircled{P}$ = phosphate in (*1.3*) and subsequent structures, cf (*1.10*).

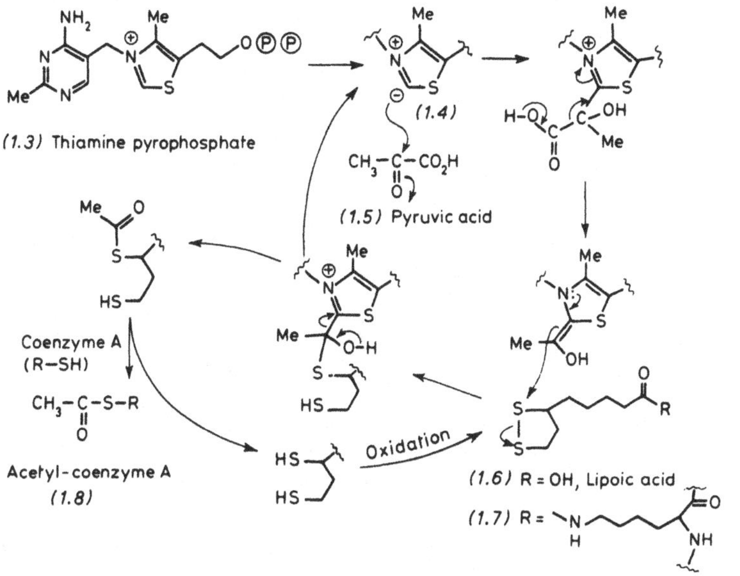

**Scheme 1.1**

acetyl group [5]. In the sequence shown, pyruvic acid (*1.5*) loses carbon dioxide giving coenzyme-bound acetaldehyde, which is oxidized to the CoA ester of acetic acid.

A similar long arm is apparent in biotin (*1.9*), again enzyme bound through a lysine residue. The coenzyme (*1.9*) assists in the carboxylation of acetyl-CoA (*1.8*) with carbon dioxide yielding malonyl-CoA (*1.14*). Exchange of both acetyl-CoA and malonyl-CoA occurs with acyl carrier proteins (ACP) having free thiol groupings. Condensation then occurs between acetyl-S-ACP and malonyl-S-ACP with simultaneous decarboxylation; the

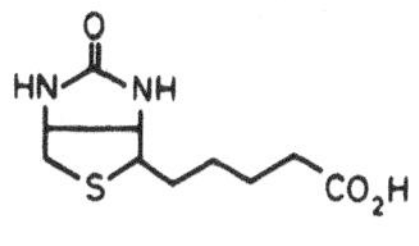

(*1.9*) Biotin

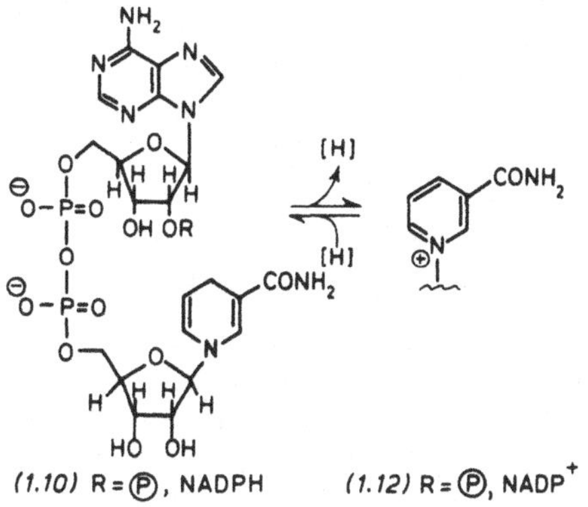

(*1.10*) R = $\textcircled{P}$, NADPH　　(*1.12*) R = $\textcircled{P}$, NADP$^+$
(*1.11*) R = H, NADH　　(*1.13*) R = H, NAD$^+$

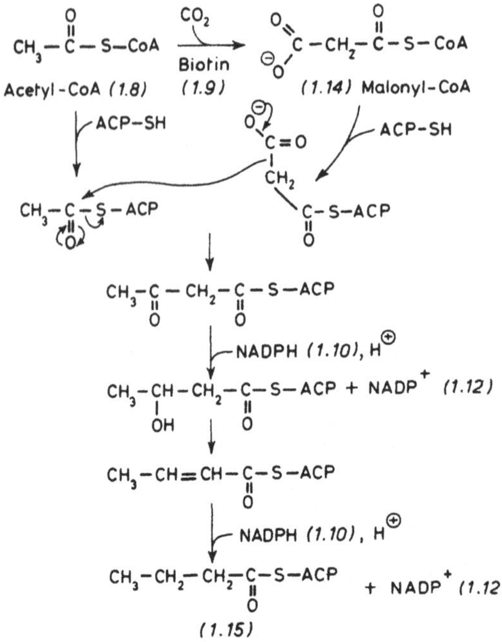

**Scheme 1.2**

carboxylate anion is transferred into the new bond (Scheme 1.2) [6]. Subsequent steps involve reduction, dehydration and double-bond saturation. They implicate in part the widely utilized reducing coenzyme, NADPH (*1.10*) (Scheme 1.2). The first sequence gives butyryl-S-ACP (*1.15*) and for the generation of longer chains, as (*1.1*), the sequence, of malonyl-CoA addition, reduction, dehydration and reduction, is repeated the requisite number of times.

It has been most elegantly demonstrated that the carboxylation of acetyl-CoA (*1.8*) to give malonyl-CoA (*1.14*) proceeds with retention of configuration; subsequent condensation of malonyl-CoA with acetyl-CoA proceeds with inversion of configuration at the malonyl methylene group. Further, elimination of water (Scheme 1.2) takes place (in yeast) in a *syn* stereochemical sense [8].

In examining the overall conversion of pyruvate into a fatty acid (Schemes 1.1. and 1.2) it is interesting to note the exploitation of particular chemical properties of sulphur:

1. as an easily reduced disulphide (*1.6*)
2. as an easily oxidized dithiol
3. in reactive thioesters which aid the Claisen-type condensation reactions.

Also of crucial importance for the condensation is the use of a malonic acid derivative (*1.14*) as a source of a stable anion. (Further discussion of fatty acid biosynthesis in relation to polyketide formation is taken up in Chapter 3.)

### 1.1.3 The biosynthesis of polyacetylenes and prostaglandins

The formation of unsaturated fatty acids, e.g. oleic acid (*1.2*), which are also primary metabolites, may occur by at least two routes, one aerobic the other anaerobic. Essentially though, both involve desaturation of a fully saturated fatty acid [2]. Polyacetylenes, e.g. crepenynic acid (*1.16*), which are secondary metabolites, also apparently derive by step-wise desaturation of a saturated fatty acid [9]. The path to crepenynic acid (*1.16*) is illustrated in Scheme 1.3.

$$(1.1) \rightarrow (1.2) \rightarrow CH_3(CH_2)_4 \, CH = CHCH_2CH = CH(CH_2)_7 \, CO_2H$$

$$\downarrow$$

$$CH_3(CH_2)_4 \, C \equiv C \, CH_2CH = CH(CH_2)_7 \, CO_2H$$

(*1.16*) Crepenynic acid

**Scheme 1.3**

Prostaglandins, e.g. prostaglandin $E_2$ (*1.19*), are physiologically active secondary metabolites found in mammals. The biosynthesis of these compounds also involves an unsaturated fatty acid, e.g. arachidonic acid (*1.17*). Formation of the characteristic prostaglandin skeleton involves the formation of an intermediate endoperoxide (*1.18*). Some of the succeeding steps are indicated in Scheme 1.4 [10, 11].

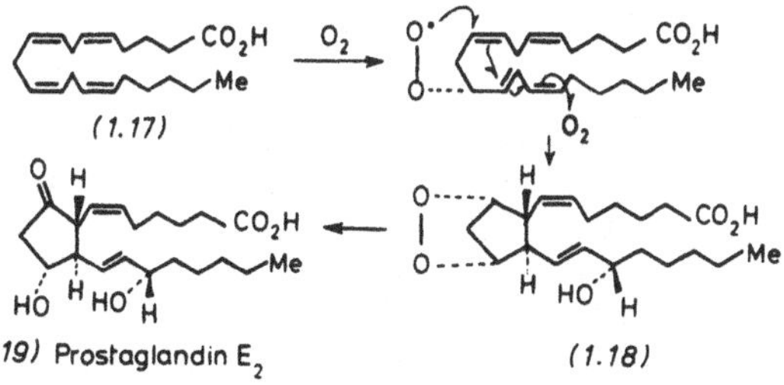

(*1.17*)

(*1.19*) Prostaglandin $E_2$          (*1.18*)

**Scheme 1.4**

It is interesting to note the formation of polyacetylenes (and prostaglandins) from fatty acids, since it is unusual for secondary metabolites to be formed from a fatty acid. The route that dominates the formation of these metabolites is the polyketide one (Chapter 3).

## 1.2 STEREOCHEMISTRY AND BIOSYNTHESIS

### 1.2.1 Chirality and prochirality [12]

It is a common observation that enzymes deal stereospecifically with their substrates; the acceptable substrates must have a particular stereochemistry and the products in turn are formed with a particular stereochemistry (for an illustration of this see particularly section 5.1). This stereospecificity is associated in many instances with reactions involving chiral centres.

In addition to the stereospecificity of reactions associated with chiral centres there is further stereospecificity to be found in reactions at prochiral centres. The conversion of ethanol (*1.20*) into acetaldehyde (*1.21*) by the enzyme, alcohol dehydrogenase, with $NAD^+$ as co-enzyme provides a simple illustration. The methylene group in (*1.20*) is prochiral*. Oxidation of the ethanol proceeds stereospecifically with removal of the *pro-R* proton from

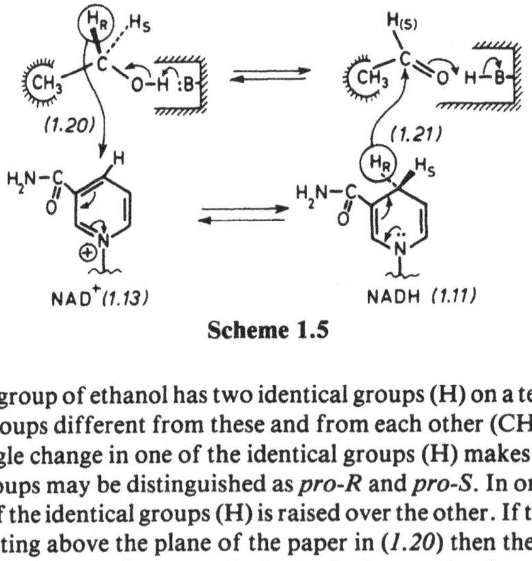

**Scheme 1.5**

*The methylene group of ethanol has two identical groups (H) on a tetrahedral carbon atom and two groups different from these and from each other ($CH_3$, OH), and so is *pro*-chiral: a single change in one of the identical groups (H) makes the centre **chiral**. The identical groups may be distinguished as *pro-R* and *pro-S*. In order to do this the priority of one of the identical groups (H) is raised over the other. If this is done for the hydrogen projecting above the plane of the paper in (*1.20*) then the configuration at the methylene carbon atom becomes *R*. So the hydrogen that is raised in priority is termed *pro-R*. [The reader can try labelling the carboxy-methyl groups in (*1.22*) similarly; answer: (*1.64*).] The acetaldehyde double bond is also prochiral: the two faces of the double bond are not identical (addition may give a chiral product). These faces may be labelled *re* and *si* by following the normal priority rules for chirality (*re* = *R*, *si* = *S*). The face of (*1.21*) viewed from above is *re* (for further discussion see [13]). An alternative way of defining prochirality is as follows. A prochiral molecule is one which has a plane of symmetry [for both (*1.20*) and (*1.21*) this is in the plane of the paper] but no axis of symmetry in the plane [rotation of e.g. (*1.21*) in the plane of the paper around an axis running through the carbonyl group does not give a molecule identical with (*1.21*), until, of course, a rotation of 360° has been carried out].

this group. The reverse reaction involves stereospecific addition of a proton to the *re*-face (in this case, the top face as seen by the reader) of the acetaldehyde carbonyl group (i.e. proton removal and addition to the same side of the two molecules). The reaction is, moreover, stereospecific with regard to co-enzyme. The interconversion of NADH and $NAD^+$ involves, respectively, removal of a proton from a prochiral centre and proton addition to form one. Removal and addition again involves the same face of the molecules concerned.

The stereospecific proton removal from (*1.20*) to give (*1.21*) can be understood simply as follows: imagine that at the active site of the enzyme there is one binding site specific for the ethanol OH and one specific for $CH_3$. This uniquely locates the molecules on the enzyme surface as shown in (*1.20*). Now imagine that the enzyme/co-enzyme proton removal can only occur physically from above the plane of the paper. This results in unique removal of the *pro-R* proton. There is no way on this model that the *pro-S* proton can be removed. A similar argument can be applied to proton addition to (*1.21*) and to proton removal from, and addition to, co-enzyme. [This argument was first applied some thirty years ago to citric acid (*1.22*) in relation to its place in the citric acid cycle. Citric acid has a prochiral centre (*) [14].]

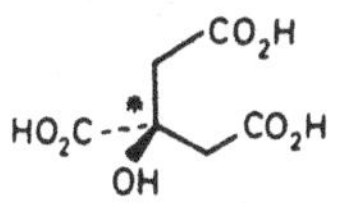

(1.22) Citric acid

## 1.2.2 Chiral methyl groups [15, 16]

Enzyme reactions involving methyl groups show none of the stereochemistry associated with prochiral centres as, e.g., the methylene group in ethanol (see above). The hydrogen atoms are indistinguishable (provided they are all one hydrogen isotope, see below) but enzyme-catalysed reactions do occur which involve methyl groups in various ways. Since enzymes are involved the reactions are expected to proceed with a particular stereochemistry. Three main classes of reaction can be identified [15]. By way of illustration we shall briefly examine one of these (Scheme 1.6) (for further discussion see

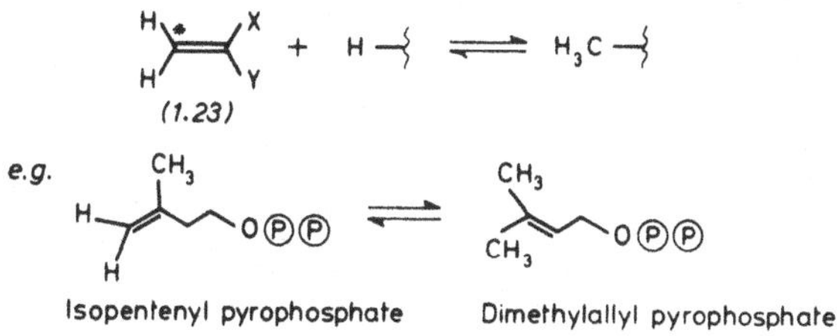

(1.23)

*e.g.*

Isopentenyl pyrophosphate          Dimethylallyl pyrophosphate

**Scheme 1.6**

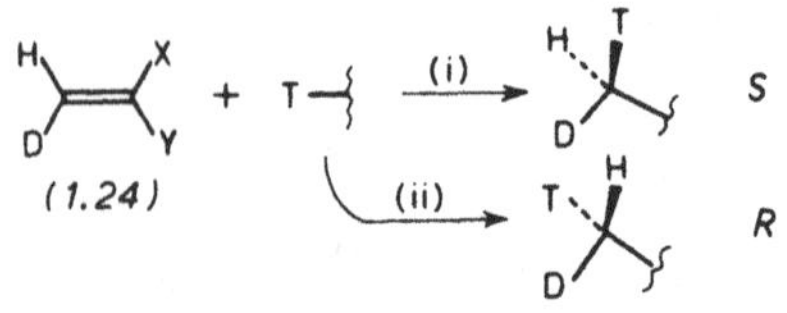

**Scheme 1.7**

section 4.4). Proton addition to C\* of the double bond in (*1.23*), with formation of a methyl group, can, in principle, occur from above or below the plane of the double bond. If the three protons involved in the generation of the methyl group are labelled with the three isotopes of hydrogen [¹H, ²H, and ³H, indicated in Scheme 1.7 as, respectively, H, D, and T] then the methyl group generated will be chiral. The stereochemical course of the reaction can be deduced, provided that:

1. the chirality (*R* or *S*) of the asymmetric methyl group thus generated can be determined
2. the substitution of hydrogen isotope around the double bond in (*1.24*) is known, e.g. as shown.

The central analytical problem is to determine the chirality of methyl groups generated or modified in biochemical reactions. The solution is dazzlingly ingenious. First, samples of chiral acetic acid [as (*1.26*)] of known absolute configuration were synthesized. All molecules of acetic acid contained deuterium (and hydrogen) but, as is customary, only very few molecules were also labelled with tritium. Only very few molecules were therefore chiral. Since the analysis is for tritium, however, this does not matter. [Subsequently an economical and supremely elegant synthesis of chiral acetic acid has been developed (Scheme 1.8), which involves two stereospecific concerted reactions and transfer of the chirality in (*1.25*) into (*1.26*) [17].]

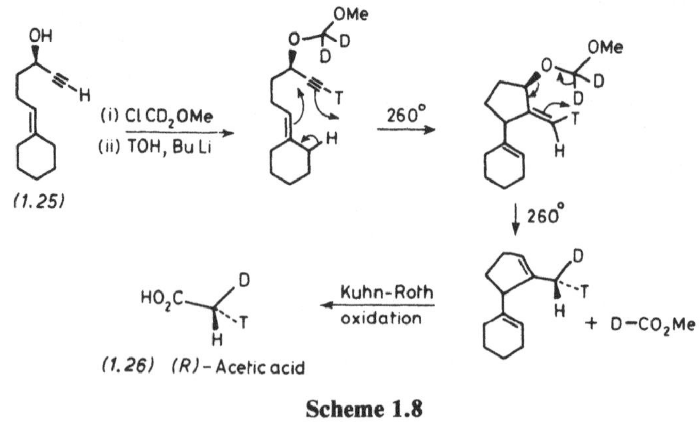

**Scheme 1.8**

The method of analysis developed is for acetic acid and involves the use of two enzymes. Chiral acetic acid [as (*1.26*)] is irreversibly condensed as its CoA-derivative with glyoxylic acid (*1.27*), using the enzyme malate synthase, to give malic acid (*1.28*). The condensation occurs with loss of hydrogen isotope by a primary kinetic isotope effect ($k_H > k_D > k_T$). This means that loss of H is favoured over loss of D which is in turn favoured over loss of T. The result is a high retention of tritium.

By loss of deuterium ($^2$H) or protium ($^1$H) two samples of *radioactive* malate are formed: (*1.28*) as major product by loss of H, and (*1.29*) as minor product from (*R*)-acetic acid by loss of $^2$H (D); (*1.33*) and (*1.34*) are, respectively, the major and minor radioactive products from (*S*)-acetic acid (*1.32*). (The malate synthase reaction is now known to proceed with inversion of configuration. Since the method of analysis depends on simply getting one result for *R*-acetic acid and the opposite for the *S*-isomer this does not matter for the purposes of assay.)

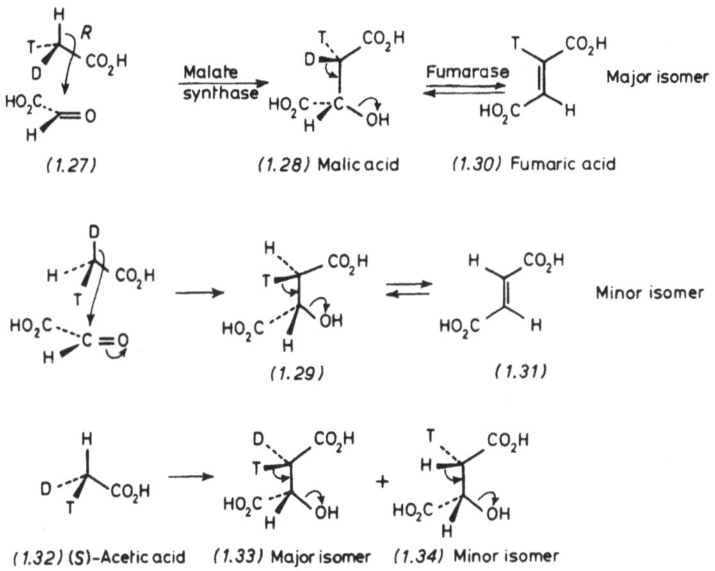

Scheme 1.9

The malic acid is isolated and incubated with fumarase until no further loss of carbon-bound tritium results. [The enzyme catalyses *trans* removal of water, as (*1.28*) ⇌ (1.30).] For the samples of malate derived from (*R*)-acetic acid, the minor product (*1.29*) loses TOH. For the material derived from (*S*)-acetic acid it is the major product (*1.33*) which loses TOH. So for (*R*)-acetic acid a higher retention of tritium (ca. 75%) is observed in the fumarase equilibration and for (*S*)-acetic acid a lower value (ca. 25%). With the method of analysis established for samples of acetic acid of known chirality it can be used to determine the configuration of samples of acetic acid of unknown chirality derived from biological reactions (for examples see section 4.2 and 4.4, and [15]).

### 1.2.3 Hydroxylation at saturated carbon atoms [2]

An oft encountered reaction in secondary metabolite biosynthesis is hydroxylation at saturated carbon atoms. These reactions are mediated by a mixed-function oxidase (or mono-oxygenase) i.e. an oxidase which uses molecular oxygen for hydroxylation, according to the equation: substrate-H + $O_2$ + $XH_2 \rightarrow$ substrate-OH + $H_2O$ + X.

These hydroxylations are stereospecific proceeding with retention of configuration [12, 18] (examples are cited in following Chapters). This is the expected result for an electrophilic substitution at saturated carbon, i.e. attack at the point of highest electron density which is the sigma bond itself [19].

## 1.3 SOME REACTIONS OF GENERAL IMPORTANCE IN SECONDARY METABOLISM

It will be convenient to discuss under this heading three quite different reaction types which will be mentioned again in succeeding chapters. Two concern aromatic compounds only, the third is of widespread significance.

### 1.3.1 Oxidative coupling of phenols [20–22]

Phenols are biosynthesized essentially in two ways. One is along the polyketide pathway starting with acetyl-CoA (Chapter 3), the other along the shikimic acid pathway (Chapter 5). A phenol, or indeed any benzenoid compound, so formed may be at a terminus in biosynthesis or be involved in the formation of other metabolites. Of general importance in this regard is the coupling together of two phenolic residues, e.g. (*1.35*) → (*1.38*). A firm mechanistic base was given to this process by the quite brilliant recognition that this bond forming could occur by inter- or intra-molecular coupling of two mesomeric radicals [as (*1.36*)] formed by the one-electron oxidation of each of a pair of phenols. Carbon–carbon bond formation can only occur, according to this hypothesis, *ortho* or *para* to the phenolic hydroxy groups. Numerous studies on the biosynthesis of various phenolic compounds (sections 6.2.2, 6.3, 6.4 and 3.3) have demonstrated the overall correctness of this hypothesis: coupling is always *ortho* or *para* to phenolic hydroxy groups (occasionally C—O—C bonds may be formed); a hydroxy group must always be present on each of the aromatic rings (*O*-alkylation, for instance, blocks the coupling reaction). In one exceptional case (section 6.3.5) it has been found that coupling only occurs if one of the aromatic rings bears *two* hydroxy functions.

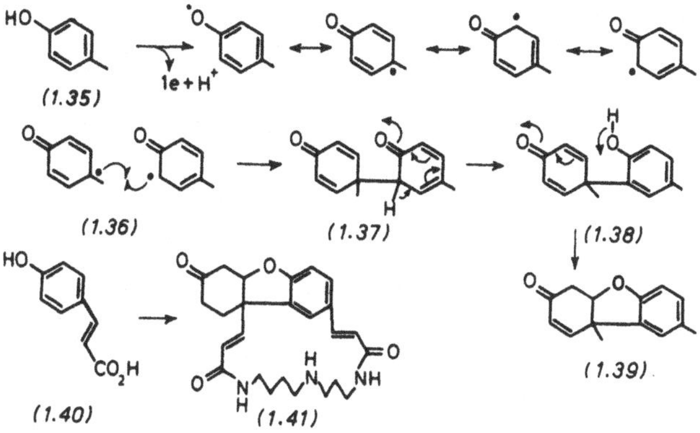

**Scheme 1.10**

Chemical conversion of *p*-cresol into the ketone (*1.39*) provides a simple illustration of the hypothesis (Scheme 1.10), in which coupling occurs between the position *ortho* to one phenolic hydroxy group and the position *para* to the other. Aromaticity is regained by proton loss from the coupling sites(s) [as (*1.37*) → (*1.38*) and (*1.42*) → (*1.43*)]. In the formation of (*1.38*)

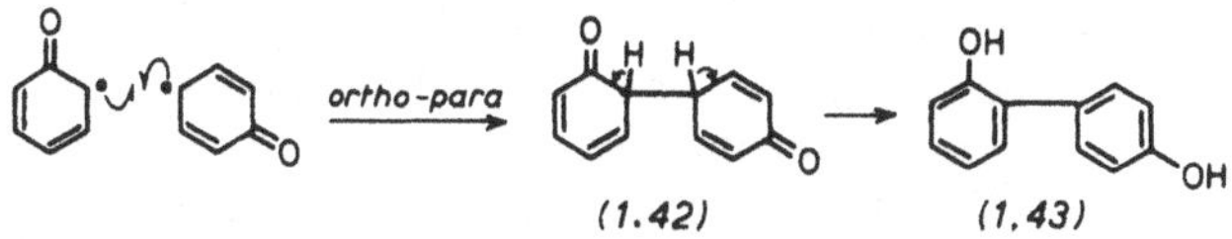

only one ring can become aromatic in this way; the methyl group in the other ring secures it in the dienone form (*1.38*). Such dienones have been isolated from living things. They may react in the way shown for (*1.38*) → (*1.39*). A strikingly close analogue of (*1.39*) is the plant alkaloid lunarine (*1.41*). It arises plausibly from two molecules of *p*-hydroxycinnamic acid (*1.40*) along a pathway similar to that shown for (*1.39*). [In support, phenylalanine, a precursor for (*1.40*), is incorporated into lunarine (*1.41*) [23].]

Instead of undergoing a ring-closure reaction of the type just described, dienones may rearrange *in vivo* [as (*1.44*) → (*1.45*)] and thus regain aromaticity (the 'dienone-phenol' rearrangement). Reduction gives dienols [as (*1.46*)] which either rearrange [as (*1.46*) → (*1.48*) ('dienol-benzene' rearrangement)] or undergo ring-closure reactions of the type (*1.46*) → (*1.47*). An outstanding example of the latter reaction involving a dienol is found in the biosynthesis of morphine (section 6.3.4), and of the former in the biosynthesis of isothebaine (section 6.3.2). (The various possibilities are summarized in Schemes 1.10 and 1.11; *ortho-para* coupling is illustrated; similar schemes can be written for *ortho-ortho* and *para-para* coupling.)

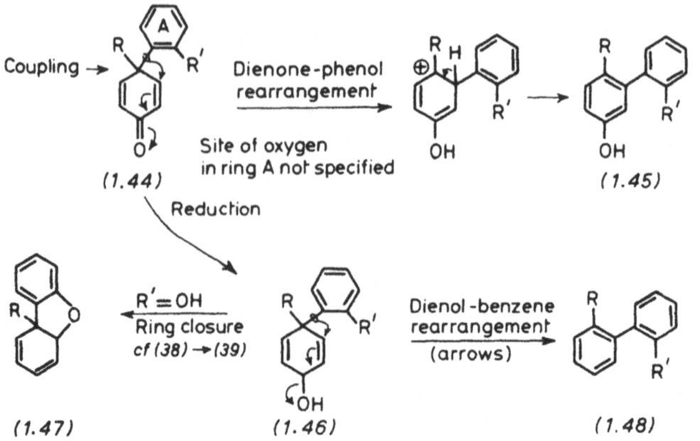

**Scheme 1.11**

## 1.3.2 Hydroxylation of aromatic substrates [24, 25]

It has been found generally in a variety of living systems that hydroxylation of aromatic substrates involves utilization of molecular oxygen and proceeds *via* an arene oxide [as (*1.49*)]. Collapse of this arene oxide to give (*1.51*) occurs with a 1,2 shift of the substituent X (e.g. X = $^1$H, $^2$H, $^3$H, Cl). In the hydroxylation of 4′-chlorophenylalanine (*1.52*) proton loss occurs from (*1.50*) to give 3′-chloro-4′-hydroxyphenylalanine (*1.54*) as the major product. For X = $^3$H (T) or $^2$H (D) loss of hydrogen isotope occurs from (*1.50*) in accord with a primary isotope effect ($k_H > k_D > k_T$). Thus hydroxylation of [4′-$^3$H]phenylalanine (*1.53*) gives the α-amino acid tyrosine (*1.55*) with a high retention (ca. 90%) of tritium at C-3′. A lower retention is observed for [4′-$^2$H]phenylalanine (ca. 70%) as a reflection of the different isotope effects. This 1,2-shift in a aromatic hydroxylation is called the NIH shift (named after the National Institutes of Health, USA, where the rearrangement was discovered).

Subsequent hydroxylation of tyrosine [as (*1.55*)] to give dopa [as (*1.56*)] occurs without NIH shift, i.e. loss of substituent X, although another arene oxide is probably implicated. For the conversion of [4′-$^3$H]phenylalanine (*1.53*) → (*1.55*) → (*1.56*), ca. 90% tritium is retained after the first hydroxylation. Half of this is lost in the next reaction because there are two equivalent (C-3′ and C-5′) sites, each bearing half of the residual tritium, which are available in (*1.55*) for oxygenation and consequent loss of tritium.

## 1.3.3 Methylation [2]

*C*-, *O*-, and *N*-methylations are encountered frequently in the biosynthesis of secondary metabolites. All appear to involve nucleophilic substitution on the *S*-methyl group of *S*-adenosyl-L-methionine (*1.59*) (Scheme 1.12). In

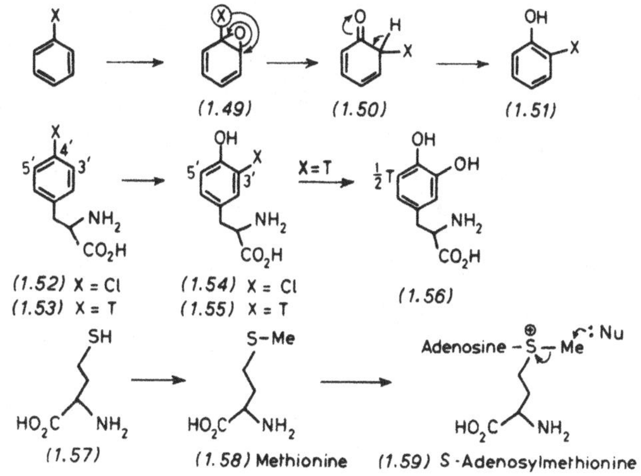

**Scheme 1.12**

biosynthetic feeding experiments it is normal to use [*methyl*-$^{14}$C]methionine [as (*1.58*)] as a precursor for methyl groups; reaction *in vivo* with ATP and Mg$^{2+}$ affords (*1.59*). Particularly in early studies formic acid was used to test for C$_1$ units in biosynthesis. It can serve as a source for methyl groups (as well as other C$_1$ units) via the tetrahydrofolate-linked intermediates, N$^{10}$-formyltetrahydrofolate, N$^5$-formyltetrahydrofolate, methyltetrahydrofolate (*1.62*) and methylenetetrahydrofolate (*1.60*). The conversion of (*1.61*)

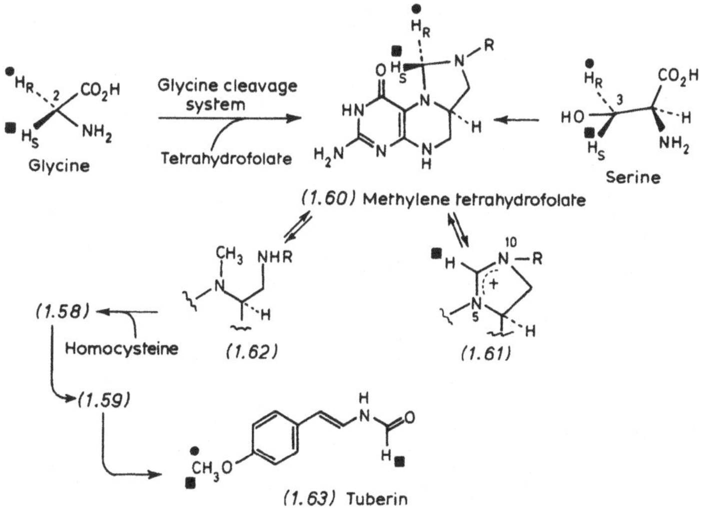

**Scheme 1.13**

into (*1.60*) involves a reduction; further reduction affords (*1.62*) from which the methyl group may be transferred to homocysteine to yield methionine (*1.58*). These tetrahydrofolate intermediates, however generated, can provide not only $C_1$ units at the methanol level of oxidation (*1.62*) = (*1.58*) but also at the formaldehyde (*1.60*) and formic acid (*1.61*) levels of oxidation. The extremes are found, for example, in the metabolite, tuberin (*1.63*) which is formed as indicated in Scheme 1.13 where C-2 of glycine and C-3 of serine act as sources for both $C_1$ units. The implication of these amino acids in such reactions involving tetrahydrofolate is well established (other examples appear in Chapter 7) and the detailed stereochemistry indicated in Scheme 1.13 has been deduced for tuberin (*1.63*) and in related cases [26].

$$\begin{array}{c} R\diagup^{CO_2H} \\ HO_2C^{\cdots}\diagup^{S}\diagdown_{CO_2H} \\ OH \end{array}$$

(*1.64*)

## REFERENCES

Further reading: Rétey, J. and Robinson, J.A. (1982) *Stereospecificity in Organic Chemistry and Enzymology*, Verlag-Chemie, Weinheim; Stryer, L. (1988) *Biochemistry*, 3rd edn, Freeman, New York; [2] and [3]

1. Swan, G.A. (1967) *An Introduction to the Alkaloids*, Blackwell Scientific Publications, Oxford.
2. Mahler, H.R. and Cordes, E.H. (1971) *Biological Chemistry*, 2nd edn, Harper and Row, New York.
3. Staunton, J. (1978) *Primary Metabolism: a mechanistic approach*, Clarendon Press, Oxford.
4. Dagley, S. and Nicholson, D.E. (1970) *An Introduction to Metabolic Pathways*, Blackwell Scientific Publications, Oxford.
5. Reed, L.J. (1974) *Acc. Chem. Res.*, **7**, 40–6.
6. Arnstadt, K.-I., Schindlbeck, G. and Lynen, F. (1975) *Eur. J. Biochem.*, **55**, 561–71.
7. Lynen. F. (1980) *Eur. J. Biochem.*, **112**, 431–42.
8. Sedgwick, B., Morris, C. and French, S.J. (1978) *J. Chem. Soc., Chem. Comm.*, 193–4.
9. Bu'Lock, J.D. and Smith, G.N. (1967) *J. Chem. Soc. (C)*, 332–6.
10. Nicolaou, K.C., Gasie, G.P. and Barnette, W.E. (1978) *Angew. Chem. Int. Ed. Engl.*, **17**, 293–378.
11. Gibson, K.H. (1977) *Chem. Soc. Rev.*, **6**, 489–510.
12. Bentley, R. (1969) *Molecular Asymmetry in Biology*, Academic Press, Vol. 1; (1970), vol. 2.
13. Hanson, K.R. (1966) *J. Am. Chem. Soc.*, **88**, 2731–42.

14. Ogston, A.G. (1948) *Nature,* **162**, 963.
15. Cornforth, J.W. (1970) *Chem. in Brit.,* **6**, 431–6.
16. Floss, H.G. and Tsai, M.-D. (1979) *Advances in Enzymology,* **50**, 243–302.
17. Townsend, C.A., Scholl, T. and Arigoni, D. (1975) *J. Chem. Soc. Chem. Comm.,* 921–2.
18. Battersby, A.R., Staunton, J., Wiltshire, H.R., Bircher, B.J. and Fuganti, C. (1975) *J. Chem. Soc. Perkin 1,* 1162–71; and their ref. 11.
19. Olah, G. (1972) *Chem. in Brit.,* **8**, 281–7.
20. Barton, D.H.R. and Cohen, T. (1957) In *Festschrift Dr A. Stoll*, Birkhaüser, Basle, pp. 117–43.
21. Erdtman, H. and Wachtmeister, C.A. (1957) In *Festschrift Dr A. Stoll*, Birkhaüser, Basle, pp. 144–65.
22. Taylor, W.I. and Battersby, A.R. (eds) (1967) *Oxidative Coupling of Phenols*, Arnold, London.
23. Poupat, C. and Kunesch, G. (1971) *Compt. rend.,* **273C**, 433–6.
24. Auret, B.J., Boyd, D.R., Robinson, P.M. *et al.* (1971) *J. Chem. Soc. Chem. Comm.,* 1585–7.
25. Guroff, G., Daly, J.W., Jerina, D.M. *et al.* (1967) *Science,* **157**, 1524–30.
26. Cable, K.M., Herbert, R.B., Bertram, V. and Young, D.W. (1987) *Tetrahedron Lett.,* **28**, 4101–4; and refs cited.

# 2 Techniques for biosynthesis

## 2.1 INTRODUCTION

In the study of secondary metabolism the ongoing problem is twofold: first, to identify the source(s) in primary metabolism from which a secondary metabolite has its genesis; and secondly, to identify by what mechanisms and manner of intermediates it is thus fashioned. It is structure that first hints at a solution made complete by a battery of special techniques.

The biosynthetic pathway by which a primary metabolite is formed may be an intricate one involving complex chemistry. Thus the structure of a primary metabolite does not necessarily yield any immediate clues to its mode of biosynthesis (e.g. the biosynthesis of the $\alpha$-amino acid, tyrosine, in section 5.1). By contrast the structure of a secondary metabolite often allows accurate speculation about its origins and even its mechanism of formation. Such speculation has long been the hand-maid of· structure determination, e.g. consideration of the possible biogenesis of the alkaloid, morphine, indicated the correct structure (2.2) for it in the midst of confusing chemical evidence. Both the isoprene rule (Chapter 4) and the polyketide hypothesis (Chapter 3), underpinned by coherent biogenetic speculation, have been, and continue to be, invaluable in the structure determination of natural products. They are based on the fact that large numbers of secondary metabolites are formed from one or two simple repeating units.

It is because, in the event, secondary metabolites derive from but a handful of primary metabolites along pathways involving generally the simplest reactions in organic chemistry and little rearrangement, that speculation about their biosynthesis can be so accurate. Also invaluable in this regard, is the identification of metabolites of similar structure in the same, or related, species. Thus benzylisoquinoline alkaloids [as (2.1)] and morphine (2.2) are both found in *Papaver* species. It follows reasonably that morphine derives from a compound of type (2.1) (section 6.3.4).

The wealth of intelligent speculation about the biosynthesis of secondary metabolites has provided a firm base from which to mount wide-ranging experimental forays into secondary metabolism in search of origins and

Tyrosine    (2.1) Reticuline    (2.2) Morphine

mechanisms of formation. The experiments are carried out using a variety of techniques. The dominant technique involves the administration of likely precursors to an organism, and examination of the secondary metabolite produced to see if the compounds 'fed' were used in the formation of the metabolite. In this way, tyrosine and (*2.1*) have been shown to be involved in morphine biosynthesis. To track the likely precursor through to the metabolite, the precursor must be labelled, or marked, in some way. A variety of 'labels' are used. First, there are radioactive isotopic labels, e.g. $^{14}C$ and $^{3}H$ (tritium) which are very sensitively assayed by scintillation counting (section 2.2.1) (tyrosine was examined as a morphine precursor with a $^{14}C$ label at C-2). Second, there are stable isotopic labels, e.g. $^{13}C$, $^{15}N$, $^{18}O$, $^{2}H$ (deuterium) which are very much less sensitively assayed by mass spectrometry and/or n.m.r. spectroscopy (section 2.2.2).

In addition to the results of feeding experiments with labelled precursors, clear information on the sequence of biosynthesis of several related metabolites produced by an organism may be obtained. This is done by feeding a very early precursor (usually $^{14}CO_2$ in plants) and noting the order in which the metabolites are labelled. This order gives the sequence of biosynthesis.

Experiments with purified enzymes involved in biosynthesis, or even experiments with crude enzyme preparations, can provide important definition of a pathway. Studies with enzymes are briefly referred to, by way of introduction, in section 2.3, as is the use of mutants.

It is usually a simple matter to feed a micro-organism, growing in liquid culture, with a labelled precursor. Commonly aqueous solutions are added at predetermined time(s) to the culture and the metabolite is isolated and purified some time later. It is more difficult to feed precursors to plants. A variety of methods are used:

1. a wick is threaded through the lower part of the plant stem and dipped into an aqueous solution of the precursor
2. assimilation through the roots by standing the plant in an aqueous solution of the precursor
3. injection into a hollow stem or seed capsule
4. standing excised parts of the plant in an aqueous solution of the precursor
5. painting a solution of the precursor on the leaves.

In feeding experiments with labelled precursors common problems arise.

1. A very reasonable precursor may be utilized in metabolite biosynthesis at a very low level or not at all. This may be:
   (a) because of the difficulty in getting the precursor to the site of biosynthesis (a perennial problem)
   (b) because it is genuinely not involved in the biosynthesis of the chosen metabolite
   (c) because it is used much more efficiently for the formation of another secondary, or primary, metabolite
   (d) because, in studies with plants, the particular metabolite might not be undergoing biosynthesis at the time of the experiment.
   Thus a negative result with a labelled precursor is always ambiguous, never certain;
2. Even if a labelled compound, [(C in (2.3)] is shown to be an efficient pre-

$$A \longrightarrow B \longrightarrow C \longrightarrow D$$
$$\qquad \updownarrow \qquad \text{Metabolite}$$
$$\qquad E$$
$$\qquad (2.3)$$

cursor* for a metabolite, D, it does not mean it is an obligatory intermediate in the biosynthesis of D. It may just have been a convenient foreign substrate dealt with economically by the living system. However, if the precursor, C, can be shown to occur in the organism under examination and be formed like the metabolite, D, from the same earlier precursor(s), e.g. A, then it is most likely to be a true intermediate. Its status as an intermediate can be further strengthened by the appropriate experiments with enzymes isolated from the living system;
3. The precursor may be at a shunt (E) from the main pathway (A → B → C → D) in (2.3). It is difficult to resolve whether or not this is so and final conclusions depend on the balance of the evidence.

## 2.2 ISOTOPIC LABELLING

Simple compounds bearing isotopic labels can be bought. More complex compounds must be synthesized [9, 10].

### 2.2.1 Radioactive isotopes

Tritium ($^3$H) and $^{14}$C, with half-lives of 12.26 years and 5600 years respec-

---

*The efficiency with which a precursor is utilized is given either by the percentage of label incorporated from the precursor into the metabolite or by the extent to which the precursor label is diluted in the derived metabolite. It is normal to compare the values for several precursors: efficiency is only really significant in a relative sense. A precursor earlier in a pathway will normally be incorporated less efficiently and with higher dilution (passing through more metabolic pools) than one which appears later. Acceptable incorporations in plants are often below 1%, but experiments with microorganisms usually return acceptable incorporations above 1%.

tively, are the most commonly used radioactive isotopes. They are used extensively to monitor the uptake of precursors into metabolites and to determine the fates of individual hydrogens or carbons in a precursor during biosynthesis. Tritium finds particular application in this latter regard because any number of proton additions/removals and rearrangements may occur in the course of the biotransformation of precursor into metabolite. (Tritium loss is normally measured by reference to a $^{14}C$ label in the precursor which is known not to be lost during biosynthesis.)

The isotopes, $^{14}C$ and $^{3}H$, are both $\beta$-emitters. They are assayed by liquid scintillation counting. In simple terms, the method involves dissolving the radioactive compound in a scintillant 'cocktail' containing phosphors which convert the $\beta$-radiation into light. The light emitted is measured within the scintillation counter by photomultiplier tubes and is expressed as counts against time, a normal way of expressing radioactivity. In practice radioactivity is measured within the counter at less than 100% efficiency. The actual efficiency must be determined and allowance made for this in calculating the true activity of the sample [in disintegrations $sec^{-1}$ $mmol^{-1}$; 1 Curie = $3.7 \times 10^{10}$ d $sec^{-1}$; 1 m Ci = 37 MBq (Bq = Becquerels)]. Reliable radioactivity measurements depend on:

1. recrystallization of the sample, or a derivative, to constant radioactivity
2. obtaining counts for the radioactive compound at above twice the value for an inactive (background) sample (this is the lower limit for radioactivity measurement).

The fate of radioactive labels, particularly carbon, is determined by unambiguous degradation of the radioactive metabolite to isolate the labelling site from all others. Specific labelling of a particular site, or sites, indicates specific utilization of the precursor, e.g. [1-$^{14}$C]acetic acid (2.4) specifically labels the sites indicated in orsellinic acid (2.5), thus showing that

acetic acid is specifically a source for (2.5) (specific incorporation), and that it is used in the way shown. Spread of acetate radioactivity, say, over all the carbon atoms in (2.5) would clearly indicate fragmentation of the precursor before utilization in the biosynthesis of (2.5). Without determination of the site of labelling the real significance of the measured incorporation of a precursor, which could be due either to specific, or to random, incorporation, would remain unknown.

It is possible for a fairly complex precursor to suffer partial fragmentation prior to use. Use of a single label and even showing that the label is specifically located at the expected site may well not reveal this. So intact incorporation is more accurately established by using a precursor with labels in two sites as distant as possible from each other. The precursor can only be incorporated into the metabolite intact if the ratio between the two labels in precursor and metabolite is maintained. If once two fragments are formed, each bearing a single label, the chances are remote that they will recombine from the metabolic pools in the living system to give another precursor with the same ratio of labels. Two $^{14}$C labels may be used in this sort of experiment but it is simpler to use a $^{14}$C and a $^{3}$H label. A notable example is to be found in the biosynthesis of mesembrine alkaloids where the use of doubly labelled precursors indicated their fragmentation and where singly labelled precursors indicated intact incorporation (section 6.4.2).

Although $^{14}$C and $^{3}$H are both $\beta$-emitters, $^{3}$H is fortunately a lower energy $\beta$-emitter than $^{14}$C and there are sufficient differences in their energy spectra to enable one to obtain a separate estimate of the amount of $^{14}$C and $^{3}$H a molecule contains.

Unlike stable isotopes radioactive ones are generally only present in compounds at very low enrichment, i.e. the bulk of the material is non-radioactive.

When using a mixture of labels, e.g. $^{14}$C and $^{3}$H, it is normal to prepare the compounds with single labels and then mix them. Because of the low enrichment of isotopic label in a compound it is most unlikely that a *single molecule* of e.g. a multiply labelled species will, in any case, have more than one site labelled, thus (*2.6*) (labels as shown) is made up of (*2.7*), (*2.8*) and (*2.9*). So also, for example, ethanol tritiated on C-1 is made up of (*2.10*), (*2.11*) and (*2.12*). Normal chemical removal of a hydrogen atom from C-1 will result in

loss of $^{1}$H rather than $^{3}$H, by a primary kinetic isotope effect, i.e. high tritium retention: ca. 90%. Stereospecific hydrogen removal from the methylene group will not be influenced by the isotope effect and will result in removal of $^{3}$H from (*2.11*) and $^{1}$H from (*2.12*), or *vice versa*, i.e. 50% retention of tritium. A 50% tritium retention, i.e. stereospecificity, in such a system during biosynthesis indicates an enzyme-catalysed reaction. [For further discussion of reactions at centres like that of the methylene group in (*2.10*) and the associated stereochemistry see section 1.2.1.] The actual stereochemistry of this and other reactions can be probed by employing the individual, chirally labelled, molecules [as (*2.11*) and (*2.12*)]. One species will suffer complete loss of tritium in a stereospecific (enzyme-catalysed) reaction, the other complete retention. If the chirality of the fed material is known then the stereochemical course of the reaction follows. Numerous applications are to be found in following chapters. The most outstanding are in the complex area of steroid biosynthesis (Chapter 4) where brilliant use has been made of chirally tritiated mevalonic acid samples to determine the intricate detail of biosynthesis.

Use of the three isotopes of hydrogen to fabricate chiral methyl groups for study of the stereochemistry of enzyme-catalysed reactions involving methyl groups is to be noted (sections 1.2.2, 4.2 and 4.4).

Although radioactive isotopes are used commonly in trace amounts, tritium if present in high concentration can be measured as an n.m.r. signal [11], e.g. in the biosynthesis of cycloartenol (section 4.2). Information about the location and stereochemistry of a tritium label was immediately apparent from the n.m.r. spectrum which would not have been the case with scintillation counting.

### 2.2.2 Stable isotopes [3–8]

A striking feature of the recent exploration of the biosynthesis of secondary metabolites is the ever increasing use of stable isotopes and the ever increasing application of sophisticated n.m.r. techniques in analysis. The stable isotopes which are used are $^{13}C$ (nuclear spin $=\frac{1}{2}$), $^2H$ (nuclear spin $= 1$), $^{15}N$ (nuclear spin $= \frac{1}{2}$) and $^{18}O$. Of these, the first three give rise to n.m.r. signals and may be analysed for by this method or by mass spectrometry. $^{18}O$ (natural abundance 0.204%) does not give rise to n.m.r. signals. It may be analysed for by mass spectrometry or by $^{13}C$ n.m.r. spectroscopy: the presence of $^{18}O$ on a particular carbon atom causes an upfield shift of the resonance for that carbon atom compared to when $^{16}O$ is present. (The magnitude of these shifts range from ca. 0.01–0.035 ppm for singly bonded oxygens to 0.03–0.55 ppm for doubly bonded oxygens (see below).) $^{17}O$ (natural abundance 0.037%; nuclear spin $= 5/2$) can be observed directly by n.m.r. spectroscopy but its use is limited by low sensitivity (3% of that of $^1H$ at 100% isotopic purity), very broad lines and also expense. (For an example of where it has been used in a labelling study see [12].)

Stable isotopes can yield more information more quickly than radioactive ones. Although rapidly applied, the methods of analysis (n.m.r. spectroscopy and mass spectrometry) are much less sensitive than the method used to assay radioactivity. So a high enrichment is necessary and much more precursor needs to be fed. There is thus a danger of so much being administered that the normal metabolism of the organism under study is disturbed, resulting in false conclusions. In practice, however, this does not seem to have been a serious problem (for an interesting example involving $^{15}N$ see [13]). The risk of overloading the living system is not found with radioactive precursors because they are almost always fed in trace amounts with minimum disturbance of normal metabolism.

All organic molecules contain a significant amount of $^{13}C$ (natural abundance: 1.1% of $^{12}C$) and much lesser amounts of $^{2}H$ (natural abundance: 0.015%) and, where relevant, $^{15}N$ (natural abundance: 0.36%) and $^{18}O$. The presence of naturally abundant stable isotopes within a molecule ultimately limits the amount of precursor label which can be detected in a metabolite. This is particularly true where $^{13}C$, the most abundant isotope, is involved. Enhanced signals in the $^{13}C$ n.m.r. spectrum of a metabolite formed from a labelled precursor indicates the location of the label. A lower practical limit for accurate estimation is a signal enriched to the extent of 20% over the natural abundance signal; in practice much higher enrichments are sought for accuracy. Similar problems in estimating low enrichment are encountered in mass spectrometric analysis. Any single $^{13}C$, $^{2}H$ or $^{15}N$ label has to be measured on the M + 1 peak. The natural abundance peak can be quite intense, e.g. $C_{10}H_{12}$ shows M + 1 of 11% of $M^{+}$. Thus accurate estimation of small increments in intensity will be difficult. If more than one $^{13}C/^{2}H/^{15}N$ is present this difficulty is alleviated since for, e.g. $C_{10}H_{12}$, M + 2, the peak to be measured if two stable isotopes are present, has an intensity of only 0.6%.

There are particular problems associated with the direct measurement by n.m.r. spectroscopy of $^{2}H$ [low sensitivity, poor dispersion of chemical shifts (ca. 15% of those in $^{1}H$ n.m.r.) and broad signals ($^{2}H$ nuclear quadrupole); the low natural abundance, however, allows detection at very low levels] and $^{15}N$ [sensitivity (0.2% of $^{1}H$), relaxation and n.O.e. problems and chemical shift values often dependent on solvent]. Both $^{2}H$ and $^{15}N$ are often measured by reference to a $^{13}C$ label in $^{13}C$ n.m.r. spectroscopy. $^{2}H$ is also commonly analysed for by $^{2}H$ n.m.r. (for examples see below).

A significant advantage of stable isotopes over radioactive ones resides in the high enrichment (approaching 100%) of stable isotopes used. It is a simple thing to have two or more labels within the same precursor molecule (cf by contrast, radioactive labels, above). The advantages will be apparent in the ensuing discussion (see also following chapters).

The first requirement in using $^{13}C$ labelling for studying the biosynthesis of a particular compound is to assign the resonances in the natural abundance $^{13}C$ n.m.r. spectrum of the metabolite. Any mistakes in assignment can lead to false conclusions about biosynthesis. Incorporation of $^{13}C$ label from a precursor leads to enhanced n.m.r. signals in the metabolite derived from it. Multicolic acid (*2.15*) biosynthesized from [$^{13}C$]acetate provides one example [14] among many [3–6, 8]. Enhanced n.m.r. signals in (*2.16*) for C-1, C-3, C-5, C-7, C-9, and C-10 from [2-$^{13}C$]acetate and complementary results with [1-$^{13}C$]acetate outline the precursor labelling pattern.

More powerfully still, as mentioned above, one may use precursors with two or more labels within the same molecule. Along with application of $^{13}C$ in association with $^{2}H$ and $^{18}O$ the use of [1,2-$^{13}C_2$]acetate has revolutionized the study of polyketide biosynthesis (Chapter 3). [1,2-$^{13}C_2$]Acetate is normally assimilated into a metabolite under conditions where it will be formed from the labelled precursor after dilution by the organism's own unlabelled acetate. The metabolite will then be formed very largely, if not completely, so that no two labelled acetate units are adjacent to one another. In the n.m.r. spectrum of the metabolite, coupling will now be seen between the signals for the $^{13}C$ labels of intact acetate units, and these only. So in the spectrum of (2.16), derived from [$^{13}C_2$]acetate, coupling may be discerned between C-8 and C-9, C-7 and C-6, C-5 and C-2, C-10 and C-4, but C-1, C-3, and C-11 appear as singlets. So C-10, C-4, C-2, and C-5 through C-9 represent intact acetate units (see thickened bonds) and C-1, C-3, and C-11 fragmented ones. The likely biosynthesis of multicolic acid (2.15) follows as shown in Scheme 2.1.

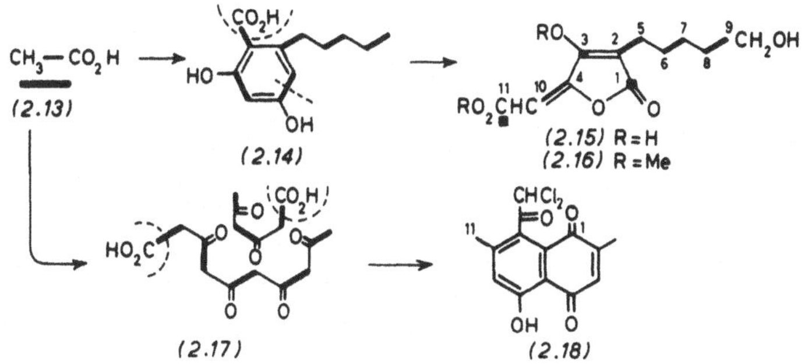

**Scheme 2.1**

Another example using [1,2-$^{13}C_2$]acetate is provided by mollisin (2.18) where uncertainty prevailed about its pattern of biosynthesis from multiples of acetic acid arising from available $^{14}C$ data. [1,2-$^{13}C_2$]Acetate (2.13) gave mollisin (2.18), the $^{13}C$ n.m.r. spectrum of which showed doublets superimposed on natural abundance singlets for all signals except those for C-1 and C-11 which were enhanced singlets. C-1 and C-11 thus derive from fragmented acetate units and the pattern of biosynthesis follows as shown in (2.17) [15]. (For further discussion see section 3.7.)

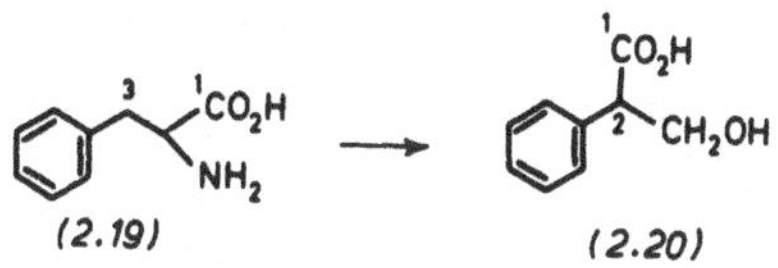

A different approach to the use of a precursor with two adjacent $^{13}C$ atoms in this way, is one in which a precursor is labelled on two atoms expected to come together in the course of biosynthesis. Coupling in the n.m.r. spectrum of the derived metabolite indicates that these atoms have indeed become contiguous. For example, a rearrangement occurs in the conversion of phenylalanine (*2.19*) into tropic acid (*2.20*) (section 6.2.2). [1,3-$^{13}C_2$]Phenyl-alanine [as (*2.19*), labels in the same molecule] gave tropic acid (*2.20*) in which coupling was observed between C-1 and C-2, thus proving that rearrangement occurs by an *intra*-molecular 1,2-shift of the phenylalanine carboxy group. (Inter-molecular rearrangement would have given species labelled on either C-1 or C-2 but not on both as a result of dilution in meta-bolic pools [16].

[U-$^{13}$C]Glucose (*2.21*) (U = universally labelled, i.e. each of the six carbon atoms is highly enriched with $^{13}$C) has found most useful application in the study of the biosynthesis of a number of metabolites (e.g. section 7.6.2). The antibiotic streptonigrin (*2.23*) will be taken as an example here. Rings C and D of (*2.23*) (see dotted lines) could be identified as arising from the amino acid tryptophan (*2.24*) but the origins of rings A and B were obscure.

(2.21)  (2.22)  (2.23) Streptonigrin

(2.24)  (2.25)

(2.26)  (2.27) Anabasine

Aromatic amino acids are biosynthesized by way of the shikimate path-way (Chapter 5) and the carbon atoms originate through the metabolism (fragmentation) of glucose in defined ways. Thus the incorporation of [U-$^{13}$C]glucose into the tryptophan-derived part of streptonigrin gave the expected pattern for rings C and D which is shown with thickened bonds in (*2.23*); the pattern was deduced from the $^{13}$C–$^{13}$C couplings observed. The $C_4$ moiety in ring D is a unit of erythrose-4-phosphate (*2.22*).

Two further $C_4$ (erythrose) units could be discerned for (*2.23*) [* = C-1 of (*2.22*)]. Notably, the pattern of ring A is the same as for ring D, i.e. one which corresponds to an aromatic moiety/intermediate of the shikimate pathway.

This provided the essential clue which led to the identification of aromatic intermediates which are involved in the elaboration of rings A and B. One of these is (*2.25*) (section 7.5.4.) [17].

In the study of the biosynthesis of porphyrins and vitamin $B_{12}$ many ingenious experiments involving compounds labelled with $^{13}$C and other isotopes have been carried out. Since discussion of these metabolites is beyond the scope of this book the reader is referred elsewhere to masterly and authoritative accounts of their biosynthesis [18, 19].

It is generally true that precursors are incorporated into plant metabolites at a sufficiently low level to preclude the use of stable isotopes. Of course, there are exceptions, most notably in the study of the biosynthesis of pyrrolizidine and quinolizidine alkaloids (section 6.2). Because of the incorporations observed for these metabolites it has been possible to use a range of stable isotopes to probe the course of biosynthesis. Where incorporations are low there is, however, an ingenious solution and that is to use a precursor bearing two contiguous labels, e.g. $^{13}$C-$^{13}$C (also $^{13}$C-$^{15}$N). The $^{13}$C n.m.r. signals from these contiguous atoms in the metabolite [as (*2.27*)] formed from the labelled precursor [as (*2.26*)] will be doublets, due to one-bond $^{13}$C-$^{13}$C coupling, which will flank the natural abundance singlets (and be dwarfed by them) in the $^{13}$C n.m.r. spectra. Incorporations can be measured at very low enrichments since the natural abundance of $^{13}$C-$^{13}$C coupled signals is very low (0.0123%). One example of this is found for the incorporation of [4,5-$^{13}$C$_2$]lysine (*2.26*) into anabasine (*2.27*) labelled as shown [20].

In appropriate cases $^{15}$N, the stable isotope of nitrogen, has found important application. Its incorporation may usefully be monitored by mass spectrometry and n.m.r. spectroscopy. Although analysis may be by $^{15}$N n.m.r. spectroscopy [the incorporation of (*2.25*) into (*2.23*) was measured in this way; ● = $^{15}$N] [22] it is more convenient and usual to use $^{15}$N in association with $^{13}$C and to analyse the metabolite by $^{13}$C n.m.r.; this also provides for the testing for intact $^{13}$C-$^{15}$N bonds during biosynthesis. Generally analysis depends on the observation of $^{13}$C-$^{15}$N coupling [23] but isotope shifts may also be observed. In the biosynthesis of streptonigrin (*2.23*) from tryptophan (*2.24*) precursor bond cleavage could, *a priori*, have occurred on either side (a or b) of the ring nitrogen atom in (*2.24*). Maintenance of $^{15}$N-$^{13}$C coupling in the $^{13}$C n.m.r. spectrum of streptonigrin derived from [1′-$^{15}$N, 2′-$^{13}$C]tryptophan indicated clearly that bond cleavage is along the dotted line b in (*2.24*) [24].

The oxygen atoms found in a secondary metabolite may originate from oxygen in the precursor that diverts from primary metabolism, from aerial oxygen (enzymic oxidation), or from water. Tracking the origin of the oxygen atoms, very commonly through the use of $^{18}$O, can provide important clues to the mechanism of biosynthesis of a secondary metabolite. This has found particular and powerful application in the study of polyketide biosynthesis (Chapter 3). The microbial metabolite polivione (*2.29*) provides an

example. The use of [1,2-$^{13}C_2$]acetate gave results which established that
(*2.29*) is a polyketide (Scheme 2.2) with the labelling pattern shown.
[1-$^{13}$C, $^{18}$O]Acetate (*2.28*) gave polivione (*2.29*) in which the $^{13}$C n.m.r.
resonances for C-7, -9, and -11 showed upfield isotope shifts due to bonding
to $^{18}$O. The complementary experiment was done with $^{18}O_2$ gas: the $^{13}$C
n.m.r. signals for C-4, -12, and -14 were now shifted (the presence of up to
two $^{18}$O atoms on C-14 gave rise to two shifted $^{13}$C resonances). From these
labelling results the biosynthetic pathway shown in Scheme 2.2 was deduced
for polivione (*2.29*) [25].

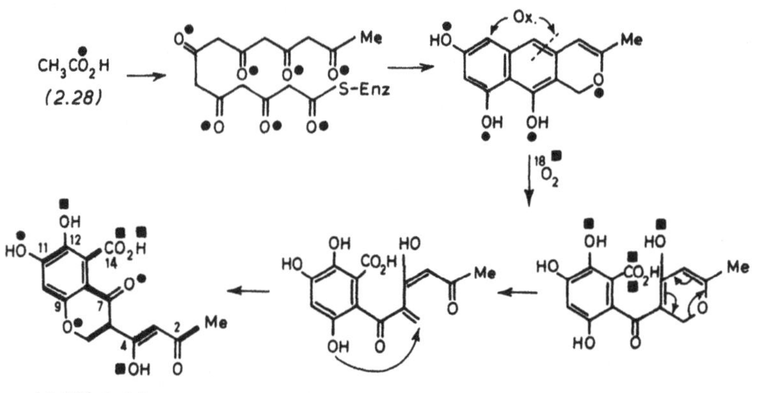

**Scheme 2.2**

The incorporation of oxygen from water into a metabolite is normally
deduced from the absence of labelling from aerial or precursor $^{18}$O. A
recurring problem with such deductions, however, is that $^{18}$O may be lost at
some stage by exchange. C-2, -4, and -7 in (*2.29*) are subject to exchange in
water; C-2 was found not to be labelled by $^{18}$O-acetate contrary to
expectations.

Deuterium ($^2$H) is both an inexpensive stable isotope and also a powerful
probe for events taking place in secondary metabolism. It is commonly
assayed by $^2$H n.m.r., by $^{13}$C n.m.r. in association with a $^{13}$C label, and by
mass spectrometry. A particular method of analysis (section 4.4) involves the
measurement of optical activity in deuteriated succinic acid samples as a
means of probing the stereochemistry of a biosynthetic sequence [26].

Precursors with stable isotopic labels, e.g. $^2$H, can provide further
information not available with radioactive labels. Thus in the biosynthesis of
microbial phenazines [as (*2.31*)], [2-$^2$H]shikimic acid (*2.30*) gave iodinin [as
(*2.31*)] some molecules of which were dilabelled as shown in (*2.31*) (measured
by mass spectrometry). This proved that these symmetrical heterocycles are
derived from two molecules of shikimic acid. Because of the symmetry in the
metabolite, and the normal low enrichment of radioactive labels, $^{14}$C and $^3$H

labels would not have allowed distinction between one and two molecules of precursor. The labelling sites, and thus orientation of (*2.30*) in (*2.31*), were established by base-catalysed exchange but, in principle, could have been determined by $^2$H n.m.r. spectroscopy [27].

(2.30) → (2.31)

The biosynthesis of the microbial metabolite tuberin (*2.33*) is from the α-amino acid tyrosine (*2.32*). The formation of the double bond in (2.33) has been examined with deuteriated samples of tyrosine. Analysis was by $^2$H n.m.r. and mass spectrometry. L-[2-$^2$H]Tyrosine [as (*2.32*)] but not D-[2-$^2$H]tyrosine gave labelled tuberin. Incorporation of the former precursor was lower than that of other tyrosine samples. This is, however, expected since amino acids readily undergo enzyme-catalysed transamination reactions which result in the loss of hydrogen and nitrogen from C-2.

Tyrosine samples chirally deuteriated at C-3 were next examined as precursors. (3S)-[3-$^2$H]Tyrosine [as (*2.32*)] gave tuberin (*2.33*) the $^2$H n.m.r. spectrum of which gave a single resonance at δ 6.35 ppm due to deuterium on C-3.

(2.32)        (2.33) Tuberin

There are two useful things to be noted about $^2$H n.m.r. The first is that chemical shift values for $^2$H and $^1$H n.m.r. spectroscopy are closely similar: the $^1$H signal for the proton on C-3 in tuberin is at δ 6.15 ppm. The second useful thing is that broadening of $^2$H n.m.r. signals can be substantially reduced by running the spectrum above room temperature (this is something everyone does but hardly anyone mentions).

(3R)-[3-$^2$H]Tyrosine [as (*2.32*)] was, unlike the (3S)-isomer, not incorporated into (*2.33*). The combined results indicate that double bond formation occurs by loss of the 3-*pro-R* proton in tyrosine (*2.32*) and, associated with it, loss of the carboxyl group. Formation of this *E*-double bond occurs, as shown, in a formal anti-periplanar sense [28].

$^2$H n.m.r. spectroscopy has been used on the one hand to detect a $^{13}$C nucleus which is neighbour to a deuterium atom. On the other hand $^{13}$C n.m.r. provides a powerful way of analysing for deuterium which is directly attached to a $^{13}$C nucleus [7]. Each deuterium atom shifts the position of resonance for the $^{13}$C to which it is attached by 0.3 to 0.6 ppm upfield ($\alpha$ shift). The nuclear spin on deuterium ($= 1$) also increases the multiplicity of the signal which in the absence of $^2$H as well as normal $^1$H decoupling can lead to very complex spectra. For an example from polyketide biosynthesis, see [29].

An ingenious alternative approach is to make use of a molecule in which the deuterium label is two bonds away from the $^{13}$C label, e.g. [1-$^{13}$C, 2-$^2$H$_3$]acetate (*2.34*) [30]. The effect of the deuterium is to shift the

(2.34)                    (2.35)

$^{13}$C resonance ($\beta$-shift). These $\beta$-isotope shifts are usefully additive and are most frequently upfield (ca. 0.01 to 0.1 ppm per $^2$H) but they can also be downfield or zero particularly if the 'reporter carbon' is a carbonyl group [31]. There are distinct advantages in this technique though, for problems of line broadening and of reduced n.O.e. are largely avoided and $^2$H–$^{13}$C coupling over two bonds is negligible. Very usefully it is possible to detect *two* adjacent bonds which have remained intact during biosynthesis. The first of many examples where the technique has been applied is in the conversion of [1-$^{13}$C, 2-$^2$H]acetate (*2.34*) into 6-methyl-salicylic acid (*2.35*) which was labelled as shown [30].

## 2.3 ENZYMES AND MUTANTS

Powerful support for any biosynthetic pathway as well as detailed information on the reactions involved may be gained by isolation, purification and characterization of enzymes which will catalyse individual steps of biosynthesis. A number of examples are quoted in succeeding chapters.

An enzyme does not have to be pure to yield valuable information. In many cases simple cell-free preparations containing a mixture of many enzymes may yield most useful results. These preparations are simply made from bacterial cultures in various ways including ultrasonic disruption. For plants resort is commonly made to cell-free preparations from plant tissue cultures [32]. An impressive example is found in work on the biosynthesis of terpenoid indole alkaloids (section 6.6.2) and benzylisoquinoline alkaloids (section 6.3). Also to be noted is the application here of radioimmunoassay [33] to the analysis of radioactive alkaloids.

A biosynthetic sequence can be summarized as A → B → C → D, where D is a known metabolite which is normally accumulated by the organism and A, B, and C are unknown intermediates. If the conversion of, e.g. B into C can be prevented through the absence of the necessary enzyme then B will accumulate instead of D. B may be isolated from this blocked, or mutant, organism and its structure determined. A second mutant may be blocked between C and D, so C accumulates and may be identified.

In summary:  mutant 1:  A → B ↛ C → D
B accumulates
mutant 2:  A → B → C ↛ D
C accumulates.

If C is fed to mutant 1 then D will be produced. If B is fed to mutant 2, C will be produced. From this very simple example it can be concluded with reasonable security that the biosynthetic sequence to D involves B → C → D. In practice the situation is more complex. For applications in secondary metabolism see section 3.8 and [2]. Good examples are to be found in the biosynthesis of the aminoglycoside antibiotics streptomycin (*2.36*) [34] and astromycin [35]. For other work on aminoglycoside antibiotics see [36].

(*2.36*) Streptomycin

# REFERENCES

Further reading: [2]–[8].

1. Robinson, R. (1955) *The Structural Relations of Natural Products*, Oxford University Press, Oxford.
2. Brown, S.A. (1972) In *Biosynthesis (Specialist Periodical Reports)* (ed. T.A. Geissman), The Chemical Society, London, vol. 1, pp. 1–40.
3. Simpson, T.J. (1975) *Chem. Soc. Rev.*, **4**, 497–522.
4. Tanabe, M. (1973) In *Biosynthesis (Specialist Periodical Reports)* (ed. T.A. Geissman), The Chemical Society, London, vol. 2, pp. 241–99.

5. Tanabe, M. (1975) In *Biosynthesis (Specialist Periodical Reports)* (ed. T.A. Geissman), The Chemical Society, vol. 3, pp. 247–85.
6. Tanabe, M. (1976) In *Biosynthesis (Specialist Periodical Reports)* (ed. J.D. Bu'Lock), The Chemical Society, London, vol. 4, pp. 204–47.
7. Garson, M.J. and Staunton, J. (1979) *Chem. Soc. Rev.*, **8**, 539–61.
8. Vederas, J.C. (1987) *Nat. Prod. Rep.*, **4**, 277–337.
9. e.g. Murray, III, A. and Williams, D.L. (1958) *Organic Syntheses with Isotopes*, Interscience, New York, 2 vols.
10. Ott, D.G. (1981) *Syntheses with Stable Isotopes*, Wiley, New York.
11. Evans, E.A., Warrell, D.C., Elvidge, J.A. and Jones, J.R. (1985) *Handbook of Tritium NMR Spectroscopy and Applications*, Wiley, New York.
12. Sankawa, U., Ebizuka, Y., Noguchi, H. *et al.* (1983) *Tetrahedron*, **39**, 3583–91.
13. Römer, A. and Herbert, R.B. (1982) *Z. Naturforsch.*, **37c**, 1070–4.
14. Gudgeon, J.A., Holker, J.S.E. and Simpson, T.J. (1974) *J. Chem. Soc. Chem. Comm.*, 636–8.
15. Casey, M.L., Paulick, R.C. and Whitlock, jun., H.W. (1976) *J. Amer. Chem. Soc.*, **98**, 2636–40.
16. Leete, E. (1987) *Can. J. Chem.*, **65**, 226–8.
17. Gould, S.J. and Cane, D.E. (1982) *J. Amer. Chem. Soc.*, **104**, 343–6.
18. Battersby, A.R. (1987) *Nat. Prod. Rep.*, **4**, 77–87.
19. Battersby, A.R. and McDonald, E. (1979) *Acc. Chem. Res.*, **12**, 14–22.
20. Leete, E. (1982) *J. Nat. Prod.*, **45**, 197–205.
21. Hedges, S.H., Herbert, R.B. and Wormald, P.C. (1983) *J. Chem. Soc. Chem. Comm.*, 145–7.
22. Erickson, W.R. and Gould, S.J. (1987) *J. Amer. Chem. Soc.*, **109**, 620–1.
23. Levy, G.C. and Lichter, R.L. (1979) *Nitrogen-15 Nuclear Magnetic Resonance Spectroscopy*, Wiley, New York.
24. Gould, S.J., Chang, C.-C., Darling, D.S. *et al.* (1980) *J. Amer. Chem. Soc.*, **102**, 1707–12.
25. Demetriadou, A.K., Laue, E.D. and Staunton, J. (1985) *J. Chem. Soc. Chem. Comm.*, 764–6.
26. Popjak, G. and Cornforth, J.W. (1966) *Biochem. J.*, **101**, 553–68.
27. Etherington, T., Herbert, R.B., Holliman, F.G. and Sheridan, J.B. (1979) *J. Chem. Soc. Perkin 1*, 2416–9.
28. Herbert, R.B. and Mann, J. (1984) *Tetrahedron Lett.*, **25**, 4263–6.
29. Brereton, R.G., Garson M.J. and Staunton, J. (1984) *J. Chem. Soc. Perkin 1*, 1027–33.
30. Abell, C. and Staunton, J. (1981) *J. Chem. Soc. Chem. Comm.*, 856–8.
31. Simpson, T.J. and Stenzel, D.J. (1982) *J. Chem. Soc. Chem. Comm.*, 1074–6.
32. Butcher, D.N. and Ingram, D.S. (1976) *Plant Tissue Culture*, Arnold, London.
33. Chapman, D.I. (1979) *Chem. in Brit.*, **15**, 439–47.
34. Ohnuki, T., Imanaka, T. and Aiba, S. (1985) *Antimicrobial Agents Chemotherapy*, **27**, 367–74.
35. Itoh, S., Odakura, Y., Kase, H. *et al.* (1984) *J. Antibiot.*, **37**, 1664–9; and following paper.
36. Goda, S.K. and Akhtar, M. (1987) *J. Chem. Soc. Chem. Comm.*, 12–14; and refs. cited.

# 3 *Polyketides*

## 3.1 INTRODUCTION

Acetic acid is found in living systems as its coenzyme A ester (*3.1*). This is a reactive thioester and is a pivotal compound in biosynthesis. On the one hand, it is involved in the formation of the important long-chain fatty acids and their transformation products (section 1.1.2). On the other hand, (*3.1*) is

*(3.1)* Acetyl ¦ coenzyme A

the source of small fragments in the biosynthesis of numerous secondary metabolites whose origins chiefly lie elsewhere. There is, though, a large group of metabolites whose structures are legion [5] which is simply based, like the fatty acids, on the almost exclusive use of linear chains of repeating acetate units, generically the polyketides. This group of metabolites is the subject of discussion in this chapter but examples are to be found too in Chapters 5, 6 and 7. It should also be noted that the terpenes and steroids formed by repeating $C_5$ isoprene units derive ultimately from acetyl-coenzyme A (*3.1*) (Chapter 4). It seems that in nature, little opportunity has been missed to use acetyl-coenzyme A to construct the backbone of many secondary metabolites or to embellish metabolites formed principally from other sources.

For those engaged in the structure determination of natural products there is a security to be found in structures which can be correlated with simple repeating units. So the repeating isoprene unit found in the terpenes (Chapter 4) has been invaluable in structure assignment. At the turn of the century Collie [6, 7] recognized that many natural products contained within them the $[CH_2\text{-}CO]_n$ unit and that this could be exploited in chemical

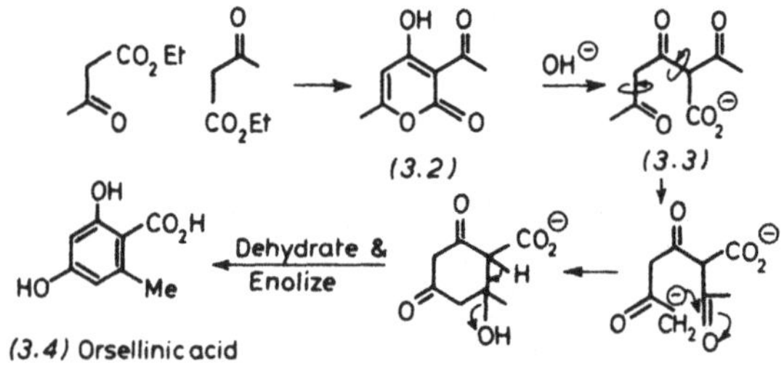

**Scheme 3.1**

synthesis, as in the conversion of dehydroacetic acid (3.2) via (3.3) into orsellinic acid (3.4), a naturally occurring polyketide (Scheme 3.1). These ideas lay dormant until the middle of this century when Birch independently hypothesized that the biosynthesis of numerous secondary metabolites could be correlated with repeating $CH_2$-CO units based on acetate [8, 9]. Thus orsellinic acid (3.4) can be thought of as arising through the poly-$\beta$-keto-acyl-CoA derivative (3.5) [and not (3.3) which has an irregular $[CH_2\text{-}CO]_n$ sequence] derived by linear combination of acetate units (Scheme 3.2). The same triketo-intermediate (3.5) could lead to a number of other $C_8$ metabolites. Detailed discussion of the biosynthesis of these compounds is taken up in section 3.3. It may be noted at this point though that the appropriate [1-$^{14}$C]acetate labelling of (3.4) (Scheme 3.2) and many other metabolites has been observed, thus validating the hypothesis [1].

Comparison of the illustrated pathway to (3.5) (Scheme 3.2; cf Scheme 3.4) with that of fatty acid biosynthesis (section 1.1.2; Scheme 1.2) indicates that a clear distinction between them is achieved by the absence of a reductase in polyketide formation, i.e. oxygen atoms of acetate are retained in polyketide metabolites. In the case of orsellinic acid (3.4), in particular, results of experiments with [$^{18}$O]acetate established this to be correct [10]. It should be noted, however, that universal retention of acetate oxygen through the poly-$\beta$-keto-acyl-CoA to the final metabolite is not observed. One is missing, for example, in 6-methylsalicyclic acid (3.14) and several in curvularin (3.109). (See, especially, sections 3.8 and 3.9 for what appear to be polyketides formed with varying levels of oxidation and dehydration corresponding to the steps illustrated in Scheme 1.2, section 1.1.2.)

The Birch hypothesis initiated many studies on polyketide biosynthesis

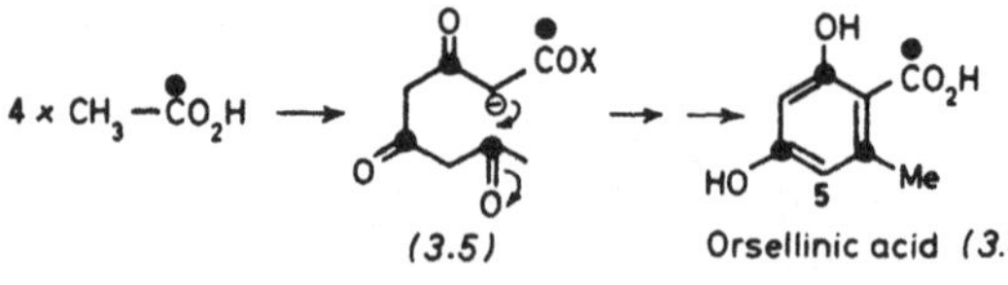

**Scheme 3.2**

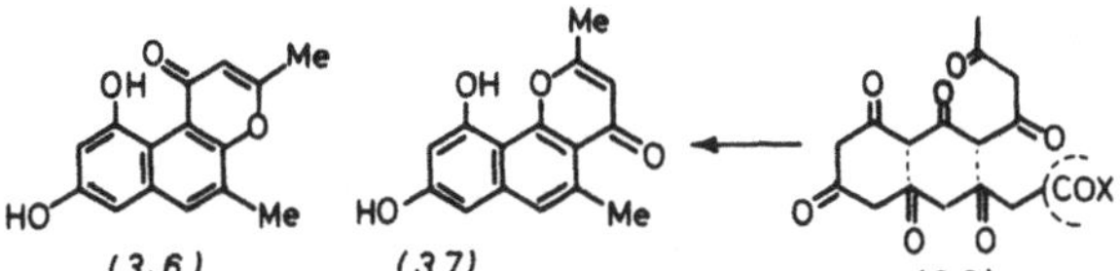

using [$^{14}$C]acetate, studies which have been extended in the impressive and powerful use of acetate labelled with $^{13}$C [11] and also $^{18}$O and $^{2}$H. A second important consequence of the hypothesis has been in structure determination of natural products, thus returning to the original ideas of Collie. For example, at an early stage, the hypothesis indicated (correctly) that the structure of eleutherinol is (3.7) rather than (3.6) as originally thought [(3.7) may arise simply from the poly-β-keto-acyl-CoA (3.8); derivation of (3.6) would be more complex].

## 3.2 FORMATION OF POLY-β-KETO-ACYL-CoA'S

### 3.2.1 Acetate and malonate [2, 12, 13]

Formation of a poly-β-keto-acyl-CoA [as (3.9)] occurs as for fatty acid biosynthesis by condensation of acetyl-coenzyme A with malonyl-CoA. Malonyl-CoA is generally derived by carboxylation of (3.1) (Scheme 3.3). An alternative path to malonyl-CoA is via oxaloacetate, an intermediate in the citric acid cycle.

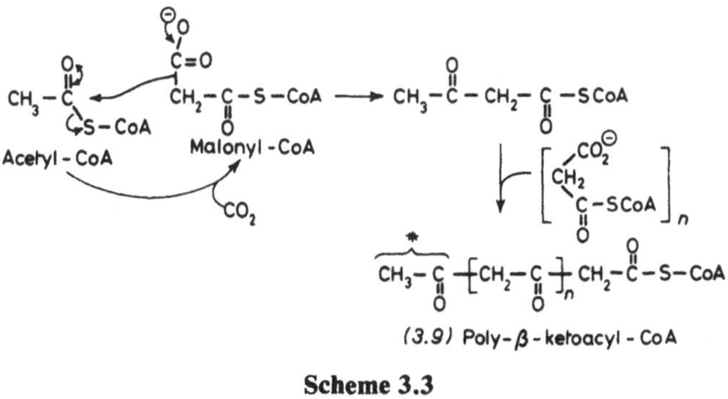

(3.9) Poly-β-ketoacyl - CoA

**Scheme 3.3**

Incorporation of [1-$^{14}$C]acetate into a polyketide metabolite gives labelling throughout the chain, of alternate carbon atoms, which are of course the normally oxygenated ones. Labelled malonate, on the other hand, tends to label the second and successive acetate-derived units more heavily than the first [*unit in (3.9)] which derives directly from acetate and not malonate. Because of this, the first acetate unit, called 'starter' acetate, tends to be more heavily labelled by labelled acetate than the other units. This differential labelling may be used to detect starter units, i.e. the acetate unit at the beginning of a polyketide chain. Incorporation of up to three deuterium atoms from $[^2H_3]$acetate on a methyl group in a metabolite similarly indicates starter acetate; malonate units can only have a maximum of two deuterium atoms.

Assimilation of $[1-^{14}C]$acetate into an organism may be followed by direct conversion into a polyketide metabolite, or the labelled acid may be turned through the citric acid cycle before use in polyketide biosynthesis. The result is, however, that $[1-^{14}C]$acetate is again generated. On the other hand, $[2-^{14}C]$acetate which enters the citric acid cycle exits with some label on C-1 in addition to that on C-2. Thus, $[2-^{14}C]$acetate can lead to a polyketide with label distributed over all its carbon atoms.

Some polyketides include extra carbon atoms as $[CHRCO]_n$, R = H and Me. The extra $C_1$ units may arise from *S*-adenosylmethionine with methylation of an intermediate enol [as *(3.25)* → *(3.26)*]. Alternatively introduction is by substitution in biosynthesis of propionyl-CoA *(3.10)* and methylmalonyl-CoA *(3.11)* for, respectively, acetyl-CoA and malonyl-CoA. Although *(3.11)* may derive by carboxylation of *(3.10)* as for the lower homologues, it is believed that normal biosynthesis is the reverse: *(3.10)* is formed by decarboxylation of methylmalonyl-CoA *(3.11)*. This compound, *(3.11)*, is derived from the citric acid-cycle intermediate, succinyl-CoA in a $B_{12}$ mediated reaction. For examples on the utilization of *(3.10)* and *(3.11)* in

$$CH_3-CH_2-\underset{\underset{O}{\|}}{C}-S-CoA \qquad CH_3-\underset{\underset{CO_2H}{|}}{CH}-\underset{\underset{O}{\|}}{C}-S-CoA$$

*(3.10)* Propionyl-CoA          *(3.11)* Methylmalonyl-CoA

biosynthesis see sections 3.8, 3.9 and 7.6.2. The use of a malonamyl residue [see *(3.123)*] as a starter unit in tetracycline biosynthesis (section 3.8) is to be noted as is the important one of f-coumaryl-CoA in flavone biosynthesis (section 5.4).

### 3.2.2 Assembly of poly-β-keto-acyl-CoA's

Formation of a polyketide involves condensation of acetyl-CoA with the appropriate number of malonyl-CoA units, modification of the completed poly-β-ketone where required, and release of the product in stable form, as, for example, in 6-methylsalicylic acid biosynthesis (Scheme 3.4). The whole sequence occurs enzyme-bound, without release, or acceptance, of intermediates externally. Thus probing of the biosynthetic sequence by normal feeding experiments with possible intermediates fails. Instead, very properly, information is gained in different ways by working with the actual enzymes involved. The biosynthesis of several polyketides has been explored in this way, and most extensively that of 6-methylsalicylic acid *(3.14)* [14–17].

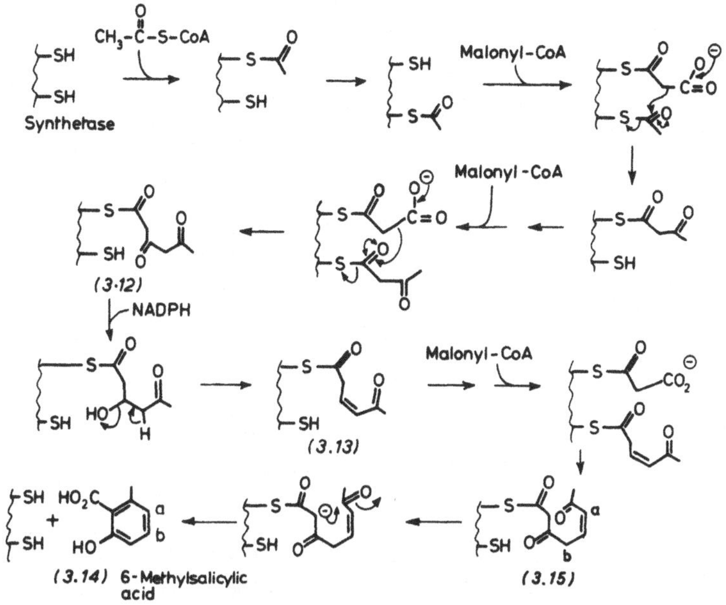

CH₃–C–S–CoA

Synthetase

Malonyl–CoA

Malonyl–CoA

(3·12)

NADPH

Malonyl–CoA

(3.13)

Malonyl–CoA

(3.14) 6-Methylsalicylic acid

(3.15)

**Scheme 3.4**

Extensive purification of 6-methylsalicylate synthetase from *Penicillium patulum* has been carried out. It is distinct from fatty acid synthetase, separable from it, and of half the molecular weight; both enzymes are complexes of several enzymes. One molecule of 6-methylsalicylic acid is generated by 6-methylsalicylate synthetase from one molecule of acetyl-CoA and three of malonyl-CoA in the presence of one molecule of NADPH as coenzyme; no free intermediates can be detected. In the absence of NADPH, triacetic acid lactone (*3.16*) is formed as the sole product. The same lactone is formed when coenzyme is omitted from reactions of fatty acid synthetases derived from a variety of sources. So there are similarities in the way the two enzymes function. There are differences, however. Most obviously in 6-methylsalicylate formation reduction occurs only once (with eventual formation of a *cis*-double bond) and at a late stage, whereas for fatty acids reduction occurs after each malonate addition (cf Scheme 1.2; section 1.1.2). Also, it has been observed that the rate of (*3.16*) formation is higher with 6-methylsalicylate synthetase than with fatty acid synthetase, both isolated from *P. patulum*.

The similarities in the function of the two enzymes and of course in the substrates has suggested that the mechanisms of enzyme action are similar. Also, the 6-methylsalicylate synthetase has two different types of thiol grouping (deduced from inhibition studies) which are analogous to the ACP carrier and condensing sites in fatty acid synthetase. A possible mechanism of (*3.14*) formation is illustrated in Scheme 3.4. The different receptors for the

(3.16)    (3.17) Atranorin     (3.18) R=H   (3.20) Gliorosein   (3.21) Barnol   (3.22) R=H
                               (3.19) R=Me                                       (3.23) R=

malonate electron pair through the sequence contrasts with a uniformity of receptor type in fatty acid biosynthesis (see Scheme 1.2, section 1.1.2). The *cis*-double bond in (*3.13*) can be thought of as conferring on the growing chain the requisite conformation for eventual cyclization. Where reduction does not occur, as in orsellinic acid (*3.4*) synthesis, conformational rigidity can be achieved via the enol forms of the keto-groups. In some cases *C*-methyl groups may also affect the conformation of the folded chain (section 3.9).

> The incorporation of $^2$H-labelled acetate has been used to test these ideas in the biosynthesis of metabolites of the (*3.14*)-type (double bond formation as in Scheme 3.4) and those of the (*3.4*)-type (double bonds by enolization only) [18]. Double-bond formation by enolization is expected to be reversible. Thus deuterium at e.g. b in (*3.15*) could be washed out by exchange by ketone–enol interconversion. Double bond formation by reduction and elimination, on the other hand, is irreversible and so should result in higher retention of deuterium, e.g. a in (*3.15*) relative to b. The appropriate pattern was observed for several examples of both the (*3.4*)- and (*3.14*)-type thus providing support for biosynthesis as summarized in Scheme 3.4. Mono- and tri-deuteriated acetate incorporation into 6-methylsalicylic acid has also provided evidence for the steps associated with aromatization leading to (*3.14*) being stereospecific and therefore under enzymic control (no isotope effect was observed in the labelling of sites a and b in (*3.14*) [19]).

*C*-Methylation occurs before release of the completed polyketide from the synthetase complex. The evidence for this conclusion comes in part from the failure of unmethylated compounds to serve as precursors, e.g. orsellenic acid (*3.4*) is not a precursor for atranorin (*3.17*) [20]. Similarly 5-methylorcylaldehyde (*3.19*), and not orcylaldehyde (*3.18*), is a precursor for gliorosein (*3.20*) and related compounds [21]. Barnol (*3.21*) is interesting in that one methyl group derives from methionine (*) and the other by reduction of a carboxy-group. The ethyl group derives from C-2 of acetate and methionine. [The polyketide skeleton, as (*3.5*), is revealed by the dotted lines in (*3.21*), see also (*3.22*)]. The results of feeding experiments with various possible precursors [e.g. orsellinic acid (*3.4*) and 5-methylorsellinic acid (*3.27*), neither of which were incorporated into (*3.21*)] provides persuasive evidence for both methylation steps occurring before aromatization of the intermediate (3.22) [22]. Apparent methylation of a methyl rather than a methylene group is most unusual and it may be that the actual intermediate for methylation is (3.23) (both methylations are then normal in occurring on methylene groups) which becomes truncated *en route* to barnol (3.21).

Incorporation of [$^2$H$_3$]methionine into gliorosein (3.20) occurs with retention of all the deuterium atoms [23] arguing for methylation by a simple nucleophilic displacement on *S*-adenosylmethionine [see (3.25)].

*Aspergillus flaviceps* normally produces 5-methylorsellinic acid (3.27) and its transformation product flavipin (3.28). A cell-free enzyme preparation used [$^{14}$CH$_3$]methionine for formation of the aromatic metabolite [in this case, 5,6-dimethylresorcinol, as (3.27)] without addition of any poly-β-keto-acyl-CoA acceptor. So this substrate was already enzyme bound. Various products derivable from (3.24) were obtained on hydrolysis of the enzyme complex thus indicating (3.24) to be the substrate for methylation (Scheme 3.5) [24]. This then provides further evidence for methylation occurring before release of the completed polyketide from the synthetase complex.

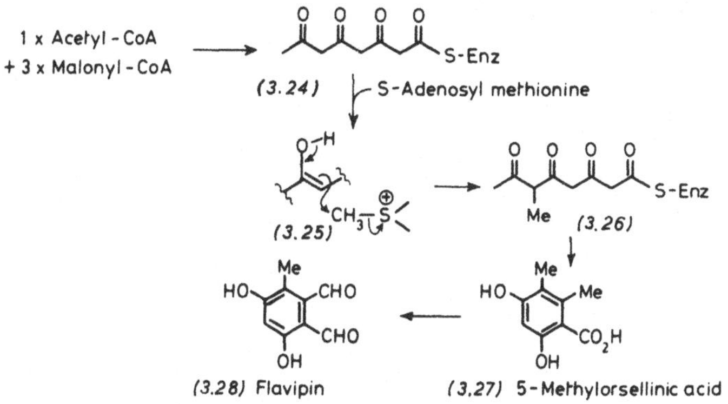

**Scheme 3.5**

Prenylation in contrast to methylation occurs on the polyketide after liberation from the enzyme (see mycophenolic acid, section 4.6). A normal prenylation occurs here, and in the formation of (3.29), *ortho* to a phenolic hydroxy-group, see Scheme 3.6; siccanin (3.30) derives from (3.29) [25].

Commonly for polyketides, $n = 4$ to 10 in the general formulation [CH$_2$-CO]$_n$ but may be as high as 19 to 20 in macrolide antibiotics (sections 3.9 and 7.6.2). Numerous examples are known where $n = 4$, 5, 7 and 8 while those where $n = 3$, 6, 9 and 10 are less common. Cyclization of a poly-β-keto-acyl-CoA chain to give a six-membered ring can take a number of courses. Almost all the possible cyclization modes are represented in naturally occurring polyketides with, apparently, a single restriction.

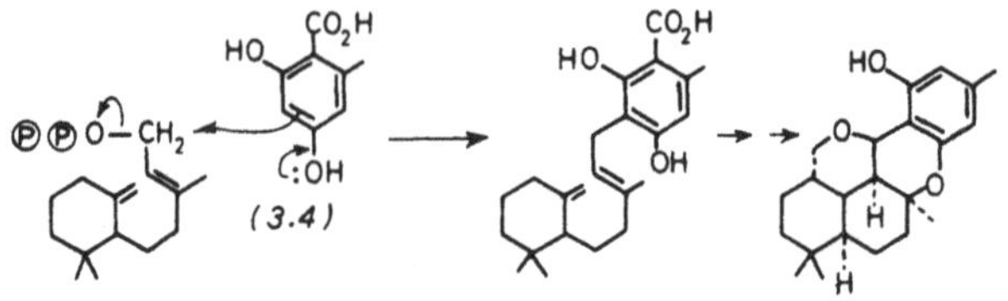

**Scheme 3.6**

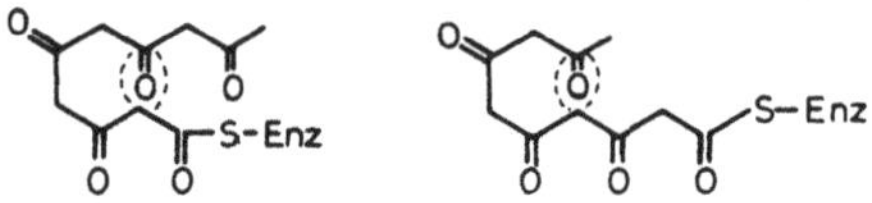

Uncyclized residues from the methyl end of a polyketide are never shorter than the residues bearing the carboxy-group. Thus cyclization of (*3.31*) may occur as shown but not as in (*3.32*). This clearly relates in some (unknown) way to the mechanism of ring closure. It is also observed that whatever oxygens are deleted in a polyketide the one next to a carboxy-group as in (*3.14*), is always retained. It may be concluded that the anion (enol) used in cyclization $\alpha$ to a thioester function requires stabilization by both the ketonic and thioester carbonyl groups.

## 3.3 TETRAKETIDES

The biosynthesis of some examples of this numerous group of polyketides is summarized in Scheme 3.7. Note that the single and particular labelling site in fumigatin (*3.36*) derived from orsellinic acid indicates the orientation of the original poly-$\beta$-keto-acyl-CoA and also that the carboxy-group is lost after introduction of the C-5 hydroxy-group; otherwise a symmetrical intermediate [(*3.43*); labelling as shown] and dilabelled (*3.36*) would have been generated. The formation of usnic acid (*3.40*) is by oxidative coupling (Section 1.3.1) of two molecules of methylphloroacetophenone (*3.39*). Further examples in tetraketide biosynthesis are (*3.14*), (*3.17*), (*3.20*), (*3.21*), (*3.28*), and (*3.30*); see also below.

Tetraketides are formed from four acetate/malonate units. This is the minimum number for formation of a common stable aromatic ring. As already seen, *inter alia*, in orsellinic acid (*3.4*) the oxygenation pattern is an alternating one. Aromatic rings formed from phenylalanine and tyrosine (Chapter 5) have a different pattern, as seen in (*3.44*). [A particularly good example, where one aromatic ring derives from acetate/malonate and a second from phenylalanine, is found in the flavonoids (section 5.4).] At its simplest then one can distinguish visually between these two alternative

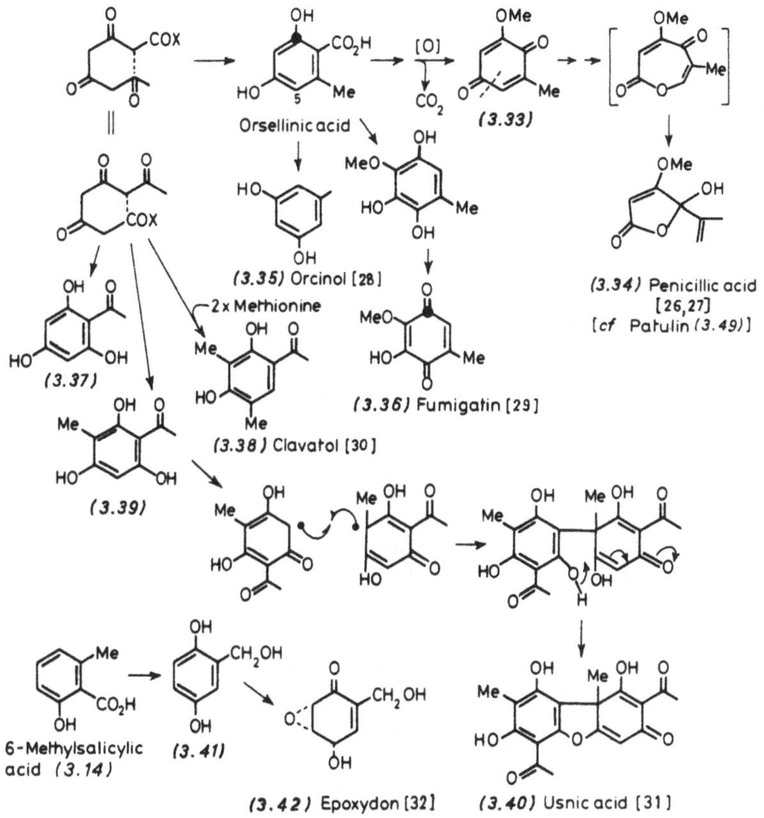

**Scheme 3.7**

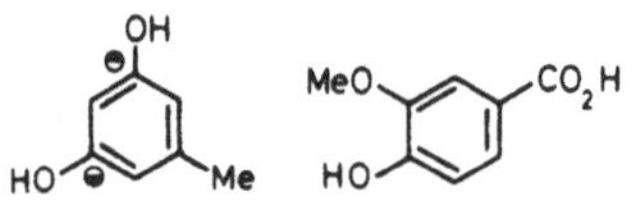

routes to aromatic rings. Further aromatic oxygenation of course blurs the visual distinction, see (3.21), as does the loss of polyketide oxygen during biosynthesis.

The biosynthesis of the aromatic metabolite, 6-methylsalicylic acid (3.14), has been discussed above. This acid is the source of a variety of metabolites in which hydroxylation and oxidation of the aromatic nucleus and side-chain methyl group occur in major pathways [4] following decarboxylation to *m*-cresol (3.45). A terminus in this set of oxidative reactions is patulin (3.49) (Scheme 3.8).

The [1-$^{14}$C]acetate labelling in patulin is shown (●). No regular polyketide

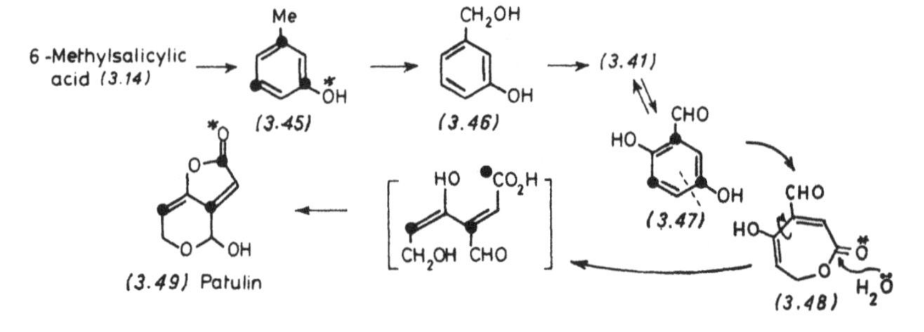

**Scheme 3.8**

chain is apparent but the incorporation of radioactive 6-methylsalicylic acid (3.14) and other related compounds indicate that (3.49) arises by cleavage of an aromatic ring [see dotted line in (3.47)].

Interesting experiments relate to the use of $^2$H and $^{18}$O as probes for the mechanism of patulin formation. m-[$^2$H]Cresol (3.50) gave dideuterio-patulin, the mass spectral fragmentation of which was consistent with location of deuterium as in (3.51), and the cleavage as in Scheme 3.8 [33]. The origin of the oxygen atoms in (3.49) was established in experiments with $^{18}$O$_2$ gas and [1-$^{13}$C, $^{18}$O$_2$]acetate [34]. From the $^{18}$O isotope-induced shifts observed in the $^{13}$C n.m.r. spectra (see discussion in section 2.2.2) of the derived patulin (3.49) it was clear that the oxygen of the carbonyl group (*) is derived from acetate whilst the remainder arise by oxidative processes from aerial oxygen. Cleavage of the aromatic ring in (3.47) is probably preceded by formation of an arene oxide leading to (3.48).

Enzyme preparations from *P. patulum* have been obtained which have been studied for their substrate specificity and products formed. For example, one preparation converted m-hydroxybenzyl alcohol (3.46) into gentisyl alcohol (3.41) but the corresponding aldehyde (3.52) was not ring-hydroxylated by any preparation from *P. patulum*. Thus (3.46) but not (3.52) is on the pathway to patulin (3.49). Additional evidence has been obtained in studies with mutants. All the various strands of evidence lead to the pathway which is summarized in Scheme 3.8 [34].

A different cleavage of an aromatic ring is apparent in penicillic acid (3.34) formation. Penicillic acid has its genesis via orsellinic acid (3.53), and labelling of (3.34) = (3.54) by [1,2-$^{13}$C$_2$]acetate, see (3.53) → (3.54), allows

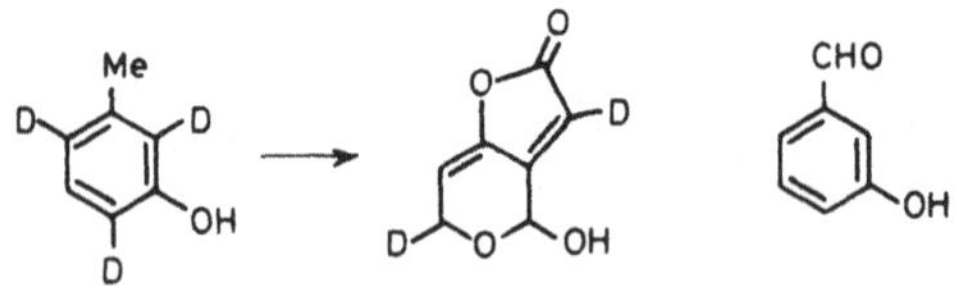

(3.53) (3.54) Penicillic acid

■ = fragmented acetate

clear definition of the site of aromatic ring cleavage as that shown [dotted line in (3.33) and (3.53)]. Results of precursor feeding experiments indicate the sequence of biosynthesis which is abbreviated in Scheme 3.7. An enzyme preparation which effects the final oxidative ring opening has been isolated; it is a mono-oxygenase [26, 27]. Similar enzymes, working on quinones [see (3.33)] as substrates for aromatic ring opening, may be involved in the biosynthesis of multicolic acid and the aflatoxins (which are discussed below).

Astepyrone (3.55) is formed, like penicillic acid, through similar cleavage of the aromatic ring in the acid (3.53) (see dotted line) or the corresponding aldehyde. In this case the carboxyl group is retained to become the carbonyl group in (3.55) [35]. Botryodiplodin (3.56) is again biosynthesized via orsellinic acid (3.53) but now cleavage occurs between C-3 and C-4 instead of C-4 and C-5; C-4 is lost at some stage [36].

(3.55) Astepyrone (3.56) Botryodiplodin

A mono-oxygenase is implicated in the biosynthesis of fungal tropolones, e.g. stipitatonic acid (3.58). Ring expansion of an aromatic ring could, as in colchicine biosynthesis, account for the formation of the tropolone ring (section 6.3.7). This is strongly supported by the appropriate specific incorporation of labelled 3-methylorsellinic acid (3.57) into stipitatonic acid (3.58), and $^{18}O_2$ labelling results [see (3.58)] [37, 38]. Another tropolone-containing metabolite, sepedonin (3.60), derives from a pentaketide with inclusion of a $C_1$ unit [*site in (3.60)]. By analogy with stipitatonic acid biosynthesis (3.59) is a likely intermediate [39, 40].

(3.57) (3.58) Stipitatonic acid

Colletodiol (3.64) is a simple non-aromatic polyketide. [$^{13}C$]Acetate labelling is consistent with its formation by union of a tetraketide with a triketide. The origin of all of the oxygen and hydrogen atoms in (3.64) has

(3.59)    (3.60) Sepedonin

been established. From the results it is possible to build up a detailed picture of the biosynthesis of this metabolite. The compound (*3.61*) was proposed as a common intermediate for both halves of (*3.64*); dehydration gives (*3.62*) and further elaboration with the addition of a further C$_2$ unit affords (*3.63*). Epoxidation of one of the double bonds followed by ring opening (see *3.65*), either before or after lactonization, affords (*3.64*) [41].

The triketide moiety found in colletodiol is rarely found naturally. One of these rare triketides is furanomycin (*3.66*) which is constructed from one molecule of propionate and two of acetate [42].

The fungal α-pyrones rosellisin (*3.67*) [43] and coarctatin (*3.68*) [44] are tetraketides each formed with the inclusion of two carbons into the skeleton which are derived from methionine (*).

## 3.4 PENTAKETIDES

Pentaketides are formed from five acetate/malonate units. A regular pattern formed by a single polyketide chain is to be seen in e.g. mellein (*3.69*) [45], also ochratoxin A which is of similar structure [46, 47]. Commonly compounds of type (*3.59*) are produced which are further modified as for the tetraketides in ways which include aromatic ring scission. A simple modifica-

(3.61)    (3.62)    (3.63)    (3.64) Colletodiol

(3.65)    (3.66)    (3.67) Rosellisin

(3.68) Coarctatin    Me—CO$_2$H ⟶    (3.69) Mellein

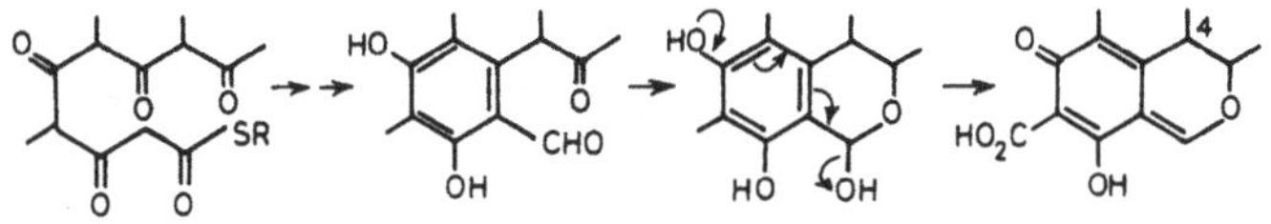

Scheme 3.9

tion is in the formation of (dihydro)isocoumarins [as (3.72)]. Citrinin (3.71) is derived by way of (3.70), i.e. from five acetate units with three $C_1$ units provided by methionine. Extensive labelling studies delineate the course of biosynthesis with clarity (Scheme 3.9). Apart from the carboxyl group in (3.71), which is formed by oxidation of a methyl group, all the oxygen atoms originate in acetate. The incorporation of deuterium from deuteriated acetate at C-4 in (3.71) excludes biosynthesis via an isocoumarin [as (3.72) $\Delta^{3,4}$] [48, 49]; austdiol has a structure similar to citrinin and its biosynthesis appears to be similar [50]. A hexaketide aschochitin is structurally related to (3.71) and it has a similar biosynthesis. Good evidence has been obtained in this case for the direct conversion of a thioester to an aldehyde (cf Scheme 3.9) without the intermediacy of the corresponding acid [51].

The dihydroisocoumarin (3.72) is produced along with terrein (3.73) by some strains of *Aspergillus terreus*; and (3.72) is a precursor for (3.73). Results of experiments with [$^{13}C_2$]acetate indicate that formation of (3.73) occurs by fracture of (3.72) in the manner shown (Scheme 3.10 path a; thickened bonds: intact acetate units) [52]. A compound, (3.74), with related structure is thought to be formed in a slightly different manner (Scheme 3.10, path b) [53]; see also [54].

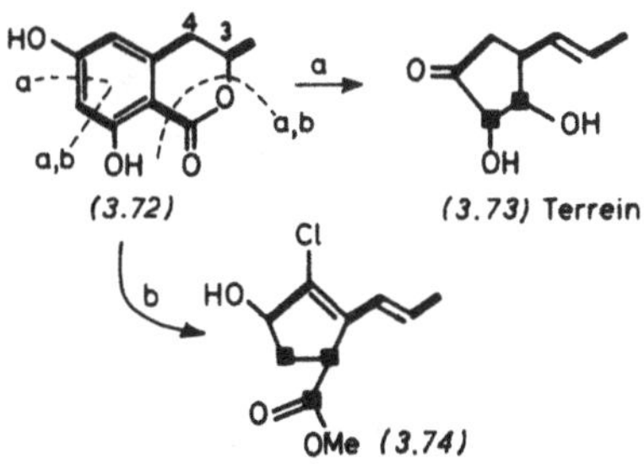

Scheme 3.10

The pattern of [$^{13}C$]-acetate and -methionine incorporation in sclerin (3.75) might be taken to indicate an unusual biosynthesis from two poly-ketide chains with the introduction of a $C_1$ unit on to a methyl rather than a methylene group (cf barnol, above). An alternative, more satisfying suggestion was that (3.75) arises by rearrangement of the cometabolite,

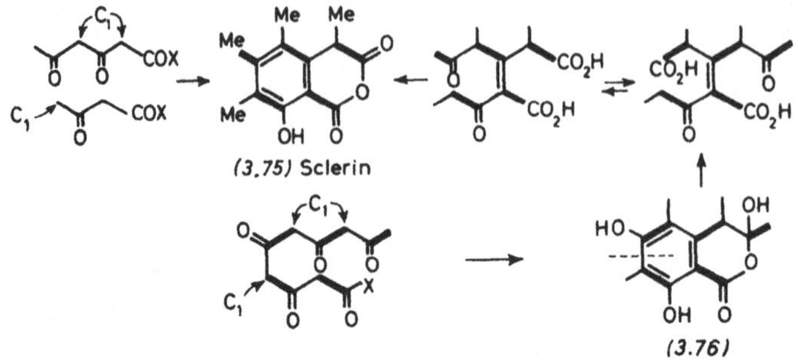

**Scheme 3.11**

sclerotinin A (*3.76*) itself formed from a single polyketide chain. The correctness of this pathway has been confirmed (Scheme 3.11) [55].

A further way in which a linear polyketide chain may be rearranged is found for aspyrone (*3.77*) and asperlactone (*3.78*) which are closely related metabolites of *Aspergillus melleus*. The results of applying the full battery of labelled acetate precursors and of feeding advanced precursors closely define the biosynthesis of these two metabolites (Scheme 3.12) [56].

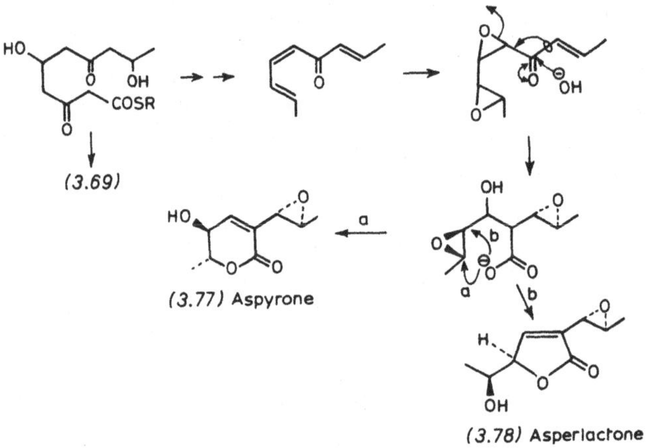

**Scheme 3.12**

[2-$^{13}$C]Acetate labelling [● in (*3.79*)] of asperlin (*3.79*) indicates a simple unfragmented polyketide chain [57].

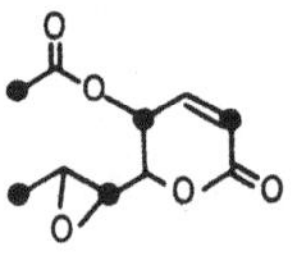

(*3.79*) Asperlin

*(3.80)*                    *(3.81)* Diplosporin

The mycotoxin diplosporin *(3.81)* is biosynthesized from five acetate units with the inclusion of two $C_1$ units (*) from methionine. Its biosynthesis may well proceed through *(3.80)*. Use of [*methyl*-$^2H_3$]methionine as a precursor gave *(3.81)* in which deuterium was present at C-2 and C-5 ($^2H$ n.m.r.), showing that, during biosynthesis, oxidation of these methionine-derived units does not go beyond the aldehyde level of oxidation. In the $^{13}C$ n.m.r. spectrum of diplosporin formed from [1-$^{13}C$, $^2H_3$]acetate the resonance for C-11 showed $\beta$-$^2H$ isotope shifts (see section 2.2.2) indicating that 1 to 3 deuterium atoms were present at C-12. This confirms that C-11 plus C-12 constitute the starter acetate [58].

## 3.5 HEXAKETIDES

Relatively few hexaketides have been isolated. Variotin *(3.82)* is an example. It derives in the expected manner from acetate (malonate) and methionine with the pyrrolidone ring having its origins in glutamic acid [59, 60]. Labelled acetate and formate (source via methionine for methyl groups) were incorporated into alternaric acid *(3.83)* in a manner consistent with condensation of a hexaketide-derived acyl derivative with dihydrotriacetic acid lactone [see dotted line in *(3.83)*] [61]. Rubropunctatin *(3.84)* can be split into two fragments on the basis of labelling results (dotted line). The right-hand part derives in the usual way; the left-hand part though seems to derive through the $\beta$-keto acid formed by condensation of hexanoic acid with an acetate unit [62, 63].

Radicin *(3.86)* is particularly notable for being the first compound to have had its biosynthesis studied by direct $^{13}C$ n.m.r. methods. The incorporation of $^{13}C$-labelled acetate into deoxyradicin *(3.85)* has also

*(3.82)* Variotin                    *(3.83)* Alternaric acid

*(3.84)* Rubropunctatin

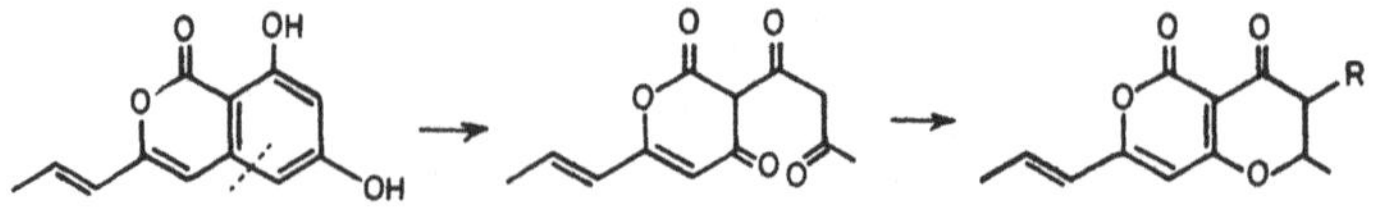

(3.85) R = H
(3.86)· R = OH, Radicin

**Scheme 3.13**

been studied. These metabolites are labelled in such a way as to suggest that they are formed from two polyketide chains; alternatively they may be formed by a pathway involving ring-cleavage as shown in Scheme 3.13 [64].

O-Methylasparvenone (3.88) is apparently formed by way of the naphthalene (3.87). Results with deuteriated acetic acid indicate that the carbonyl group at what becomes C-9 in (3.88) survives until a late stage of biosynthesis and that the proton expected at C-4 appears instead at C-3. This latter observation indicates that the C-4 hydroxyl group originates through formation of an arene oxide at C-6 and -7 in (3.87) with subsequent epoxide ring opening being accompanied by a 1,2-NIH proton shift section 1.3.2) from C-4 to C-3 [65]. (This reference includes a useful discussion of $\beta$-isotopic shifts due to deuterium; section 2.2.2.)

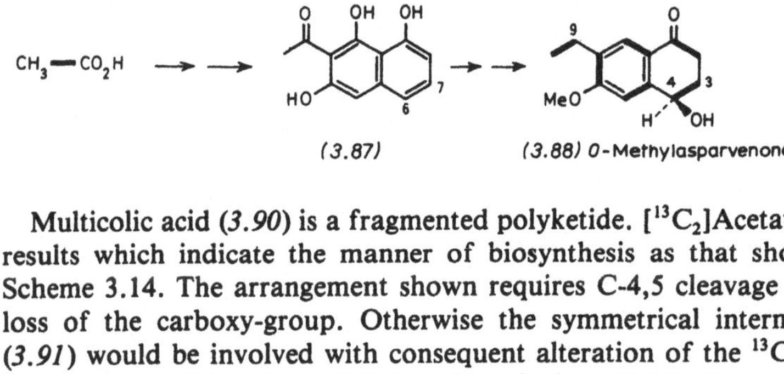

(3.87)          (3.88) 0-Methylasparvenone

Multicolic acid (3.90) is a fragmented polyketide. [$^{13}C_2$]Acetate gave results which indicate the manner of biosynthesis as that shown in Scheme 3.14. The arrangement shown requires C-4,5 cleavage before loss of the carboxy-group. Otherwise the symmetrical intermediate (3.91) would be involved with consequent alteration of the $^{13}C$ labelling pattern arising from C-4,5 and equivalent C-1,2 cleavage [66].

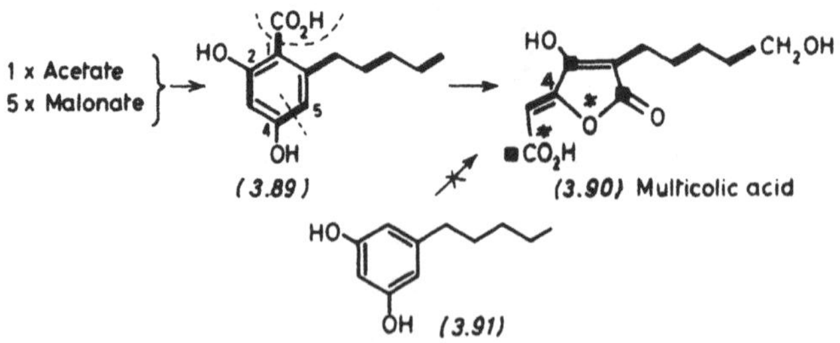

1 x Acetate
5 x Malonate

(3.89)          (3.90) Multicolic acid

(3.91)

**Scheme 3.14**

[1-$^{13}$C, $^{18}$O$_2$]Acetate labelled the oxygens indicated (*) in (*3.90*); the oxygen at C-4 remains attached to the $^{13}$C-label which appears at C-4 during the course of biosynthesis [67].

The chromanone LL-D253α (*3.93*) is biosynthesized from two poly-ketide chains (*3.92*). Results with $^{13}$C-, $^{2}$H- and $^{18}$O-labelled acetate indicate that (partial) rearrangement of the side chain occurs at some stage. This is suggested to occur as shown in Scheme 3.15 [68].

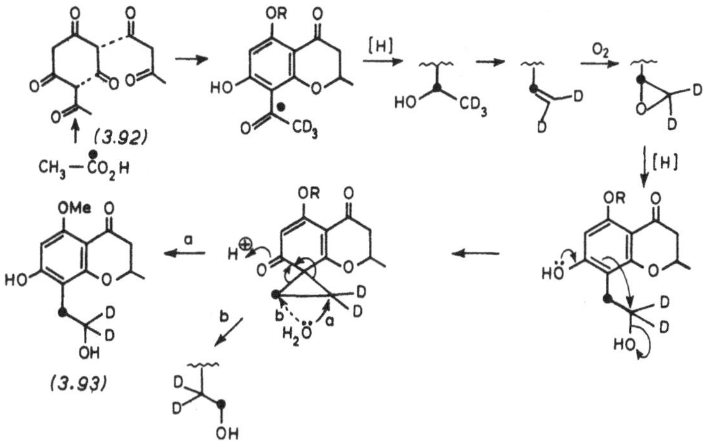

**Scheme 3.15**

## 3.6 HEPTAKETIDES

Proof that griseofulvin (*3.96*) is formed by linear combination of acetate units provided early key support for Birch's polyketide hypothesis. The benzophenone (*3.95*) is a probable intermediate. It can give griseofulvin by oxidative coupling (section 1.3.1) followed by saturation of one of the double-bonds in the resultant dienone. [2-$^{3}$H$_3$]Acetate labelled the positions indicated in (*3.96*) in accord with $^{14}$C data and the pathway shown (Scheme 3.16). Moreover, the results show that double-bond saturation in the

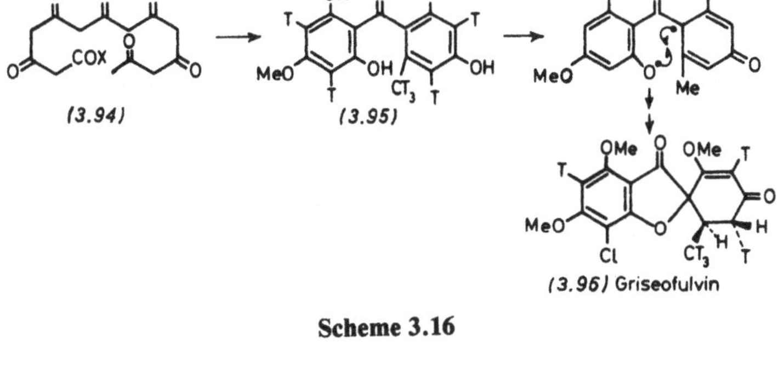

**Scheme 3.16**

intermediate dienone involves *trans* addition of hydrogen. Incorporation of [1-$^{13}$C, $^{18}$O$_2$]acetate into griseofulvin (*3.96*) and analysis by $^{13}$C n.m.r. spectroscopy gave results which prove that all the oxygen atoms in the metabolite are acetate-derived [69, 70].

Palitantin (*3.97*) is interesting because it contains a non-aromatic six-membered ring. Its biosynthesis has been explored with $^{18}$O-, $^2$H- and $^{13}$C-labelled acetate. The results are summarized in (*3.97*). In distinguishing between various mechanisms for biosynthesis the observed presence of two deuterium atoms at C-9 is crucial. Thus, for example, an aromatic intermediate is excluded. A mechanism consistent with the labelling results is given in Scheme 3.17 (C-12 was unexpectedly not labelled by $^{18}$O, but this is the consequence of ready exchange at this position) [71].

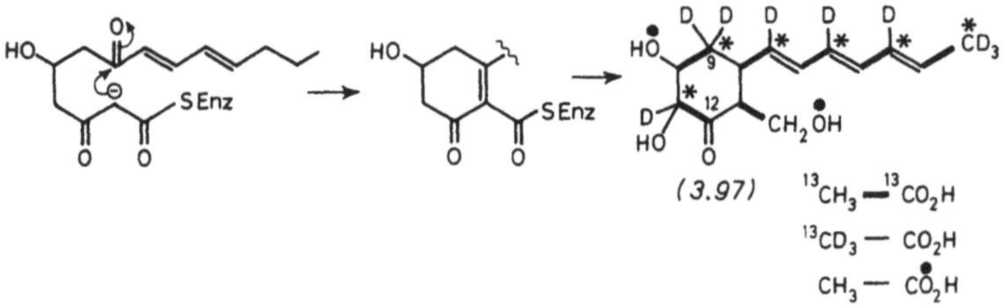

(*3.97*)   $^{13}$CH$_3$ — $^{13}$CO$_2$H

$^{13}$CD$_3$ — CO$_2$H

CH$_3$ — ĊO$_2$H

**Scheme 3.17**

As the poly-β-keto-acyl-CoA chain gets longer so the opportunity increases for widely differing cyclization reactions. A minor variation on the pattern in griseofulvin is seen in alternariol (*3.98*) formation from a single chain [72]. A quite different pattern is found in the fungal phenalenones, e.g. deoxyherquienone (*3.99*) (see dotted lines) [73]. Interestingly the plant phenalenones derive from phenylalanine, tyrosine and a single acetate (malonate) [74, 75].

Increasing complexity is seen in metabolites like cercosporin (*3.100*) which is formed by oxidative coupling of two heptaketide units [76].

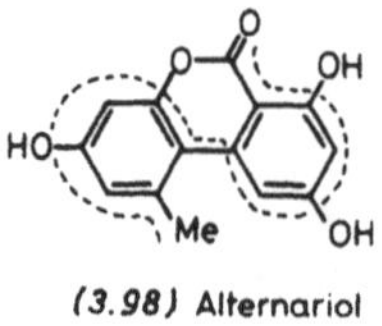

(*3.98*) Alternariol

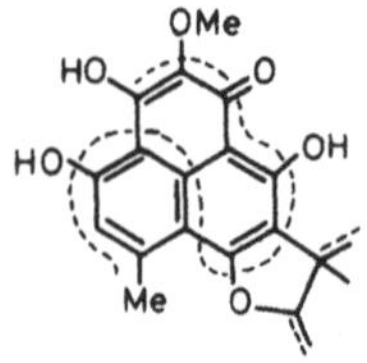

(*3.99*) Deoxyherquienone

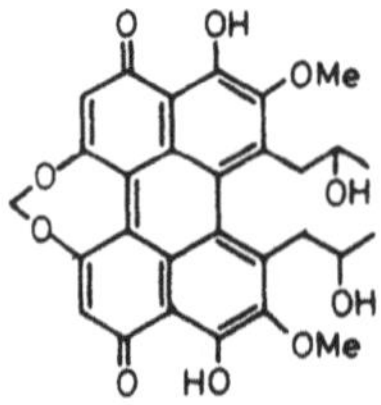

(*3.100*) Cercosporin

## 3.7 OCTAKETIDES

A common structural type is that of an anthraquinone, e.g. islandicin *(3.101)*. Ring cleavage leads on, for example, to xanthones, e.g. ravenilin *(3.103)*. The pattern of islandicin biosynthesis is indicated in Scheme 3.18 [77]. Islandicin *(3.101)* is plausibly an intermediate in ravenilin *(3.103)* biosynthesis; the [$^{13}C_2$]acetate labelling pattern is consistent with this. (The labelling pattern is a mixture of two arising as shown in Scheme 3.18 [78].) Supporting results have been obtained with [1-$^{13}C$, $^{18}O_2$]acetate [79]. The biosynthesis of sulochrin *(3.105)* has been proven to be analogously via the anthraquinone *(3.104)*, derived in turn from acetate/malonate [80].

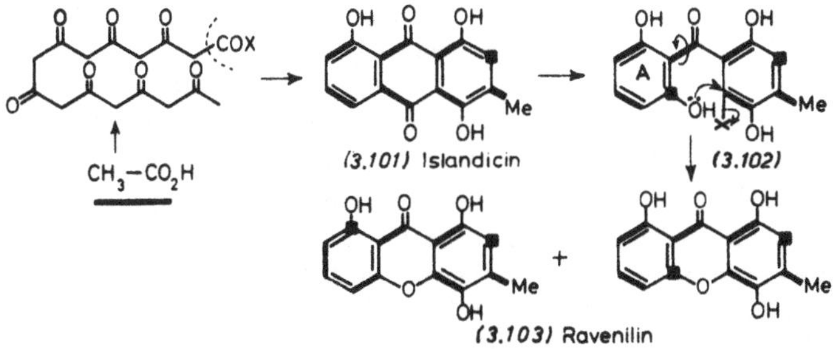

*(3.101)* Islandicin    *(3.102)*

*(3.103)* Ravenilin

**Scheme 3.18**

Tajixanthone *(3.106)* [81] and silvaticamide [82] have similar prenylated skeletons and are also derived by fragmentation of an anthraquinone. The former, like ravenilin, passes through a symmetrical intermediate for ring A [cf *(3.102)*] because two acetate patterns are observed for this ring in *(3.106)*. This means that prenylation at C-4 occurs after anthraquinone ring-cleavage. Because only one acetate pattern is observed for the corresponding ring in silvaticamide which also bears a prenyl substituent prenylation must occur before ring-cleavage.

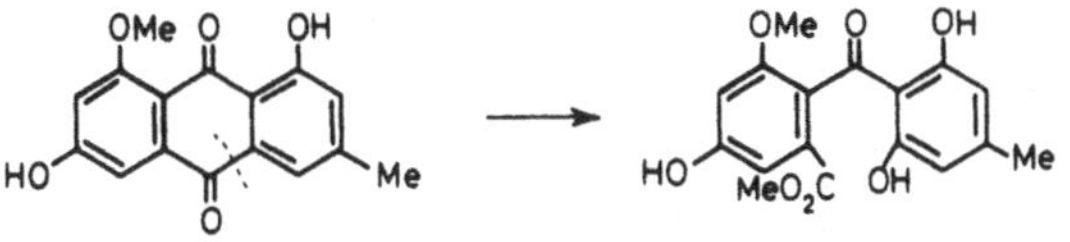

*(3.104)*     *(3.105)* Sulochrin

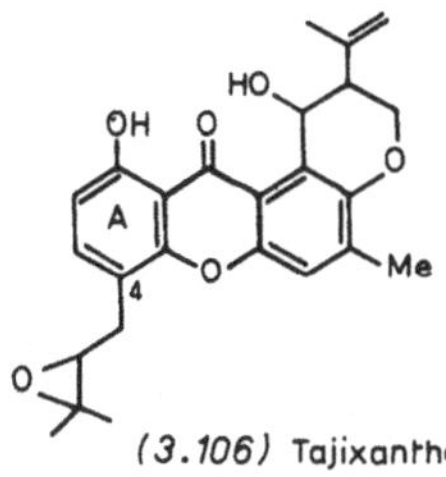

*(3.106)* Tajixanthone

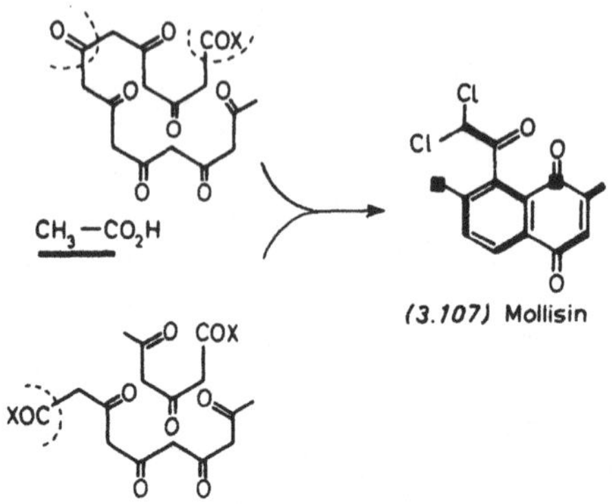

**Scheme 3.19**

Study of mollisin (*3.107*) biosynthesis provides a particularly good example where [$^{13}C_2$]acetate has cleared up uncertainties in biosynthesis arising from $^{14}C$ labelling data. The results are that the naphthoquinone (*3.107*) is formed as shown (Scheme 3.19) either via a simple fragmented octaketide chain or two separate chains [83]. (For further discussion of quinone biosynthesis see section 5.2.)

Brefeldin (*3.108*) is a quite different octaketide. It derives again by simple linear combination of eight acetate/malonate units [84] as does curvularin (*3.109*) [85]. Ochrephilone (*3.110*) biosynthesis on the other hand is more complex. This metabolite appears to originate from two polyketide chains with inclusion of three C-methyl groups (see dotted lines) [86].

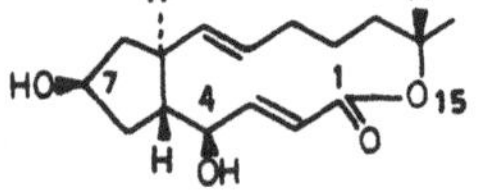

(*3.108*) Brefeldin

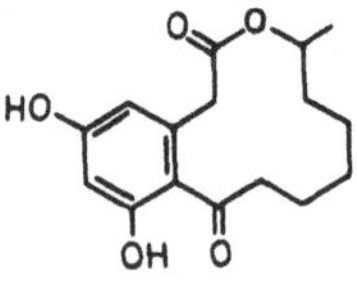

(*3.109*) Curvularin

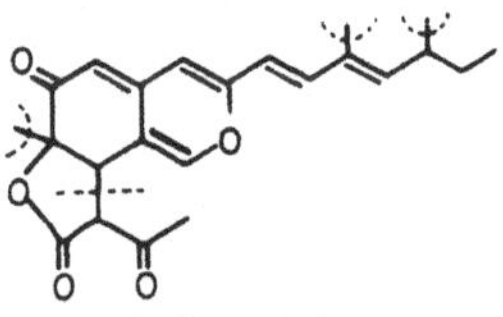

(*3.110*) Ochrephilone

Different hypotheses for the biosynthesis of brefeldin A have been considered in relation to the biosynthesis of e.g. fatty acids [87] and prostaglandins (section 1.1.3). It is useful for such considerations to know that the oxygens at C-1 and C-15 originate from acetate whilst those at C-4 and C-7 are derived from molecular oxygen. The derivation of the oxygens at these latter sites from different molecules of oxygen precludes a biosynthetic mechanism similar to that of prostaglandins [88].

Finally, a very important set of experiments involving the octaketide actinorhodin (*3.111*) needs to be noted. The genes for its biosynthesis in

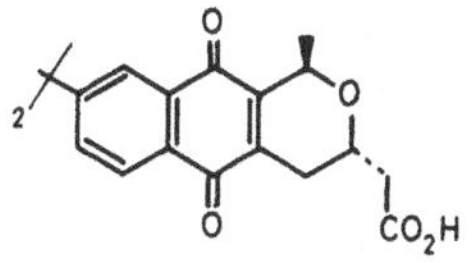

*(3.111)* Actinorhodin

one *Streptomyces* which produces it have been cloned in another organism of the same genus leading to the production of *(3.111)*. (For introductory reviews in this area see [89] and [90].) Further work has involved genetic manipulation leading to the elaboration of new or hybrid antibiotics [91].

## 3.8 NONA- AND DECA-KETIDES

Bikaverin *(3.112)* is a fairly rare example of a nonaketide. It is formed by simple folding of a single chain (see dotted line) [92].

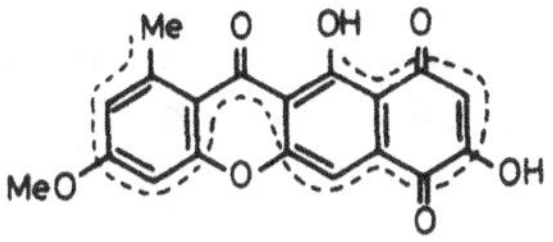

*(3.112)* Bikaverin

Asteltoxin *(3.113)* has been established as deriving from propionate (as starter) plus eight malonate units. Most curiously there is an alternative pathway operating in the same organism which begins with acetate plus methionine instead of propionate. This is found too for the structurally related fungal metabolites citreoviridin and aurovertin B. The precursor labelling of *(3.113)* is illustrated. A partial rearrangement of the polyketide

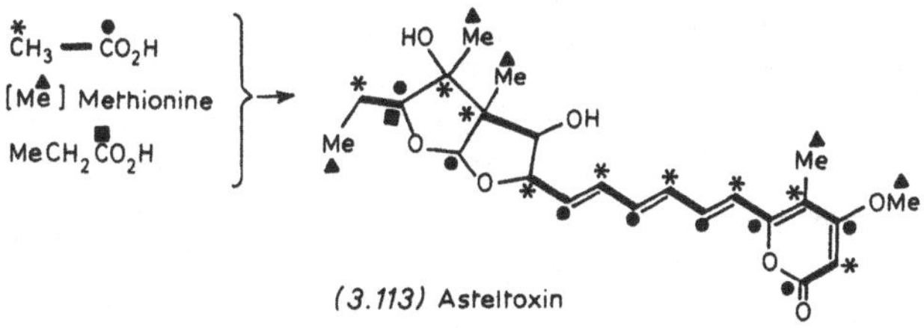

*(3.113)* Asteltoxin

chain is apparent and additional $^{18}$O-labelling studies together with these results lead to a mechanism for the rearrangement with simultaneous formation of the bisfuran ring system (Scheme 3.20) [93–95] (cf section 3.9).

The aflatoxins, e.g. aflatoxin $B_1$ *(3.121)*, are an important group of toxins. Study of the biosynthesis of these toxins has called forth the use of techniques ranging from the full armoury of stable-isotope labels to the use of mutants

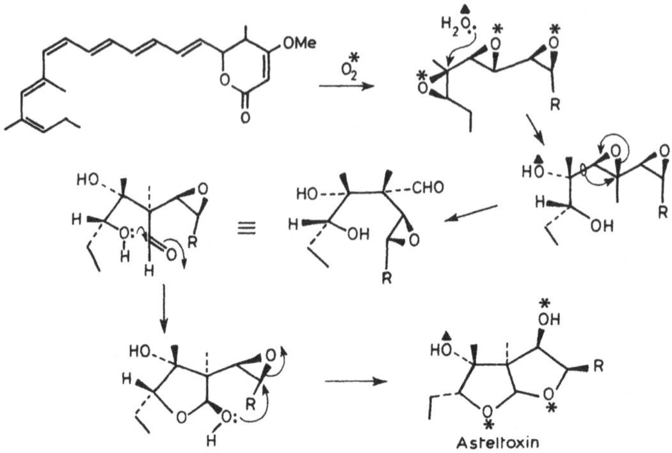

**Scheme 3.20**

and enzyme inhibitors. There have been some surprises along the way, most notable of which perhaps was the clear demonstration that instead of being formed from an acetate starter unit plus nine malonates in the usual polyketide manner these toxins begin with a hexanoate starter (*3.114*) to which malonate units are added [demonstrated for the early intermediate averufin (*3.115*)] [96]. In the light of the detailed, rigorously interlocked evidence available which has been obtained for intermediates as well as aflatoxin B₁ (*3.121*) the pathway outlined in Scheme 3.21 may be proposed [96–108]. Important intermediates are averufin (*3.115*), versiconal acetate (*3.117*), versicolorin A (*3.118*) and sterigmatocystin (*3.119*).

The demonstration that the oxygen atoms in averufin (*3.115*) derive as shown from acetate and aerial oxygen was the first study in which $^{18}$O-induced isotope shifts on $^{13}$C resonances was observed. Powerful general support was thereby provided for the formation of polyketides from molecules of poly-β-keto-acyl-CoA [105]. The mechanism illustrated for the conversion of averufin (*3.115*) into versiconal acetate (*3.117*) depends on the observations that nidurufin (*3.115*, R = OH) and its epimers are not intermediates [i.e. loss of H not OH from C-2′ in (*3.116*)] [97], that the C—O bond shown thickened in (*3.116*) is not broken, and that protons (deuterium labels) are not lost from C-1′, -4′, or -6′, nor is one of the protons on C-2′ [106, 107]. In the steps between (*3.118*) and (*3.119*) one of the phenolic oxygens must be lost (from C-6), which is rarely observed otherwise [107].

Experiments with $^{13}$C-labelled averufin (*3.115*) gave results which demonstrate that the C-8—C-11 bond in (*3.115*) becomes the C-2—C-3 bond in aflatoxin B₁ (*3.121*), whilst C-6 becomes C-5 and C-5 of averufin is lost [108]. This confirms that (*3.119*) is converted into (*3.121*) as shown.

Tetracyclines, e.g. 7-chlorotetracycline (*3.126*), are commercially impor-

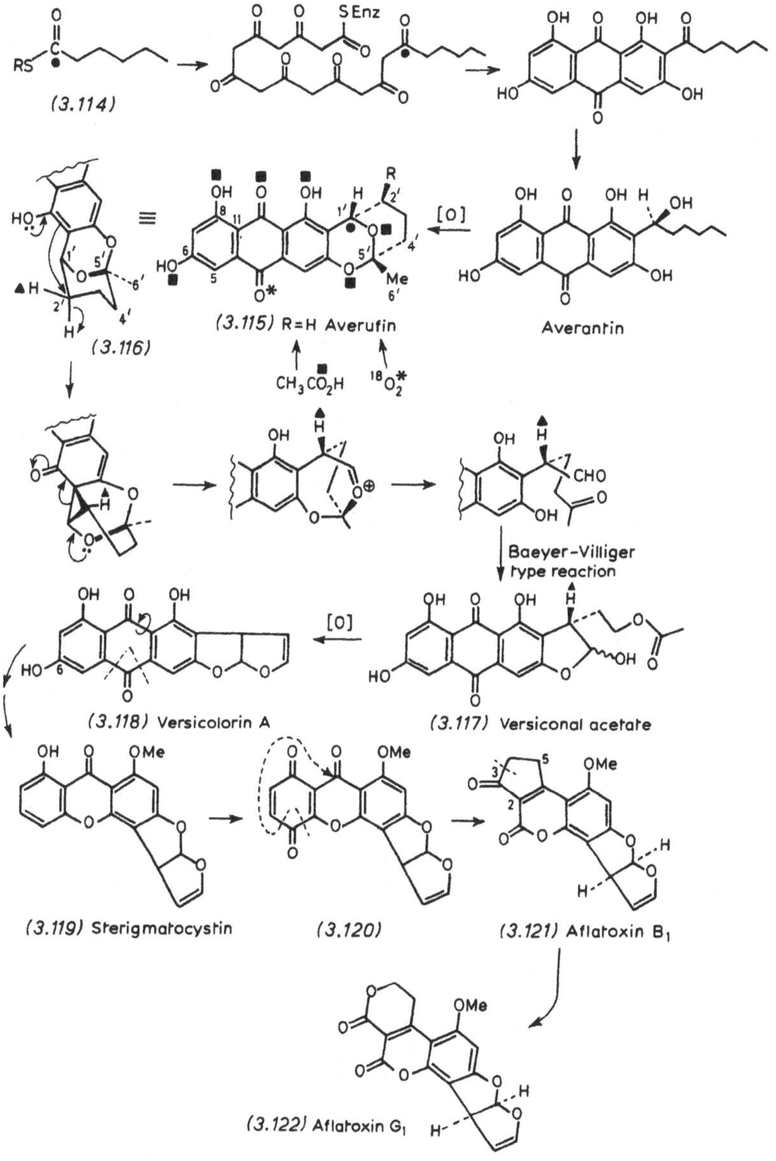

**Scheme 3.21**

tant broad-spectrum antibiotics. Detailed information is available on the course of their biosynthesis as a result of an extensive study, using mutants in particular. This involved isolation of new products from differently blocked mutants and testing them as precursors in cross-feeding experiments. The structure (*3.125*) of mutant-produced protetrone indicates the manner of folding in the polyketide as (*3.123*) (protetrone is a diversion from the main pathway, however). This polyketide is notable for its malonamyl starter [see (*3.123*)].

The deduced sequence to 7-chlorotetracycline (*3.126*) is illustrated in brief in Scheme 3.22 [109, 110]. As commonly for polyketides, methylation occurs at the enzyme-bound stage on (*3.123*). Chlorination occurs on an aromatic ring (electrophilic substitution by $Cl^+$ on an electron-rich aromatic nucleus). The labelling of oxytetracycline (*3.127*; thickened bonds indicate intact C—O bonds) by [1-$^{13}$C, $^{18}$O$_2$]acetate is consistent with the pathway illustrated in Scheme 3.22 [109]. Cetocycline has a similar structure to the tetracyclines just discussed and its labelling by [1-$^{13}$C]acetate supports a similar pathway [111].

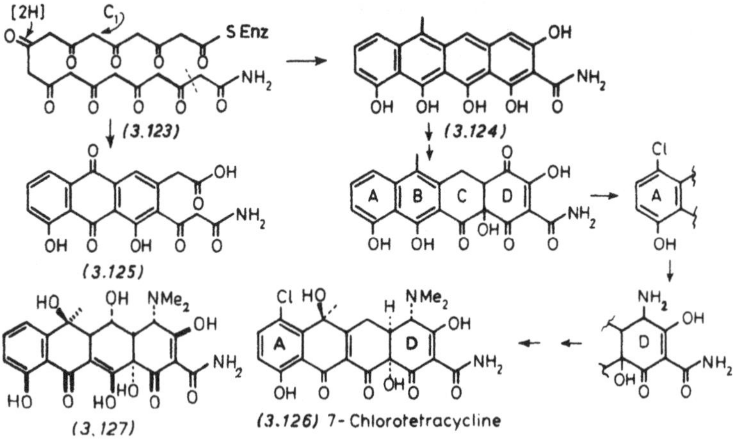

**Scheme 3.22**

A different mode of poly-$\beta$-ketone folding to that of the tetracyclines is observed for the anthracycline glycosidic antibiotics. $\epsilon$-Pyrromycinone (*3.128*), an example of the aglycone fragment in these antibiotics, is derived from nine acetate units as shown by the dotted line; the chain starter is propionate [112].

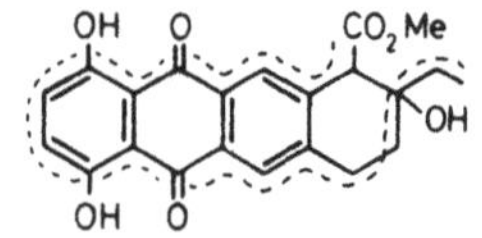

(*3.128*) $\epsilon$-Pyrromycinone

The starter unit for chrysomycin A (*3.130*) is also propionate whilst for chrysomycin B (*3.131*) it is acetate. The $^{13}$C-acetate and -propionate labelling patterns are illustrated and it seems likely that biosynthesis is from a single polyketide chain through an intermediate of type (*3.129*) which becomes fragmented [113].

(3.129)    (3.130) R= CH=CH$_2$
           (3.131) R= CH$_3$

## 3.9 POLYKETIDES WITH MIXED ORIGINS AND LARGE RING POLYKETIDES

The very largest of the polyketides are the macrolide antibiotics, e.g. nystatin (*3.136*). Further examples are the ansamycins which derive by a mixed acetate-propionate pathway (section 7.6.2). Intermediate in size are the cyto-chalasins which derive by an acetate (malonate) pathway (section 7.6.3). Where propionate units account for $C_3$ fragments in the ansamycins, methionine and acetate serve in the cytochalasins.* The macrolide antibiotics discussed below all follow the former way of generating $C_3$ units. It is clear that, if methyl groups are introduced by two different pathways, this is not adventitious. The methyl groups must have a function possibly like the double bonds in dictating the conformation of the macrocycle.

For the biosynthesis of these complex polyketides, $^{13}$C labelled precursors have been used to great advantage in plotting the origins of macrocyclic anti-biotics. Little is known, however, of the mechanism of biosynthesis. (For a discussion on the possible significance of templates see [4].)

Tylosin (*3.132*) is formed as indicated, the one $C_4$ unit deriving from ethylmalonate (= butyrate) [114, 115]. A similar origin is apparent for leucomycin (*3.133*); the origin of C-3 and C-4 is in glycerol via glycolate [116]. In both studies extensive catabolism of butyrate and ethylmalonate into propionyl fragments was observed, but by different pathways.

The platenomycins are closely related to leucomycin [117]. Picro-mycin (*3.134*) has its genesis from six propionate units and one acetate [118]. Erythromycin A (*3.135*) is wholly derived from propionate; appropriate $^{18}$O labelling by [1-$^{13}$C, $^{18}$O] propionate was observed [119]. Sixteen acetate and three propionate units account for the skeleton of the aglycone unit of nystatin (*3.136*) [120].

---

*It is a useful, general rule (with exceptions) that $C_3$ units, when they are present in polyketides, are elaborated in fungi from acetate plus methionine, whereas those in metabolites biosynthesized by Actinomycetes are formed from propionate (Snyder, W.C. and Rinehart, K.L. jun. (1984) *J. Amer. Chem. Soc.*, **106**, 787–9; and refs cited).

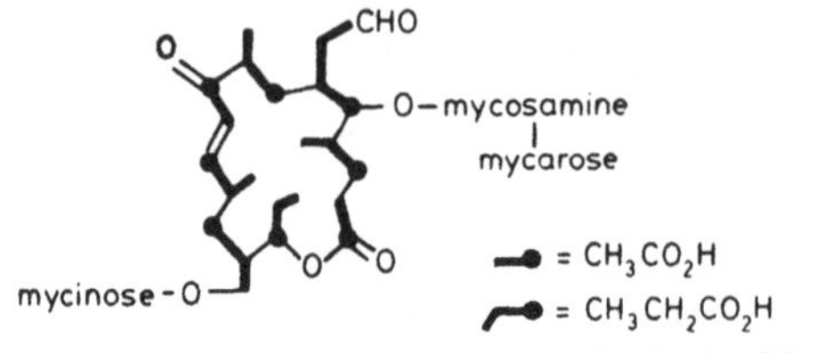

(3.132) Tylosin

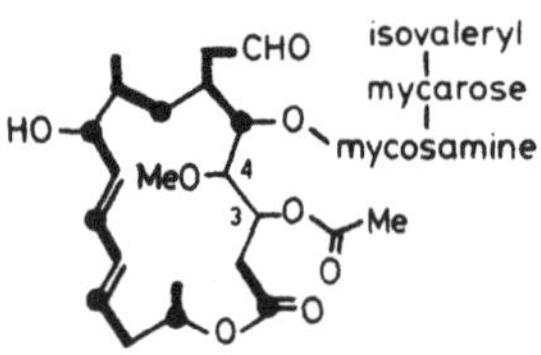

—● = $CH_3CO_2H$
↷● = $CH_3CH_2CO_2H$
⋁● = $CH_3CH_2CH_2CO_2H$

(3.133) Leucomycin

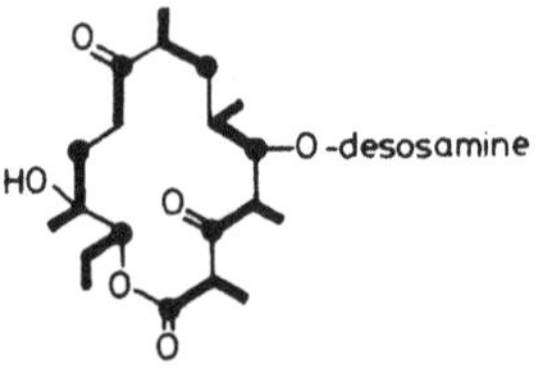

(3.134) Picromycin

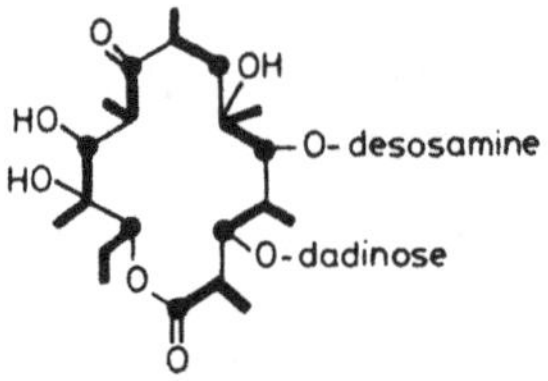

(3.135) Erythromycin A

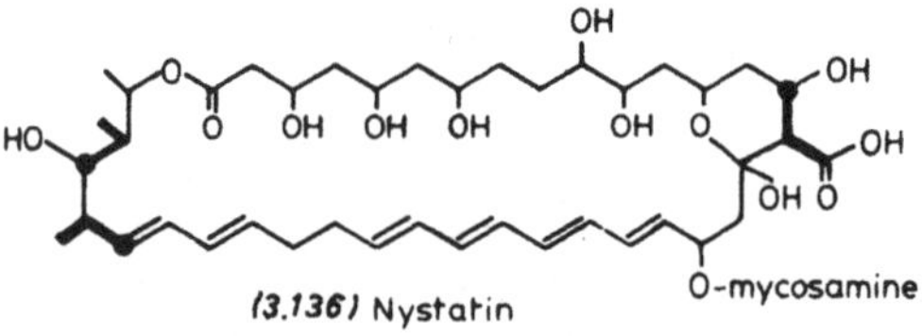

(3.136) Nystatin

Mixed origins have been deduced for aureothin (*3.137*) from propionate and (probably) a single acetate plus *p*-nitrobenzoic acid [121], and for (*3.138*) which derives from acetate, propionate, and butyrate [122]. Avenaciolide (*3.139*) derives from acetate/malonate plus succinyl-CoA (Scheme 3.23). Label from both [1-$^{13}$C]- and [2-$^{13}$C]-acetate

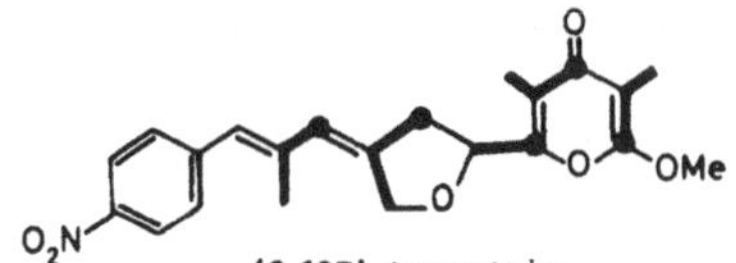

(3.137) Aureothin 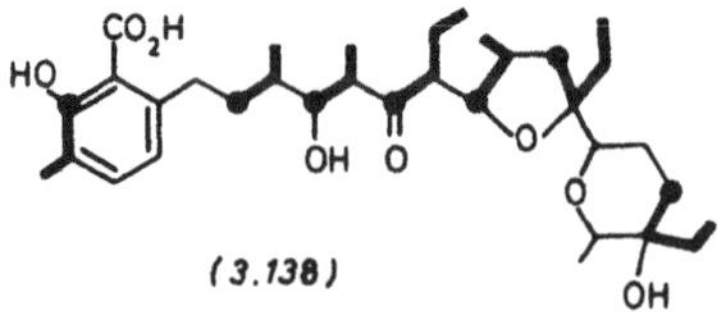          (3.138)

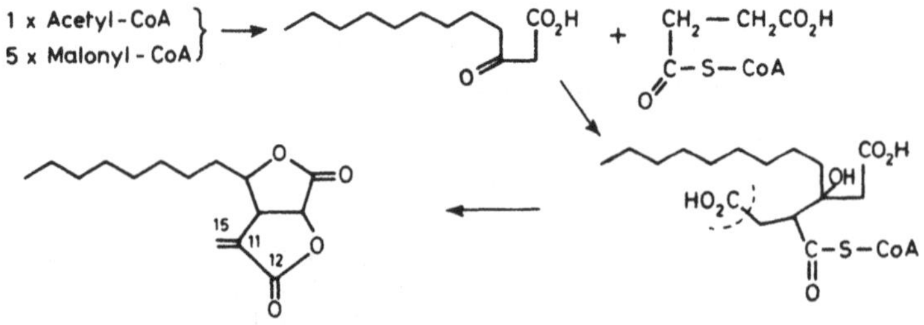

1 x Acetyl-CoA
5 x Malonyl-CoA
}

(3.139) Avenaciolide

**Scheme 3.23**

is transferred to succinyl-CoA through the citric acid cycle. (The avenaciolide was labelled directly then by acetate and indirectly through succinate.) In (3.139) derived from [2-$^{13}$C]acetate coupling was observed between C-11 and C-15 in the $^{13}$C n.m.r. spectrum consonant with a succinate origin for C-15, C-11 and C-12. (At a high enough $^{13}$C concentration two C-2-labelled acetate units will join together to give succinyl-CoA with two contiguous $^{13}$C labels [123].)

The polyether antibiotic monensin A (3.140) is constituted from

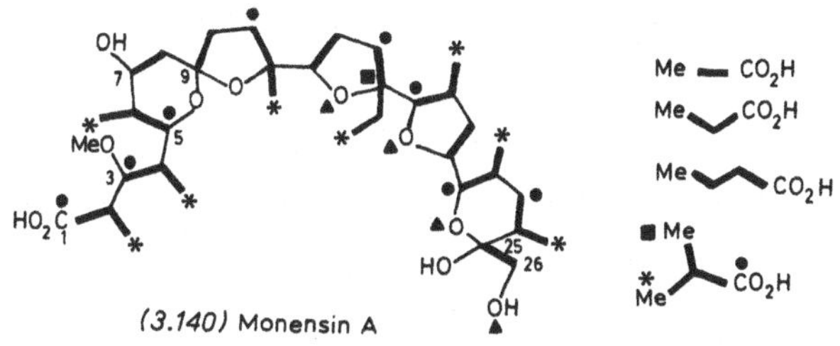

(3.140) Monensin A

acetate, propionate, butyrate, and isobutyrate as shown [124, 125]. The oxygens at C-1, C-3, and C-5 derive from propionate, those at C-7, C-9, and C-25 (the OH) originate in acetate, and the remainder are from atmospheric oxygen (▲). The results lend considerable credence to a most attractive hypothesis that this and other polyether antibiotics arise through a cascade of reactions associated with epoxide functionality (Scheme 3.24) [124]. In the course of this study the isomerization of

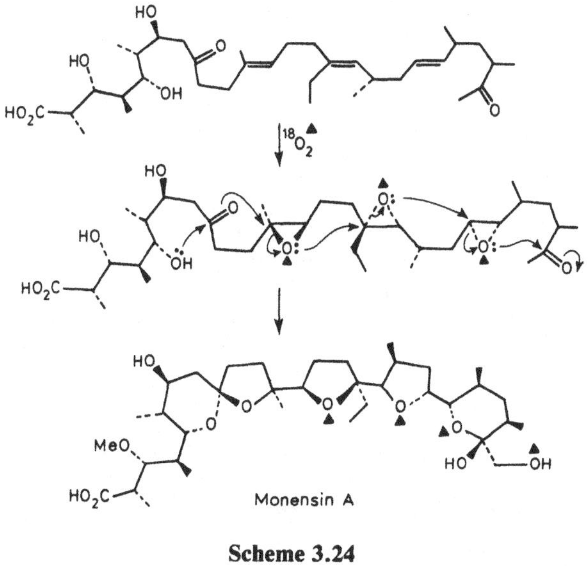

Monensin A

**Scheme 3.24**

isobutyrate to *n*-butyrate which may be degraded to methylmalonyl CoA (= propionate) was observed; this then accounts for the labelling of other units in (*3.140*) by isobutyrate (●, *). The reactions have been examined in detail [125].

Maduramycin is a polyether similar to monensin A. Maduramycin [126] and ICI 139603 [127] have origins in acetate and propionate as largely do the avermectins and milbemycins [128, 129]. Nonactin is a macrocyclic polyether constructed from four units of nonactic acid (*3.141*). The biosynthesis of (*3.141*) is as shown [130]. Acetoacetate was incorporated but only either after degradation to acetate or by rearrangement and incorporation into the propionate unit.

(3.141)

# REFERENCES

Further reading: [1]-[4]; *Biosynthesis (Specialist Periodical Reports)* and *Natural Product Reports*, The Royal Society of Chemistry, London; Steyn, P.S. (ed.) (1980) *The Biosynthesis of Mycotoxins*, Academic Press, New York.

1. Turner, W.B. (1971) *Fungal Metabolites*; Turner, W.B. and Aldridge, D.C. (1983) *Fungal Metabolites II*, Academic Press, London.
2. Packter, N.M. (1973) *Biosynthesis of Acetate-derived Compounds*, Wiley, London.
3. Packter, N.M. (1980) In *The Biochemistry of Plants* (ed. P.K. Stumpf), Academic Press, London, vol. 4, pp. 535-70.
4. Bu'Lock, J.D. (1979) In *Comprehensive Organic Chemistry* (eds D.H.R. Barton and W.D. Ollis), Pergamon, Oxford, vol. 5, pp. 927-87.
5. cf *St. Mark's Gospel*, chapter 5, v.v. 9-13.
6. Collie, J.N. (1907) *J. Chem. Soc.*, **91**, 1806-13.
7. Stewart, A.W. and Graham, H. (1948) *Recent Advances in Organic Chemistry*, 7th edn, Longmans, Green and Co., London, vol. 2, ch. 13.
8. Birch, A.J. and Donovan, F.W. (1953) *Aust. J. Chem.*, **6**, 360-8.
9. Birch, A.J. (1957) *Fortschritte der Chemie Organischer Naturstoffe*, **14**, 186-216; (1962) *Proc. Chem. Soc.*, 3-13.
10. Gatenbeck, S. and Mosbach, K. (1959) *Acta Chem. Scand.*, **12**, 1561-4.
11. Simpson, T.J. (1975) *Chem. Soc. Rev.*, **4**, 497-522.
12. Mahler, H.R. and Cordes, E.H. (1971) *Biological Chemistry*, 2nd Edn, Harper and Row, New York.
13. Spenser, I.D. (1968) In *Comprehensive Biochemistry* (eds M. Florkin and E.H. Stotz), Elsevier, Amsterdam, vol. 20, pp. 231-413 (for acetate labelling via the tricarboxylic acid cycle).

14. Dimroth, P., Ringelmann, E. and Lynen, F. (1976) *Eur. J. Biochem.*, **68**, 591–6.
15. Scott, A.I., Beadling, L.C., Georgopapadakou, N.H. and Subbarayan, C.R. (1974) *Bio-org. Chem.*, **3**, 238–48.
16. Light, R.J. and Hager, L.P. (1968) *Arch. Biochem. Biophys.*, **125**, 326–33.
17. Vogel, G. and Lynen, F. (1975) *Methods Enzymol.*, **43**, 520–30.
18. Abell, C., Garson, M.J., Leeper, F.J. and Staunton, J. (1982) *J. Chem. Soc. Chem. Comm.*, 1011–3.
19. Abell, C. and Staunton, J. (1984) *J. Chem. Soc. Chem. Comm.*, 1005–7.
20. Yamazaki, M. and Shibata, S. (1966) *Chem. Pharm. Bull. (Japan)*, **14**, 96–7.
21. Steward, M.W. and Packter, N.M. (1968) *Biochem. J.*, **109**, 1–11.
22. Better, J. and Gatenbeck, S. (1977) *Acta Chem. Scand.*, **B31**, 391–4.
23. Lenfant, M., Farrugia, G. and Lederer, M. (1969) *Compt. rend.*, **268D**, 1986–9.
24. Gatenbeck, S., Eriksson, P.O. and Hansson, Y. (1969) *Acta Chem. Scand.*, **23**, 699–701.
25. Suzuki, K.T. and Nozoe, S. (1974) *Bio-org. Chem.*, **3**, 72–80.
26. Better, J. and Gatenbeck, S. (1976) **B30**, 368.
27. Elvidge, J.A., Jaiswal, D.K., Jones, J.R. and Thomas, R. (1977) *J. Chem. Soc. Perkin I*, 1080–3.
28. Bentley, R. (1965) In *Biogenesis of Antibiotic Substances* (eds Z. Vanek and Z. Hostalek), Academic Press, New York, pp. 241–54.
29. Petterson, G. (1965) *Acta Chem. Scand.*, **19**, 543–8; 1016–7.
30. Gatenbeck, S. and Brunsberg, U. (1966) *Acta Chem. Scand.*, **20**, 2334–8.
31. Taguchi, H., Sankawa, U. and Shibata, S. (1969) *Chem. Pharm. Bull. (Japan)*, **17**, 2054–60.
32. Nabeta, K., Ichihara, A. and Sakamura, S. (1975) *Agric. Biol. Chem. (Japan)*, **39**, 409–13.
33. Scott, A.I. and Yalpani, M. (1967) *Chem. Comm.*, 945–6.
34. Iijima, H., Noguchi, H., Ebizuka, Y. *et al.* (1983) *Chem. Pharm. Bull.*, **31**, 362–5.
35. Arai, K., Yoshimura, T., Itatani, Y. and Yamamoto, Y. (1983) *Chem. Pharm. Bull.*, **31**, 925–33.
36. Renauld, F., Moreau, S. and Lablache-Combier, A. (1984) *Tetrahedron*, **40**, 1823–34.
37. Scott, A.I. and Wiesner, K.J. (1972) *J. Chem. Soc. Chem. Comm.*, 1075–7.
38. Scott, A.I. and Lee, E. (1972) *J. Chem. Soc. Chem. Comm.*, 655–6.
39. McInnes, A.G., Smith, D.G., Vining, L.C. and Johnson, L. (1971) *J. Chem. Soc. Chem. Comm.*, 325–6.
40. Takenaka, S. and Seto, S. (1971) *Agric. Biol. Chem. (Japan)*, **35**, 862–9.
41. Simpson, T.J. and Stevenson, G.I. (1985) *J. Chem. Soc. Chem. Comm.*, 1822–4.
42. Parry, R.J. and Buu, H.P. (1983) *J. Amer. Chem. Soc.*, **105**, 7446–7.
43. Nair, M.S.R. (1976) *Phytochemistry*, **15**, 1090–1.
44. Seto, H., Shibamiya, M. and Yonehara, H. (1978) *J. Antibiot.*, **31**, 926–8.
45. Holker, J.S.E. and Simpson, T.J. (1981) *J. Chem. Soc. Perkin I*, 1397–1400.
46. Weisleder, D. and Lillehoj, E. (1980) *Tetrahedron Lett.*, **21**, 993–6.
47. de Jesus, A.E., Steyn, P.S., Vleggaar, R. and Wessels, P.L. (1980) *J. Chem. Soc. Perkin I*, 52–4.
48. Barber, J., Carter, R.H., Garson, M.J. and Staunton, J. (1981) *J. Chem. Soc. Perkin I*, 2577–83.

49. Sankawa, U., Ebizuka, Y., Noguchi, H. *et al.* (1983) *Tetrahedron*, **39**, 3583–91.
50. Colombo, L., Scolastico, C., Lukacs, G. *et al.* (1983) *J. Chem. Soc. Chem. Comm.*, 1436–7.
51. Colombo, L., Gennari, C., Scolastico, C. *et al.* (1980) *J. Chem. Soc. Perkin I*, 2549–53.
52. Hill, R.A., Carter, R.H. and Staunton, J. (1975) *J. Chem. Soc. Chem. Comm.*, 380–1.
53. Holker, J.S.E. and Young, K. (1975) *J. Chem. Soc. Chem. Comm.*, 525–6.
54. Henderson, G.B. and Hill, R.A. (1982) *J. Chem. Soc. Perkin I*, 3037–9.
55. Barber, J., Garson, M.J. and Staunton, J. (1981) *J. Chem. Soc. Perkin I*, 2584–93.
56. Brereton, R.G., Garson, M.J. and Staunton, J. (1984) *J. Chem. Soc. Perkin I*, 1027–33.
57. Tanabe, M., Hamasaki, T., Thomas, D. and Johnson, L. (1971) *J. Amer. Chem. Soc.*, **93**, 273–4.
58. Gorst-Allman, C.P., Steyn, P.S. and Vleggaar, R. (1983) *J. Chem. Soc. Perkin I*, 1357–9.
59. Tanake, N. and Umezawa, H. (1962) *J. Antibiotics*, **15**, 189–9.
60. Tanake, N., Sashikata, K. and Umezawa, H. (1962) *J. Antibiotics*, **15**, 228–9.
61. Turner, W.B. (1961) *J. Chem. Soc.*, 522–4.
62. Kurono, M., Nakanishi, K., Shindo, K. and Tada, M. (1963) *Chem. Pharm. Bull. (Japan)*, **11**, 359–62.
63. Hadfield, J.R., Holker, J.S.E. and Stanway, D.N. (1967) *J. Chem. Soc. (C)*, 751–5.
64. Simpson, T.J. (1985) *Nat. Prod. Rep.*, **2**, p. 331.
65. Simpson, T.J. and Stenzel, D.J. (1982) *J. Chem. Soc. Chem. Comm.*, 1074–6.
66. Gudgeon, J.A., Holker, J.S.E. and Simpson, T.J. (1974) *J. Chem. Soc. Chem. Comm.*, 636–8.
67. Holker, J.S.E., O'Brien, E., Moore, R.N. and Vederas, J.C. (1983) *J. Chem. Soc. Chem. Comm.*, 192–4.
68. McIntyre, C.R., Simpson, T.J., Trimble, L.A. and Vederas, J.C. (1984) *J. Chem. Soc. Chem. Comm.*, 706–9.
69. Birch, A.J., Massy-Westropp, R.A., Richards, R.W. and Smith, H. (1958) *J. Chem. Soc.*, 360–5.
70. Lane, M.P., Nakashima, T.T. and Vederas, J.C. (1982) *J. Amer. Chem. Soc.*, **104**, 913–5.
71. Demetriadou, A.K., Lane, E.D. and Staunton, J. (1985) *J. Chem. Soc. Chem. Comm.*, 1125–7.
72. Sjöland, S. and Gatenbeck, S. (1966) *Acta Chem. Scand.*, **20**, 1053–9.
73. Simpson, T.J. (1979) *J. Chem. Soc. Perkin I*, 1233–8.
74. Thomas, R. (1971) *J. Chem. Soc. Chem. Comm.*, 739–40.
75. Edwards, J.M., Schmitt, R.C. and Weiss, U. (1972) *Phytochemistry*, **11**, 1717–20.
76. Okuba, A., Yamazaki, S. and Fuwa, K. (1975) *Agric. Biol. Chem. (Japan)*, **39**, 1173–5.
77. Paulick, R.C., Casey, M.L., Hillenbrand, D.F. and Whitlock, H.W. jun. (1975) *J. Amer. Chem. Soc.*, **97**, 5303–5.
78. Birch, A.J., Baldas, J., Hlubucek, R. *et al.* (1976) *J. Chem. Soc. Perkin I*, 898–904.

79. Hill, J.G., Nakashima, T.T. and Vederas, J.C. (1982) *J. Amer. Chem. Soc.*, **104**, 1745–8.
80. Curtis, R.F., Hassall, C.H. and Parry, D.R. (1972) *J. Chem. Soc. Perkin I*, 240–4.
81. Bardshiri, E., McIntyre, C.R., Simpson, T.J. *et al.* (1984) *J. Chem. Soc. Chem. Comm.*, 1404–6.
82. Maebayashi, Y. and Yamazaki, M. (1985) *Chem. Pharm. Bull.*, **33**, 4296–8.
83. Casey, M.L., Paulick, R.C. and Whitlock, H.W. jun., (1976) *J. Amer. Chem. Soc.*, **98**, 2636–40.
84. Cross, B.E. and Hendley, P. (1975) *J. Chem. Soc. Chem. Comm.*, 124–5.
85. Birch, A.J., Musgrave, O.C., Rickards, R.W. and Smith, H. (1959) *J. Chem. Soc. (C)*, 3146–52.
86. Kamiya, Y., Ikegami, S. and Tamura, S. (1974) *Tetrahedron Lett.*, 655–8.
87. Hutchinson, C.R., Shu-Wen, L., McInnes, A.G. and Walter, J.A. (1983) *Tetrahedron*, **39**, 3507–13.
88. Mabuni, C.T., Garlaschelli, L., Ellison, R.A. and Hutchinson, C.R. (1977) *J. Amer. Chem. Soc.*, **99**, 7718–20.
89. Parish, J.H. and McPherson M.J. (1987) *Nat. Prod. Rep.*, **4**, 139–56.
90. McPherson, M.J. and Parish, J.H. (1987) *Nat. Prod. Rep.*, **4**, 205–24.
91. Hopwood, D.A., Malpartida, F., Kieser, H.M. *et al.* (1985), *Nature*, **314**, 642–4.
92. McInnes, A.G., Smith, D.G., Walter, J.A. *et al.* (1975) *J. Chem. Soc. Chem. Comm.*, 66–8.
93. de Jesus, A.E., Steyn, P.S. and Vleggaar, R. (1985) *J. Chem. Soc. Chem. Comm.*, 1633–5.
94. Steyn, P.S. and Vleggaar, P. (1985) *J. Chem. Soc. Chem. Comm.*, 1796–8.
95. Steyn, P.S. and Vleggaar, R. (1985), *J. Chem. Soc. Chem. Comm.*, 1531–2.
96. Townsend, C.A. and Christensen, S.B. (1983) *Tetrahedron*, **39**, 3575–82.
97. Townsend, C.A. and Christensen, S.B. (1985) *J. Amer. Chem. Soc.*, **107**, 270–1.
98. Koreeda, M., Hulin, B., Yoshihara, M. *et al.* (1985) *J. Org. Chem.*, **50**, 5428–30; and following paper.
99. Pachler, K.G.R., Steyn, P.S., Vleggaar, R. *et al.* (1976) *J. Chem. Soc. Perkin I*, 1182–9.
100. Elsworthy, G.C., Holker, J.S.E., McKeown, J.M. *et al.* (1970) *Chem. Comm.*, 1069–70.
101. Heathcote, J.G., Dutton, M.F. and Hibbert, J.R. (1976) *Chem. Ind., (London)* 270–2.
102. Roberts, J.C. (1974) *Fortschritta der Chemie Organischer Naturstoffe*, **31**, 120–51.
103. Zamir, L.O. and Ginsburg, R. (1979) *J. Bacteriol.*, **138**, 684–90.
104. Singh, R. and Hsieh, D.P. (1977) *Arch. Biochem., Biophys.*, **178**, 285–92.
105. Vederas, J.C. and Nakashima, T.T. (1980) *J. Chem. Soc. Chem. Comm.*, 183–5.
106. Townsend, C.A., Christensen, S.B., and Davis, S.G. (1982) *J. Amer. Chem. Soc.*, **104**, 6152–3; and following paper.
107. Sankawa, U., Shimada, H., Kobayashi, T. *et al.* (1982) *Heterocycles*, **19**, 1053–8.
108. Townsend, C.A. and Davis, S.G. (1983) *J. Chem. Soc. Chem. Comm.*, 1420–2.

109. Thomas, R. and Williams, D.J. (1985) *J. Chem. Soc. Chem. Comm.*, 802-3.
110. McCormick, J.R.D. and Jensen, E.R. (1968) *J. Amer. Chem. Soc.*, **90**, 7126-7; and following paper.
111. Mitscher, L.A., Swayze, J.K., Högberg, T. *et al.* (1983) *J. Antibiotics*, **36**, 1405-7.
112. Ollis, W.D., Sutherland, I.O., Codner, R.C. *et al.* (1960) *Proc. Chem. Soc.*, 347-9.
113. Carter, G.T., Fantini, A.A., James, J.C., Borders, D.B. and White, R.J. (1985) *J. Antibiotics*, **38**, 242-8.
114. Ōmura, S., Takeshima, H., Nakagawa, A. *et al.* (1976) *Bioorg. Chem.*, **5**, 451-4.
115. O'Hagan, D., Robinson, J.A. and Turner, D.L. (1983) *J. Chem. Soc. Chem. Comm.*, 1337-40.
116. Ōmura, S., Tsuzuki, K., Nakagawa, A. and Lukacs, G. (1983) *J. Antibiotics*, **36**, 611-3; and following paper.
117. Furumai, T. and Suzuki, M. (1975) *J. Antibiotics*, **28**, 770-4; 775-82; 783-8.
118. Ōmura, S., Takeshima, H., Nakagawa, A. *et al.* (1975) *J. Antibiotics*, **29**, 316-7.
119. Cane, D.E., Hasler, H., Taylor, P.B. and Liang, T.-C. (1983) *Tetrahedron*, **39**, 3449-55.
120. Birch, A.J., Holzapfel, C.W., Rickards, R.W. *et al.* (1964) *Tetrahedron Lett.*, 1485-90.
121. Yamazaki, M., Katoh, F., Ohishi, J. and Koyama, Y. (1972) *Tetrahedron Lett.*, 2701-4.
122. Westley, J.W., Pruess, D.L. and Pitcher, R.G. (1972) *J. Chem. Soc. Chem. Comm.*, 161-2.
123. Tanabe, M., Hamasaki, T., Suzuki, Y. and Johnson, L.F. (1973) *J. Chem. Soc. Chem. Comm.*, 212-3.
124. Cane, D.E., Liang, T.-C. and Hasler, H. (1982) *J. Amer. Chem. Soc.*, **104**, 7274-81.
125. Reynolds, K. and Robinson, J.A. (1985) *J. Chem. Soc. Chem. Comm.*, 1831-2.
126. Tsou, H.-R., Rajan, S., Fiala, R. *et al.* (1984) *J. Antibiotics*, **37**, 1651-63.
127. Demetriadou, A.K., Lane, E.D., Staunton, J. *et al.* (1985) *J. Chem. Soc. Chem. Comm.*, 408-10.
128. Ono, M., Mishima, H., Takiguchi, Y. *et al.* (1983) *J. Antibiotics*, **36**, 991-1000.
129. Cane, D.E., Liang, T.-C., Kaplan, L. *et al.* (1983) *J. Amer. Chem. Soc.*, **105**, 4110-2.
130. Clark, C.A. and Robinson, J.A. (1985) *J. Chem. Soc. Chem. Comm.*, 1568-9.

# 4 *Terpenes and steroids*

## 4.1 INTRODUCTION

The polyketides form a group of metabolites which are based on patterns of repeating acetate units (Chapter 3). There is another group of metabolites, widespread in nature and rich in diverse chemistry, which is based on another repeating unit: isoprene (*4.1*). Over a hundred years ago (*4.1*) began to be

*(4.1)* Isoprene

identified as a product of thermal decomposition of rubber and other, fragrant natural compounds of low molecular weight. The idea grew that these compounds and many others, generically the terpenes, were formed by the head-to-tail linkage of isoprene units. The ideas became encapsulated in what is called the isoprene rule. This relationship of terpene to isoprene can be seen in geraniol (*4.2*), α-terpineol (*4.3*) and γ-cadinene (*4.4*) (dotted

*(4.2)* Geraniol    *(4.3)* α-Terpineol    *(4.4)* γ-Cadinene

lines). Exceptions began to appear, however, e.g. artemesia ketone (*4.5*), and notably lanosterol (*4.6*). The assignment of structure (*4.6*) to lanosterol had two fundamental consequences. One was to suggest clearly the course of cholesterol (and steroid) biosynthesis (see below). The second consequence was initiation of a reexamination of the isoprene rule and its reformulation in biogenetic terms, [7] i.e. terpenes are compounds formed as follows: (i) linear combination of isoprene units occurs to give geraniol (*4.2*) ($C_{10}$),

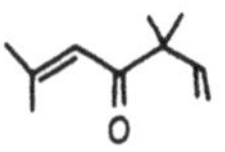

*(4.5)* Artemesia ketone

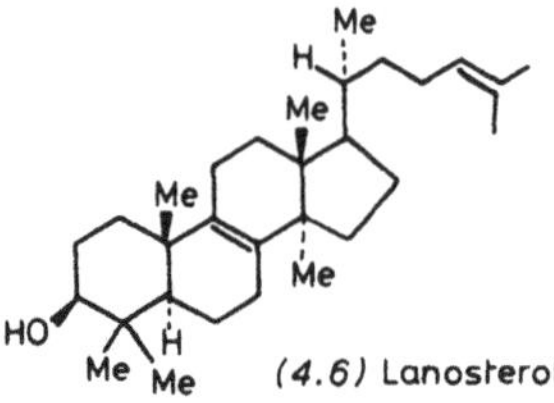

*(4.6)* Lanosterol

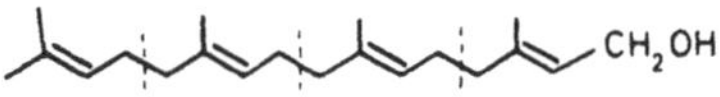

*(4.7)* Geranylgeraniol

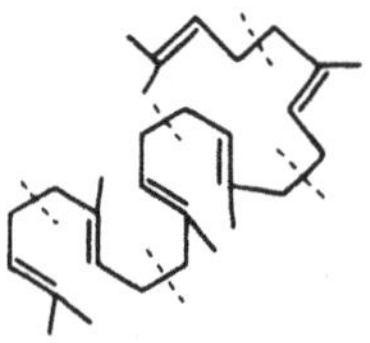

*(4.8)* Squalene

farnesol *(4.9)* (C$_{15}$), geranylgeraniol *(4.7)* (C$_{20}$), squalene *(4.8)* (C$_{30}$)* and similar compounds; (ii) other terpenes are derived from these compounds by mechanistically reasonable cyclization and, in some cases, rearrangement reactions. The derivation of α-terpineol *(4.3)* by cyclization of the geraniol *(4.2)* skeleton is mechanistically reasonable. The formation of eremorphilone *(4.10)* can be rationalized as shown in Scheme 4.1, and involves a 1,2-methyl shift (for further examples of such shifts see section 4.2 and following sections).

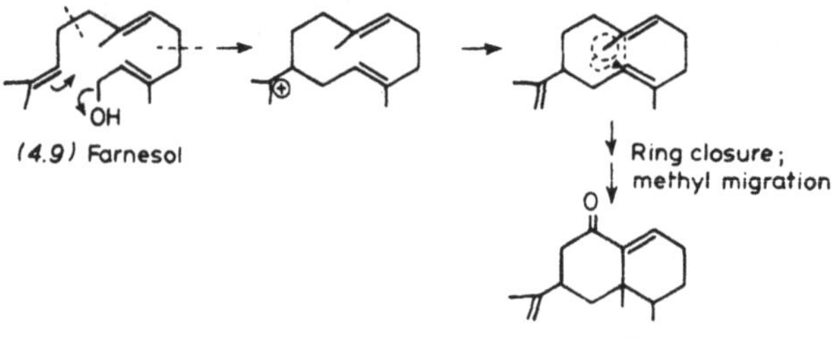

*(4.9)* Farnesol

Ring closure;
methyl migration

*(4.10)* Eremorphilone

**Scheme 4.1**

The multiples of isoprene units in a terpene serve to classify it: C$_{10}$, monoterpenes; C$_{15}$, sesquiterpenes; C$_{20}$, diterpenes; C$_{30}$, triterpenes to which the steroids are closely related; C$_{25}$, sesterpenes. It is a curious thing to note that the isoprene rule, sound and powerfully useful not least for structure elucidation, has unnatural foundations. Isoprene *(4.1)* is not a naturally occurring compound. Instead the clay from which the terpenes are fashioned in so

---

*Squalene is formed by head-on junction of two farnesol units and so obviously the head-to-tail pattern is interrupted (see Section 4.4).

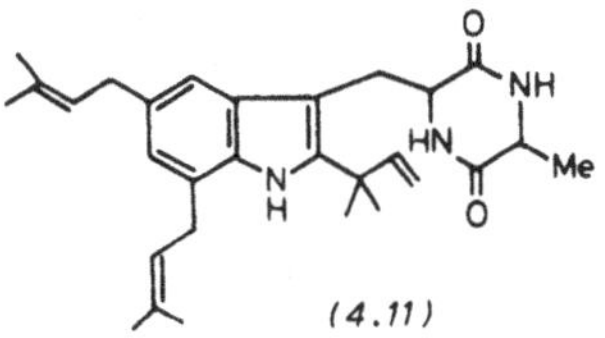

*(4.11)*

orderly a manner is isopentenyl pyrophosphate *(4.14)* and dimethylallyl pyrophosphate *(4.15)*. The latter is seen in many natural compounds e.g. *(4.11)* as a 'prenyl' unit which may also be incorporated as part of a ring system. Mevalonic acid *(4.13)* is the precursor for isopentenyl pyrophosphate *(4.14)*, and hence other terpenes. The discovery of this was in large measure a serendipitous* one. It was known that *(4.13)* was a nutrient for bacterial growth and *(4.13)* was then tried as a steroid precursor with positive results [8]. Detailed later work showed that mevalonic acid is in turn derived, as the (3R)-isomer *(4.13)* by a major pathway from acetyl-coenzyme A (Scheme 4.2; cf section 1.1.2, Scheme 1.2) [1, 2, 9, 10]. (A minor pathway is from the

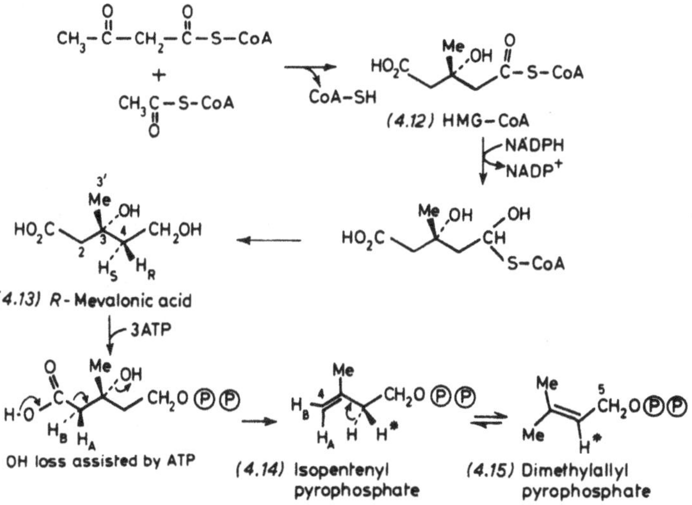

**Scheme 4.2**

$\alpha$-amino acid, leucine.) The initial steps involve reactive thio-esters (Co-A esters) and lead to $\beta$-hydroxy-$\beta$-methylglutaryl-CoA [*(4.12)* HMG-CoA]. The conversion of *(4.12)* into mevalonic acid *(4.13)* is irreversible; the mevalonic acid has no metabolic future except in the formation of terpenes and steroids.

---

*Serendipity: defined by Horace Walpole as the knack of making happy discoveries by chance (or intuition?).

Because of the biological importance of steroids, the main thrust in the investigation of terpene and steroid biosynthesis has been concerned with the latter. These investigations, surely amongst the most notable of biosynthetic studies, are marked by far-seeing intuition, intellectual and practical brilliance, and outstanding results relating to an intricate pathway (sections 4.2 and 4.4).

## 4.2 STEROIDS [1, 6, 10–13]

It was surmised fifty years ago that cholesterol (*4.18*) could be formed by cyclization of squalene (*4.8*), first isolated from a species of shark (*Squalus*). The side-chain of cholesterol, and other steroids, is isoprenoid in appearance and the labelling by [$^{14}$C]acetate was consistent with such an origin. The labelling pattern in the rings [▲ = labelling sites in (*4.18*) from [1-$^{14}$C]-labelled material] in conjunction with a consideration of possibilities raised by the structure of lanosterol (*4.6*) suggested that squalene (*4.8*) was folded as shown (in a manner different to the original guess) to give lanosterol (*4.6*) after a series of 1,2-shifts (Scheme 4.3). Modification of (*4.6*) by loss

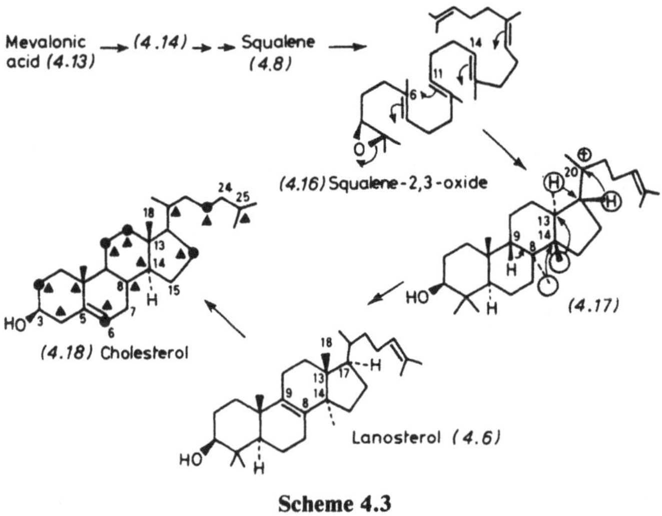

Scheme 4.3

essentially of three carbon atoms would give cholesterol (*4.18*). This was supported by showing that squalene was formed *in vivo* from acetate and converted into cholesterol [14, 15]. Importantly, mevalonic acid (*4.13*) was also shown to be a cholesterol precursor with labelling by six units in the appropriate manner. [The most recent results have been obtained with $^{13}$C-labelled materials; ● = labelling sites in (*4.18*) from [5-$^{13}$C]-labelled material. For [2-$^{14}$C]-labelling see Scheme 4.4. The reader should trace the acetate and mevalonate labels through Schemes 4.2, 4.3 and 4.4 [16].]

Cell-free extracts of liver and yeast were obtained which would synthesize squalene (*4.8*) from mevalonic acid (*4.13*) in the presence of $Mg^{2+}$, ATP, and NADPH as co-factors. In the absence of NADPH, farnesyl pyrophosphate (*4.19*) accumulated; it was transformed into squalene in the presence of NADPH. In the presence of iodoacetamide as an inhibitor a new and important intermediate, isopentenyl pyrophosphate (*4.14*), was detected [17, 18]. Thus (*4.14*) and (*4.19*) are defined as key staging posts along the route to squalene (*4.8*) and cholesterol (*4.18*).

The cyclization of squalene is aerobic and label from $^{18}O_2$ was incorporated into the C-3 hydroxy-group of cholesterol (*4.18*). The intermediate on which cyclization occurs was identified as squalene-2,3-oxide [(*4.16*), absolute stereochemistry as shown] [19–21]. The oxygenating enzyme is a monooxygenase that requires molecular oxygen and NADPH-cytochrome P-450 reductase plus free FAD to function [22]. The reasonable sequence of cyclization and migration proposed is shown in Scheme 4.3 (the squalene conformation for cyclization required by the lanosterol stereochemistry is chair-boat-chair and initially a chair for ring D). Its correctness rests on several pieces of key information. Conversion of the mevalonic acid (*4.13*) into squalene (*4.8*), involves retention of the (4-*pro-R*)-hydrogen atom [and loss of the (4-*pro-S*)-proton; for a discussion of prochirality see section 1.2.1]. Thus, in the course of biosynthesis by rat liver enzymes, [2-$^{14}$C, (4R)-$^{3}$H]-mevalonate [as (*4.13*)] gave squalene (*4.8*) with six tritium atoms ($^{14}$C:$^{3}$H, 6:6). The derived lanosterol showed a ratio of 6:5; in the mechanism shown (Scheme 4.4, cf Scheme 4.3) the mevalonoid (4-*pro-R*)-protons are shown as tritium atoms; one [at C-9 in (*4.17*)] is lost on formation of lanosterol. The cholesterol formed showed a ratio of 5:3 (one $^{14}$C label is inevitably lost as are two tritiums associated with rings A and B). A single tritium atom was located in the nucleus, at C-17$\alpha$, and two in the side-chain in accord with prediction (Schemes 4.3 and 4.4) [10]. Further evidence is that lanosterol (*4.6*) obtained from [11, 14-$^{3}$H$_2$]squalene-2,3-oxide [as (*4.16*)] in yeast retained essentially all the residual tritium at C-17 in accord with the theory [C-11 tritium becomes the hydrogen atom at C-9 in (*4.17*) and is thus, by the proposed mechanism, lost] [23]. Studies of the mechanism of the cyclization of (*4.16*) indicate that rather than the classical carbonium ion shown in (*4.17*) there may be temporary quenching of the ion by participation of a nucleophile in the cyclizing enzyme [24].

The mechanism shown (Scheme 4.3) involves two 1,2-methyl migrations [see (*4.17*)] (from C-8 to C-14, and C-14 to C-13). [3′,4-$^{13}$C$_2$]Mevalonic acid [as (*4.13*); labels in the same molecule] gave cholesterol (*4.20*) from which dilabelled acetic acid (corresponding to C-13 and attached methyl group) was obtained. This is consistent only with two 1,2-migrations because the methyl group at C-13 must derive from the same molecule of mevalonic acid as C-13 does (i.e. arise from C-14 in (*4.17*)); a 1,3-shift from C-8 to C-13 would have involved a methyl group from a different molecule of mevalonic acid terminating on C-13 [see dotted lines in (*4.20*)], and because of dilution of

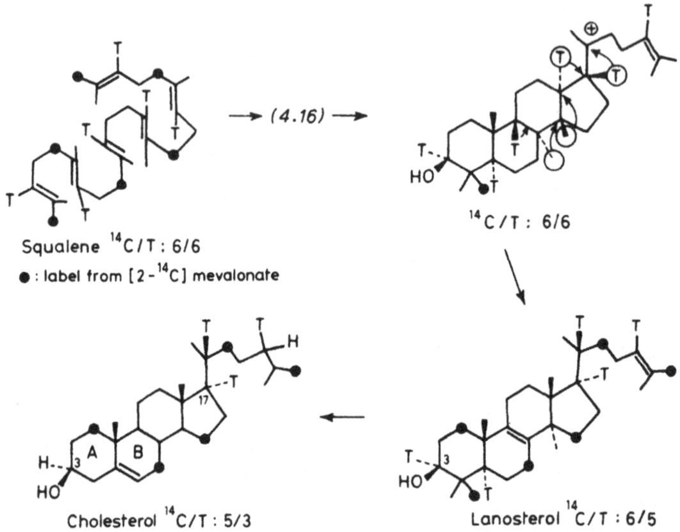

Squalene $^{14}$C/T : 6/6
● : label from [2-$^{14}$C] mevalonate

$^{14}$C/T : 6/6

Cholesterol $^{14}$C/T : 5/3

Lanosterol $^{14}$C/T : 6/5

**Scheme 4.4**

label within the system, adjacent mevalonate units were unlikely to be labelled. So a 1,3-migration would have given largely singly labelled acetic acid [25, 26].

The C-14 to C-13 methyl migration has been shown to occur with retention of configuration. (This might be expected by analogy with similar stereo-chemistry in Wagner-Meerwein rearrangements.) This was shown using mevalonic acid chirally labelled with $^3$H, $^2$H, and $^1$H at C-3′; acetic acid from C-13 and C-18 of cholesterol (*4.18*) was isolated and subjected to the usual enzyme analysis: the acetic acid obtained was of the same configuration as the mevalonate (for a detailed discussion of chiral acetic acid see section 1.2.2) [27].

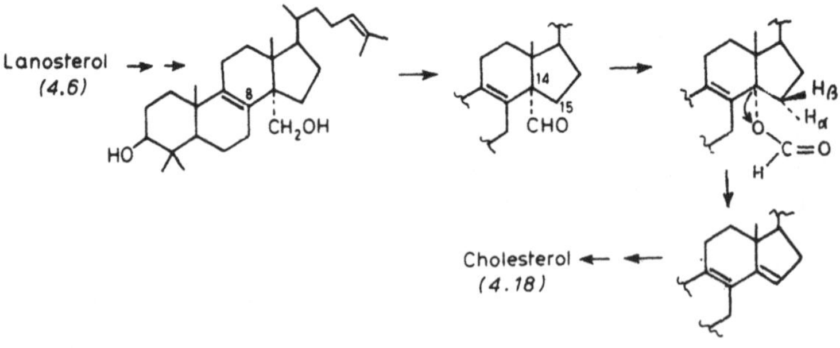

Lanosterol → → (4.6)

CH$_2$OH

HO

CHO

H$_\beta$
··H$_\alpha$
C=O
H

Cholesterol ← ← (4.18)

**Scheme 4.5**

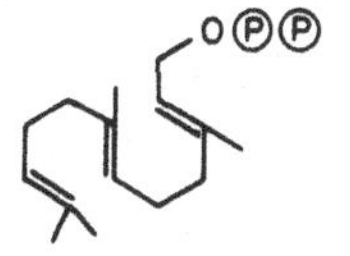

(4.19) Farnesyl
pyrophosphate

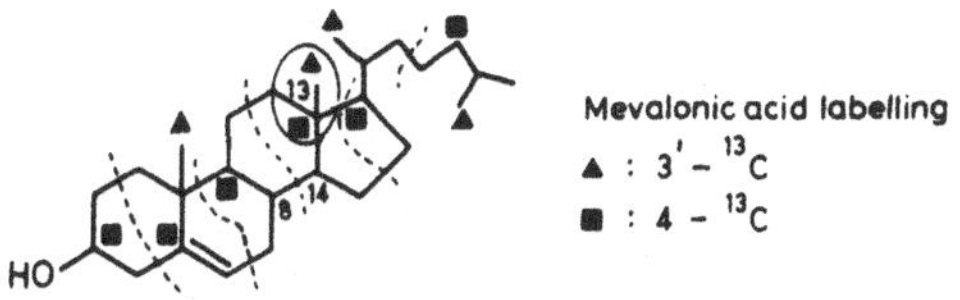

(4.20)

Mevalonic acid labelling

▲ : 3' – $^{13}$C

■ : 4 – $^{13}$C

The bioconversion of lanosterol (*4.6*) into cholesterol (*4.18*) involves most obviously loss of three methyl groups, but also saturation of the side-chain double bond and migration of the $\Delta^{8,9}$ double bond to the $\Delta^{5,6}$ position. The C-4 methyl groups are lost as carbon dioxide [28, 29], the C-14 one as formic acid. It appears that loss occurs first of the methyl group at C-14. This is associated with loss of a 15α proton; subsequent saturation of the new double bond occurs in a *trans* manner (addition of 14α and 15β protons). A possible mechanism for the elimination of the methyl group is illustrated in Scheme 4.5. In this process the postulated *O*-formyl intermediate may be formed through a Baeyer-Villiger type of rearrangement [30–33]. Stepwise oxidation of the C-4 methyl groups and loss via a C-3 ketone (loss of C-3 proton: see Scheme 4.4) has been established by detailed investigation (Scheme 4.6), and occurs with the nuclear double bond at either C-7 or C-8 [28, 29].

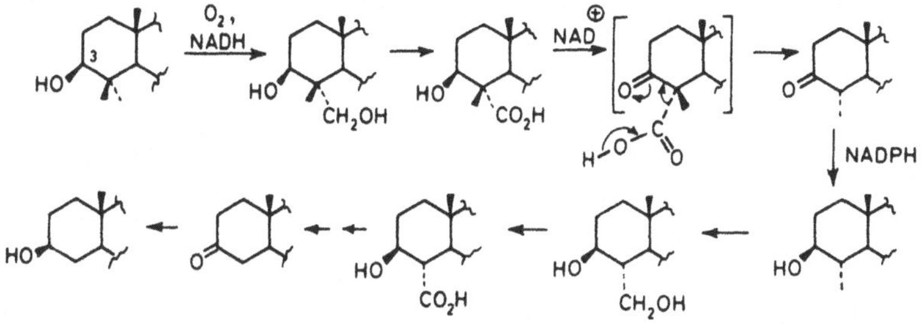

**Scheme 4.6**

Double-bond saturation in the side-chain occurs stereospecifically in the *cis* sense to the *re* face of the double bond: a proton is added to C-24 and a hydride ion from NADPH to C-25. Nuclear double-bond migration is: $\Delta^8 \rightarrow \Delta^7 \rightarrow \Delta^{5,7} \rightarrow \Delta^5$ and involves formation of 7-dehydrocholesterol (*4.23*) as an intermediate with removal of the 5α, C-6α and C-7β protons (information ingeniously gained using chirally tritiated mevalonate). [In ergosterol (*4.21*) biosynthesis the C-7α proton is lost.] Reduction of the 5,7-diene occurs in a *trans* sense but as for side-chain hydrogenation a proton and an NADPH hydrogen is added (C-8β and C-7α, respectively) [34, 35].

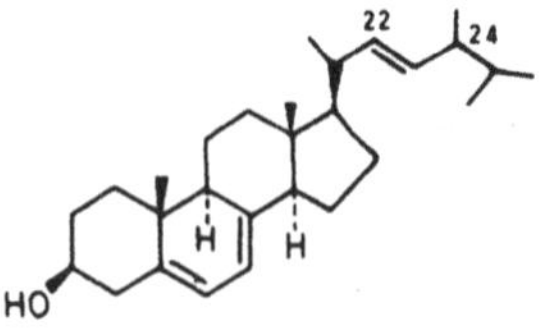

(4.21) Ergosterol 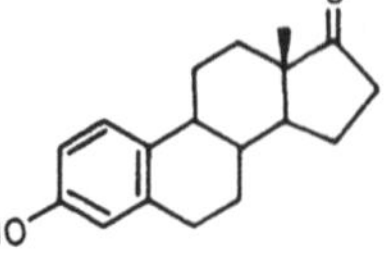        (4.22) Oestrone

From cholesterol (*4.18*) biosynthesis leads through truncation of the side-chain and other modifications to mammalian steroid hormones, e.g. oestrone (*4.22*) [11, 36]. Vitamin $D_3$ (*4.24*) is obtainable by photochemically mediated electrocyclic ring-opening of 7-dehydrocholesterol (*4.23*), followed by a thermal 1,7-sigmatropic hydrogen shift (Scheme 4.7) [37, 38]. The ecdysone insect moulting hormones, e.g. ecdysterone (*4.25*), are further metabolites of cholesterol which is derived from phytosterols taken in the insect's diet [1, 39]. Ergosterol (*4.21*) derives in yeast and fungi from lanosterol (*4.6*) [40].

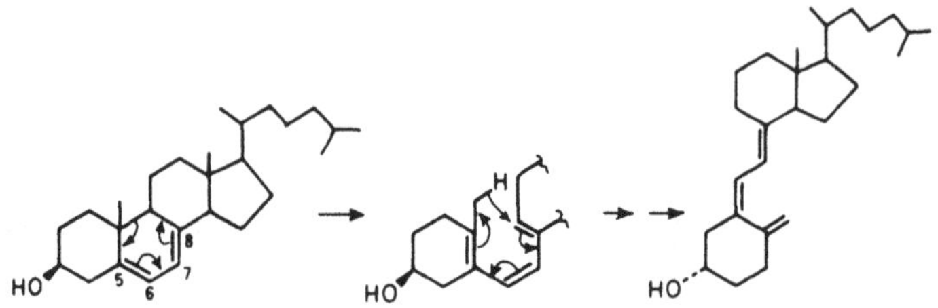

(4.23) 7-Dehydrocholesterol                    (4.24) Vitamin $D_3$

**Scheme 4.7**

The biosynthesis of steroids in photosynthetic organisms proceeds via cycloartenol (*4.27*) rather than lanosterol (*4.6*). Formation of cycloartenol (*4.27*) can be seen as an alternative way of quenching the formal carbonium ion (*4.17*) (or enzyme-stabilized equivalent) [see (*4.26*)]. Tritium n.m.r. analysis of cycloartenol (*4.27*) formed from squalene-2,3-oxide (*4.16*) with a

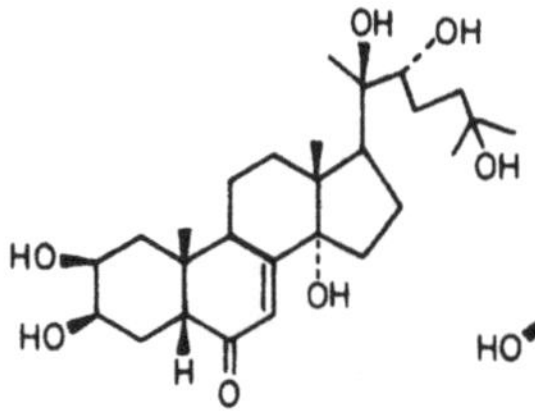

(4.25) Ecdysterone 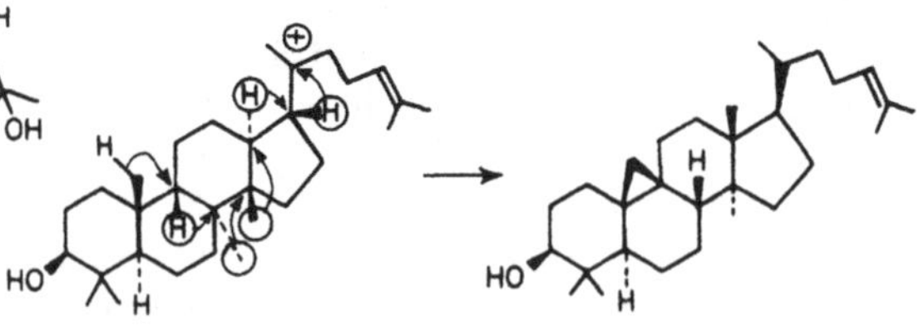        (4.26) = (4.17)                    (4.27) Cycloartenol

chiral methyl group at C-6 ($^1$H, $^2$H, and a high enrichment of $^3$H) shows elegantly that cyclopropane ring-formation occurs with retention of configuration at this carbon atom [41].

Plant sterols are characterized by alkyl substituted side-chains. These substituents and the C-24 methyl group of ergosterol (*4.21*) have been shown to arise from *S*-adenosylmethionine and by tracing the fates of individual hydrogen atoms the mechanistic rationale shown in Scheme 4.8 can be advanced; the slime mould sterol (*4.28*) retains five

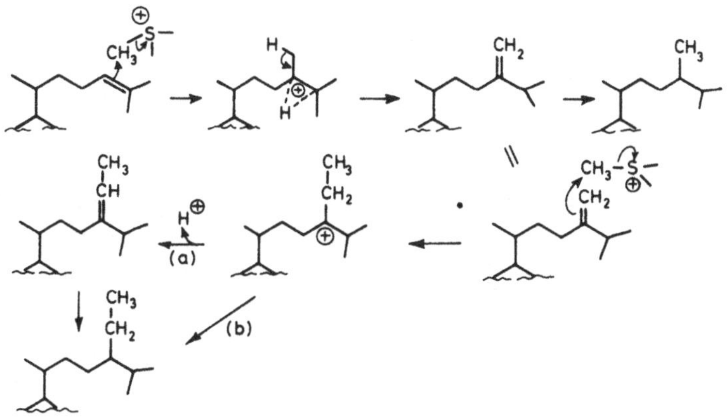

**Scheme 4.8**

methionine protons within its ethyl group, implying formation via route b, whereas poriferasterol (*4.29*) in the alga *Poterioochromonas malhamensis* retains four, thus being formed via route a [42–44]. For plant sterols side-chain alkylation is an early step in cycloartenol modification whereas for ergosterol (*4.21*) [40] in yeast it seems to be at a late stage of biosynthesis.

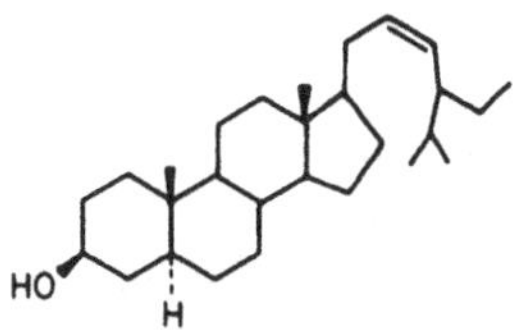

(*4.28*)

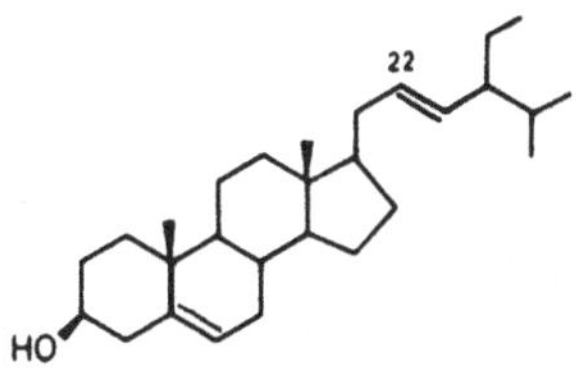

(*4.29*) Poriferasterol

Ergosterol (*4.21*) and some plant sterols, e.g. (*4.29*), contain a C-22 double bond. Generally the timing of its introduction in relation to other biosynthetic steps is unclear. Moreover, although dehydrogenation is stereospecific, which protons are removed vary with the organism type [42, 43, 45].

(4.30) Pregnenolone          (4.31) Digitoxigenin

(4.32) Tigogenin

One line of plant steroid biosynthesis leads through cholesterol (*4.18*) and pregnenolone (*4.30*) to the cardenolides, e.g. digitoxigenin (*4.31*), with inclusion of an acetate unit [46]. On the other hand, spirostanols, e.g. tigogenin (*4.32*), derive from the intact cholesterol skeleton [47, 48] and the steroidal alkaloids are close relatives.

## 4.3 PENTACYCLIC TRITERPENES

A number of groups of triterpenes can be distinguished [7] arising from different ways of discharging an intermediate carbonium ion formalized as (*4.33*) [which is generated by cyclization of squalene-2,3-oxide (*4.16*) in the conformation of a chair (ring A)-chair (B)-chair(C)-boat (initially ring D) (cf lanosterol formation above)]. This is illustrated for the formation of lupeol (*4.34*) and β-amyrin (*4.35*) in Scheme 4.9. The mechanism shown has been confirmed in the usual, powerful way with tritiated mevalonate samples, in this case $(4R)$-$[4\text{-}^3\text{H}]$mevalonic acid. This label would appear at the sites shown in [*4.33*) and, in accord with the route shown, appears as illustrated in β-amyrin (*4.35*). Supporting evidence was obtained using $[4\text{-}^{13}\text{C}]$mevalonate [49, 50].

β-Amyrin has been shown to derive from squalene-2,3-oxide [51, 52], but for fernene (*4.36*) which lacks a C-3 hydroxy-group and the protozoan meta-bolite, tetrahymanol (*4.37*), biosynthesis proceeds through direct protona-tion of squalene (*4.8*) (tetrahymanol is thus best thought of as having its hydroxy-group at C-21 rather than C-3) [53]. The structure of onocerin (*4.39*) suggests that it is formed by cyclization from both ends of squalene-diepoxide (*4.38*).

Interestingly it has been shown that the geminal methyl groups at either end of squalene may retain their identity: in soyasapogenol (*4.40*) the C-4 methyl group, but not the hydroxymethyl group, derives from C-2 of mevalonate

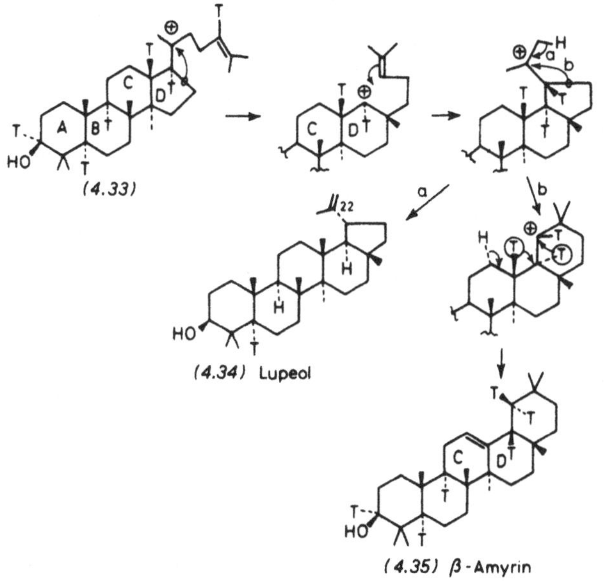

**Scheme 4.9**

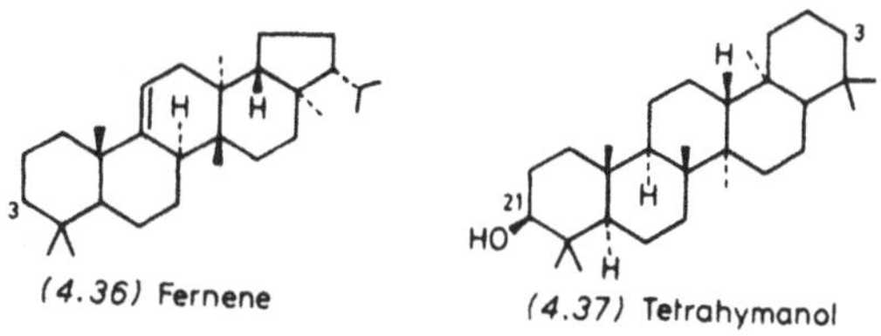

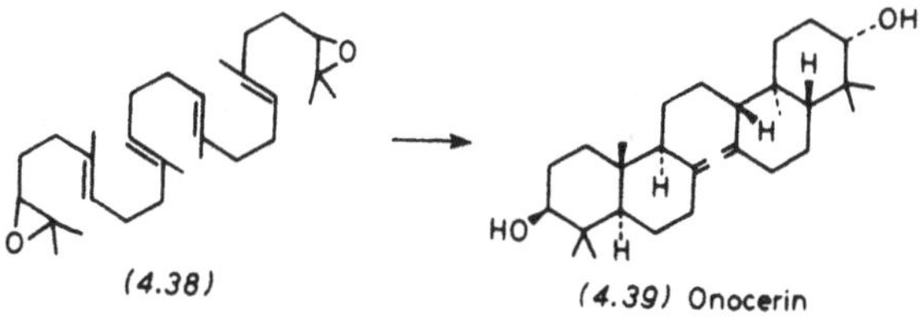

and in lupeol (*4.34*) the C-22 methyl substituent again arises from C-2 of mevalonate [54, 55].

## 4.4 SQUALENE [1, 10]

The preceding discussion has indicated the sequence of squalene (*4.8*) biosynthesis as mevalonate (*4.13*) → isopentyl pyrophosphate (*4.14*) ⇆ dimethylallyl pyrophosphate (*4.15*) → geranyl pyrophosphate (*4.41*) → farnesyl pyrophosphate (*4.19*) → squalene (*4.8*). Each stage involves stereochemical

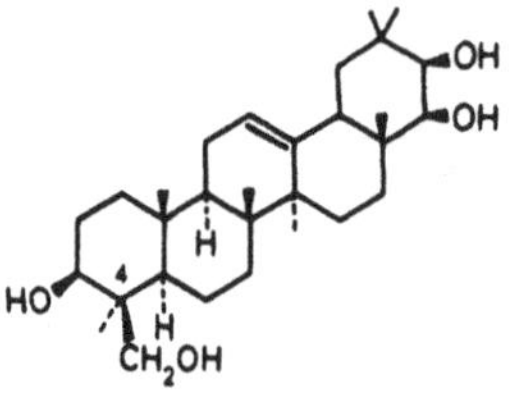

*(4.40)* Soyasapogenol 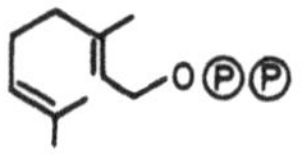          *(4.41)* Geranyl pyrophosphate

change at one or two carbon atoms. The presence of several prochiral centres in the molecules of this sequence meant that for stereochemical changes to be monitored they had to be chirally labelled with deuterium or tritium: e.g. the condensation of isopentenyl pyrophosphate *(4.14)* with dimethylallyl pyrophosphate *(4.15)* (Scheme 4.10) involves loss of one of the (mevalonoid) C-4

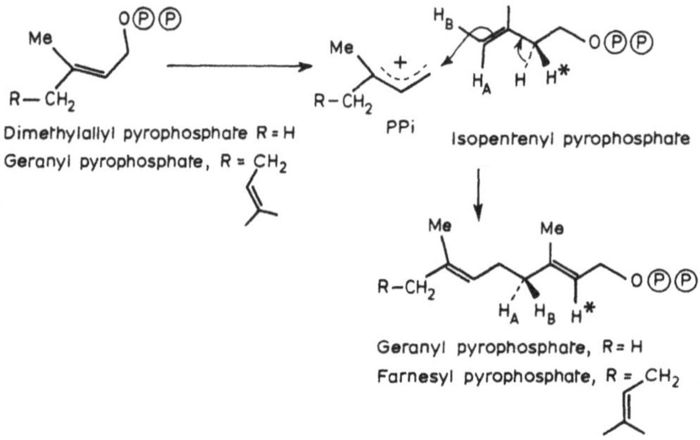

**Scheme 4.10**

protons. Conversion of labelled mevalonate samples into farnesyl pyrophosphate *(4.19)* and analysis of the *(4.19)*, established that the mevalonoid (4-*pro-S*)-proton [see *(4.13)*] was removed stereospecifically during isomerization of *(4.14)* to *(4.15)* and during successive additions of isopentenyl pyrophosphate *(4.14)* to dimethylallyl pyrophosphate *(4.15)* and *(4.41)* yielding farnesyl pyrophosphate *(4.19)*. Extensive, and quite brilliant, investigation [10] has allowed deduction of the stereochemistry, and hence mechanism, at each stage in the biosynthetic sequence to squalene. The pathway to farnesyl pyrophosphate is illustrated in Scheme 4.10; it is to be noted that reaction between *(4.14)* and *(4.15)* results in normal inversion of configuration at C-5 (mevalonate numbering) of the latter and in all successive steps. Addition to the double bond of isopentenyl pyrophosphate is to the same side as proton removal.

Results of notable, elegant experiments particularly with fluorinated analogues have provided compelling evidence that the condensation of iso-pentenyl pyrophosphate with both dimethylallyl pyrophosphate (to give geranyl pyrophosphate) and also with geranyl pyrophosphate (to give farnesyl pyrophosphate) proceeds as shown in Scheme 4.10; ionization to give an allylic carbonium ion is followed by condensation with isopentenyl pyrophosphate and then proton elimination [56].

Using pig liver enzymes the question of the stereochemistry involved at C-4 in the interconversion of dimethylallyl pyrophosphate and isopentenyl pyro-phosphate has been examined. $(2R)$-$[2$-$^3H]$Mevalonate $(4.42)$ gave iso-pentenyl pyrophosphate $(4.43)$. This, in a deuterium oxide medium, would give dimethylallyl pyrophosphate $(4.44)$ with chiral methyl group having the shown, or opposite, stereochemistry. Analysis of this methyl group as chiral acetic acid [as $(4.46)$] (section 1.2.2) derived from farnesyl pyrophosphate

$(4.42)$     $(4.43)$     $(4.44)$

$(4.46)$     $(4.45)$

$(4.45)$ demonstrated that the dimethylallyl pyrophosphate had the stereo-chemistry shown in $(4.44)$. Thus addition of the deuteron is to the *re*-face of $(4.43)$ and is of opposite stereochemistry to that observed in the association of $C_5$ units in farnesyl pyrophosphate formation. Earlier results, alluded to above, establish that the same C-4 proton is lost in isomerization and con-densation (Scheme 4.10). In the isomerization a simple one-step mechanism involving *trans* removal and addition of protons is now possible [57].

Detailed information is available on the enzymes which convert mevalonic acid $(4.13)$ through the reactions depicted in Schemes 4.2 and 4.10; prenyl-transferase catalyses the condensation of $(4.14)$ with $(4.15)$ and of $(4.14)$ with $(4.41)$ [6].

In the formation of squalene $[(4.8) = (4.47)]$ units of farnesyl pyrophos-phate come together not head-to-tail as in their formation, but 'head-on'. In this process the $(1$-*pro-S*$)$-proton in one of the farnesyl pyrophosphate units (A) (Scheme 4.11) is lost to be replaced with retention of configuration by a hydrogen from NADPH [the $(4$-*pro-S*$)$-proton] [10, 11, 58, 59]. It is interesting to note that squalene, a symmetrical molecule, is thus formed in an unsymmetrical way.

The next step in understanding squalene biosynthesis depended on the isolation, from yeast and from rat liver microsomal incubations in the absence of NADPH, of a cyclopropyl compound, presqualene pyrophos-phate $(4.48)$; it is established as an intermediate in squalene biosynthesis (in

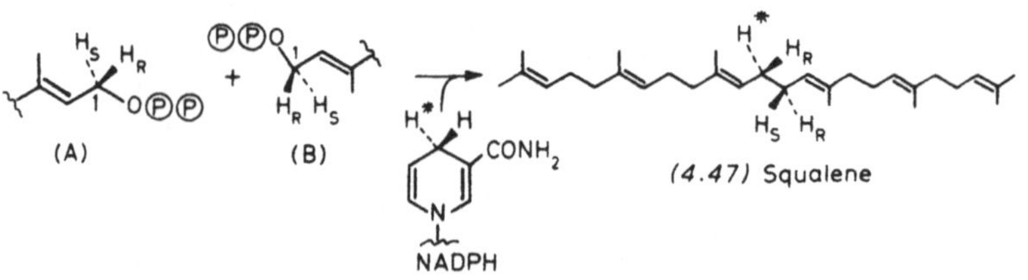

**Scheme 4.11**

microsomal preparations it is converted into squalene in the presence of NADPH) [60–63]. A reasonable mechanism for the biosynthesis of squalene (4.47) via presqualene pyrophosphate (4.48) is shown in Scheme 4.12; the condensation of the farnesyl pyrophosphate molecules may involve the initial formation of an allylic carbonium ion on the left-hand molecule (Scheme 4.12) as in Scheme 4.10. The sequence from farnesyl pyrophosphate to squalene is catalysed by a single enzyme in which there are two catalytic sites, one for the formation of (4.48) and one for the subsequent rearrangement to squalene [64, 65].

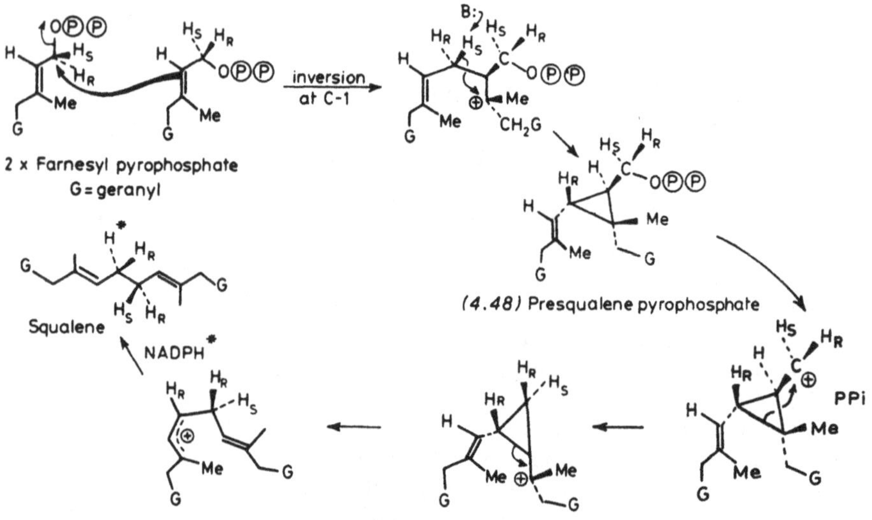

**Scheme 4.12**

## 4.5 MONOTERPENES [1, 66]

Monoterpenes, of which geraniol (4.2) is the simplest, occur very widely in higher plants. Although constituted from but ten carbon atoms the monoter-

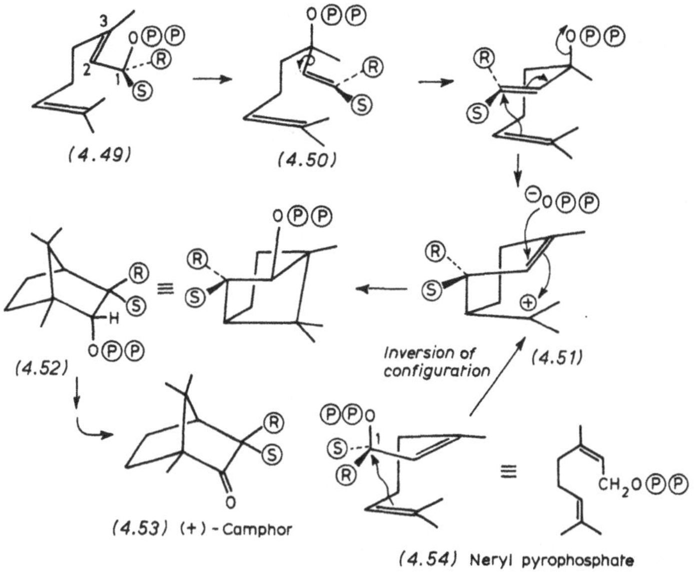

**Scheme 4.13**

penes are found with a rich diversity of acyclic and cyclic structures. The relationship of cyclic monoterpenes, e.g. (*4.3*) and (*4.53*), to geranyl pyrophosphate (*4.49*) [= geraniol (*4.2*)] is an obvious one. The *trans* double bond at C-2—C-3 in (*4.2*) would *a priori* seem to preclude its direct cyclization to give monoterpenes such as camphor (*4.53*) and neryl pyrophosphate (*4.54*) seems a more likely candidate. Debate about what the species is upon which cyclization occurs appears to have been resolved in favour of geranyl pyrophosphate which undergoes isomerization to linalyl pyrophosphate (*4.50*) prior to cyclization: (+)-camphor (*4.53*) in sage (*Salvia officianalis*) and (−)-camphor in *Tanacetum vulgare* are biosynthesized from geranyl pyrophosphate (*4.49*) by way of (+)-bornyl pyrophosphate (*4.52*) and (−)-bornyl pyrophosphate, respectively; in the formation of (*4.52*) neryl pyrophosphate (*4.54*) has been shown not to be a mandatory intermediate. Enzyme preparations of the two plants have been used to study the stereochemical fate of C-1 of (*4.49*) and (*4.54*). For both enantiomers of camphor, both precursors were incorporated without proton loss from C-1, and (*4.49*) was incorporated with retention of stereochemistry and (*4.54*) with inversion of stereochemistry at C-1 (chirally tritiated precursors were used). These results are consistent with a reaction mechanism whereby (*4.49*) is first stereospecifically isomerized to linalyl pyrophosphate [as (*4.50*)] which following rotation about C-2—C-3 to the cisoid conformer then undergoes cyclization. The path to (+)-camphor via one enantiomer of linalyl pyrophosphate, i.e. (*4.50*), is illustrated in Scheme 4.13; the other enantiomer of linalyl pyrophosphate is required for (−)-camphor. Neryl pyrophosphate is thought to cyclize either directly or via the linalyl intermediate without the attendant rotation about C-2—C-3 [67]. Stereochemical aspects of these and other cyclizations have been the subject of detailed scrutiny [68].

The cation (4.51) [≡ (4.55)] can lead to other monoterpenes [7]. An interesting observation is that ( + )-car-3-ene derives as shown in (4.56) from (4.55), which is the reverse of what was originally thought, i.e. (4.57) [● = [2-¹⁴C]mevalonate label in (4.55) and (4.56)] [69].

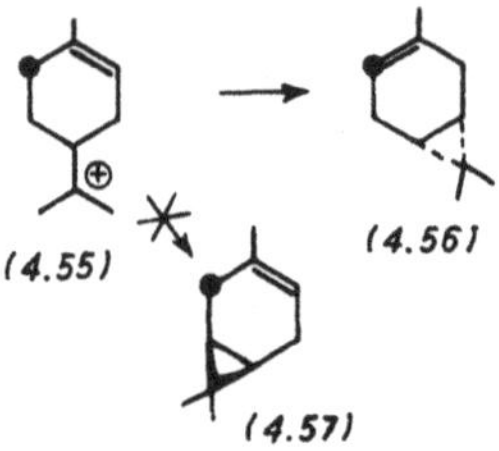

(4.55)    (4.56)

(4.57)

Monoterpenes with a rearranged skeleton are interesting. Artemesia ketone (4.58) is one such compound. It has been shown to derive from [2-¹⁴C]mevalonate consistent with biosynthesis from a compound of the chrysanthemyl type [as (4.59)] which is an analogue of presqualene alcohol [as (4.48)]. Artemesia ketone was also labelled by isopentenyl pyrophosphate (4.14) but not geraniol (4.2) [70]. [As commonly observed with mono-terpenes, mevalonate gave unequal labelling of the two $C_5$ units. This may be explained generally in terms of pools of (4.14) and (4.15) being of different sizes and accessibility. A further general problem is that of obtaining signifi-cant incorporation of mevalonate.] Chrysanthemic acid (4.60) derives from chrysanthemyl alcohol (4.59) but the mechanism of ring closure and the nature of earlier intermediates is unknown [71].

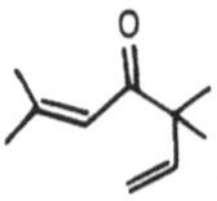

(4.58) Artemesia    (4.59) Chrysanthemyl    (4.60) Chrysanthemic
ketone                   alcohol                      acid

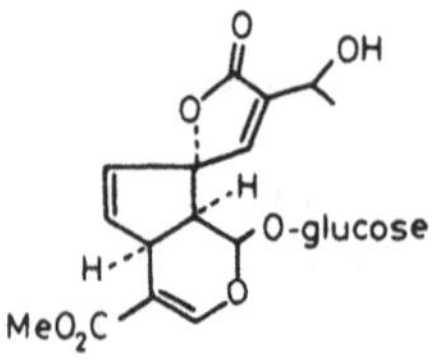

(4.61) Plumieride 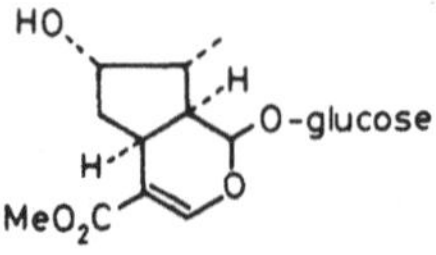    (4.62) Loganin

Iridoids, e.g. plumieride (*4.61*) [72] and loganin (*4.62*), have been the subject of extensive and fruitful study. In particular this is true for the iridoid fragment that is found in terpenoid indole alkaloids. For a discussion the reader is referred to section 6.6.2.

## 4.6 SESQUITERPENES [1, 73]

The enzyme farnesyl pyrophosphate synthetase (prenyl transferase), is responsible for the condensation of (*4.14*) with (*4.15*) to give geranyl pyrophosphate (*4.41*). The same enzyme mediates addition of a further molecule of (*4.14*) to (*4.41*) yielding 2-*trans*-farnesyl pyrophosphate (*4.19*). This is the basic unit for the elaboration of the structurally diverse sesquiterpenes. The results obtained on biosynthesis are as interesting as the sesquiterpene structures. The 2-*cis* isomer (*4.72*) of (*4.19*), required for forming some sesquiterpenes (Schemes 4.14 and 4.16) may apparently be formed by isomerization of (*4.70*). There is an alternative way in which 2-*trans*-farnesyl pyrophosphate (*4.19*) may undergo cyclization in these cases where stereochemical constraints preclude direct ring closure in the *trans*-isomer, and that is through nerolidyl pyrophosphate (*4.63*). This then involves a mechanism which is closely parallel to that discussed above for the biosynthesis of camphor (*4.53*). Detailed investigation of the biosynthesis of trichodiene (*4.64*) provides persuasive evidence for the route via nerolidyl pyrophosphate (*4.63*) illustrated in Scheme 4.14 (cf Scheme 4.13) [68, 74]. On the other hand the

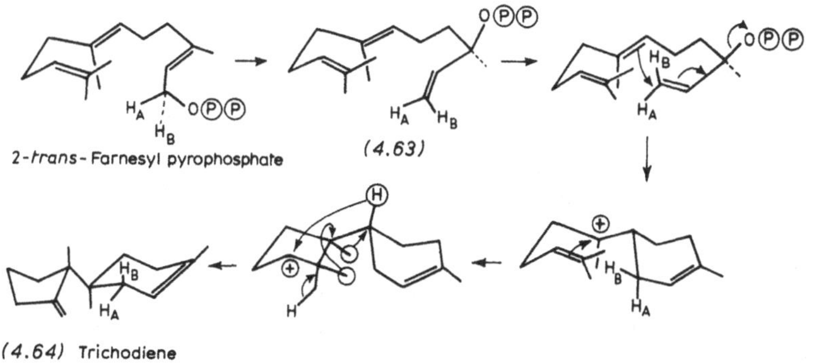

2-*trans*-Farnesyl pyrophosphate    (*4.63*)

(*4.64*) Trichodiene

**Scheme 4.14**

biosynthesis of pentalenene (*4.66*) proceeds by direct cyclization within 2-*trans*-farnesyl pyrophosphate (Scheme 4.15) [68]. Other examples of sesquiterpenes based on the humulene skeleton (*4.65*) [≡ (*4.102*)] are discussed below.

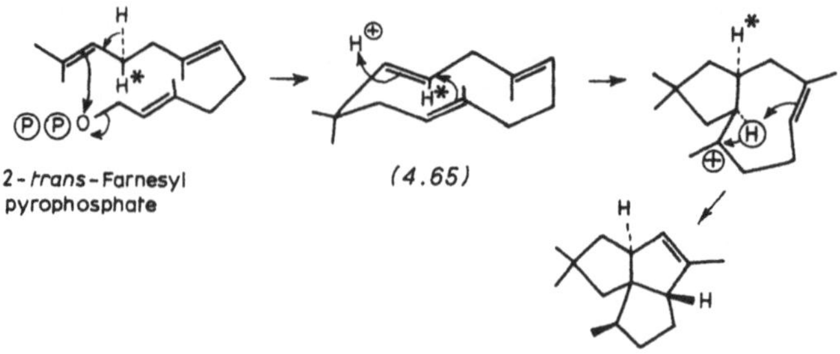

(4.66) Pentalenene

**Scheme 4.15**

Formation of abscisic acid (*4.69*), a plant growth regulator, is established to be from mevalonic acid (*4.13*) and involves loss of the (4-*pro-S*)-proton and (two-thirds) retention of the (4-*pro-R*)-proton (at C-2). This is consistent with formation of (*4.69*), with its terminal *cis* double bond, from *trans*-farnesol (*4.67*) with subsequent isomerization [terpenoid *trans* and *cis* double bonds involve loss of the mevalonoid (4-*pro-S*)-proton; the *cis* double bonds in rubber are formed with loss of the (4-*pro-R*)-proton]. Mevalonic acid samples chirally labelled with tritium at C-2 and C-5 gave results which indicate that the C-4,5 double bond introduced during biotransformation of [(*4.67*) = (*4.70*)] into abscisic acid (*4.69*) occurs with *trans* removal of hydrogen. Similar results have been observed in carotenoid biosynthesis.

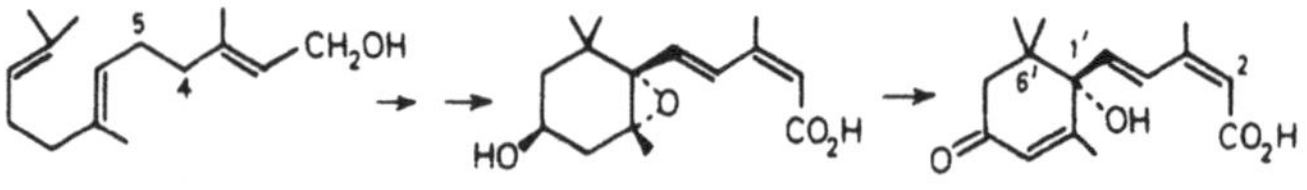

(4.67) Farnesol          (4.68)          (4.69) Abscisic acid

It is *a priori* possible that (*4.69*) is a carotenoid [cf the structure (*4.126*) for β-carotene] degradation product but it has been shown that phytoene (*4.124*) is not a precursor for (*4.69*) (in avocado fruit) although it is as usual incorporated into carotenoids (section 4.9). The observation that the abscisic acid concentration increases twenty-five fold in wilting wheat, has been cleverly exploited in examining the epoxide (*4.68*) as a precursor (fed as a mixture with the isomeric epoxide). A nine-times improvement in incorporation was observed and it was shown that the C-1′ oxygen atom was derived from the epoxide oxygen in (*4.68*). Further results confirmed that (*4.68*), with stereochemistry shown, is the precursor for abscisic acid (*4.69*). Finally, it is to be noted that formation of abscisic acid (*4.69*) and a related metabolite, phaseic acid, involves distinction being kept between C-3′ and C-2 of mevalonate in the C-6′ methyl groups [75]. The biosynthesis of abscisic acid in a fungus has also been studied in detail; the results seem to differ somewhat from those just discussed for higher plants [76].

Germacrene C (*4.71*) has been shown to derive from mevalonate and 2-*trans*-farnesol (*4.67*). Its formation can be seen to be the result of displacement of pyrophosphate from C-1 of farnesyl pyrophosphate (*4.70*) by the distal double bond (Scheme 4.16). An oxygenated relative of (*4.71*), ageratriol has been shown to derive similarly via oxygenated intermediates [77, 78]. Further cyclization of the germacrene skeleton [as (*4.71*)] would afford compounds of type (*4.4*). On the other hand caryophyllene (*4.73*), known to be derived from mevalonate, is expected to derive from 2-*cis*-farnesyl pyrophosphate (*4.72*) [or via nerolidyl pyrophosphate (*4.63*)] in a different manner (Scheme 4.16) [79]. Carotol (*4.74*) is also expected to derive from 2-*cis*-farnesol [or (*4.63*)]; the acetate labelling pattern is consistent with the path shown [80].

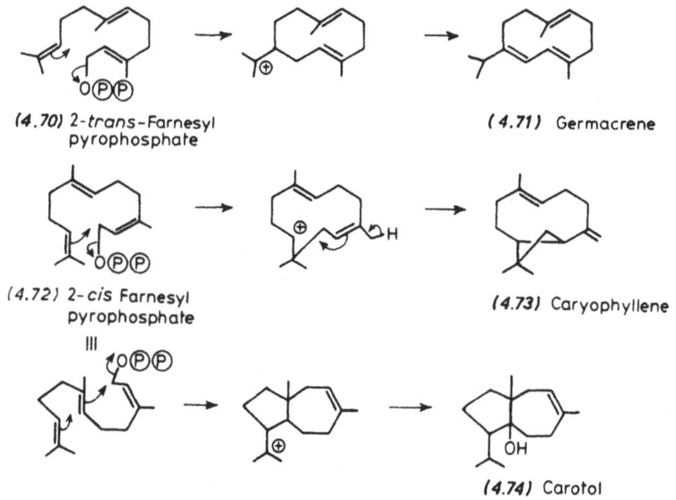

(*4.70*) 2-*trans*-Farnesyl pyrophosphate

(*4.71*) Germacrene

(*4.72*) 2-*cis* Farnesyl pyrophosphate

(*4.73*) Caryophyllene

(*4.74*) Carotol

**Scheme 4.16**

Petasin (*4.75*) although similar to γ-cadinene (*4.4*) differs from it in the siting of one methyl group. A plausible mechanism for its formation, involves a methyl shift from C-10 to C-5 as shown (Scheme 4.17, cf Scheme

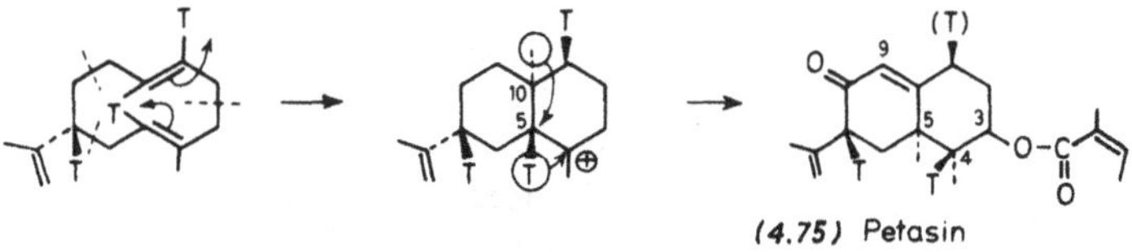

(*4.75*) Petasin

**Scheme 4.17**

(4.76) Capsidiol                    (4.77) γ-Bisabolene

4.1). Normal labelling of the skeleton of (4.75) with [2-$^{14}$C]mevalonate was observed (isopropyl group, C-3, and C-9). Tritium from (4-$R$)[4-$^3$H] mevalonate in particular appeared at C-4 indicating clearly that methyl migration is attended by a 1,2-proton shift. The proposed methyl migration is supported by results with [1,2-$^{13}$C$_2$]acetate in the biosynthesis of capsidiol (4.76) [no coupling between C-5 and C-15, therefore they are from separate mevalonate units, see (4.76)] [81]. The necessary proton shift from C-5 to C-4 in (4.76) (cf Scheme 4.17) has been established by $^2$H n.m.r. analysis of capsidiol (4.76) which had incorporated mevalonic acid (4.13) deuteriated at C-4 [81, 82]. A further point of interest with regard to petasin biosynthesis, studied in *Petasites hybridus*, was the use of an enzyme inhibitor which suppressed the biosynthesis of phytosterols and improved the mevalonate incorporation into (4.75). (A significant enhancement of mevalonate incorporation into caryophyllene in cut stems of the peppermint plant was observed with added sucrose.)

2-*cis*-Farnesol [as (4.72)] is a precursor for γ-bisabolene (4.77) in tissue cultures. A related compound paniculide B has been shown by [1,2-$^{13}$C$_2$]acetate labelling to be formed with similar folding of the farnesyl chain [83]. [$^{13}$C]Acetate and [2-$^2$H, 2-$^{13}$C]acetate labelling indicates that lubimin (4.78) is formed as shown [84, 85].

cf. Scheme 4.17              (4.78) Lubimin

The skeleton of fumagillin (4.80), obtained from *Aspergillus fumigatus*, is not classically isoprenoid, but the deduced acetate and mevalonate labelling patterns indicate it to be a sesquiterpene. Its formation from farnesol has been rationalized as shown in Scheme 4.18, with (4.79) [cf (4.77)] as a key intermediate. Powerful supplementary information has been obtained for the related metabolite, ovalicin (4.81) with [3,4-$^{13}$C$_2$]mevalonate [labelling of C-1 + C-6, C-10 + C-11, by respectively, C-4 and C-3 of single molecules of labelled mevalonate; C-7 (and C-3) were deduced to be part of a fragmented mevalonate unit: see ● positions in Scheme 4.18] and [1,2-$^{13}$C$_2$]acetate (showed incorpora-

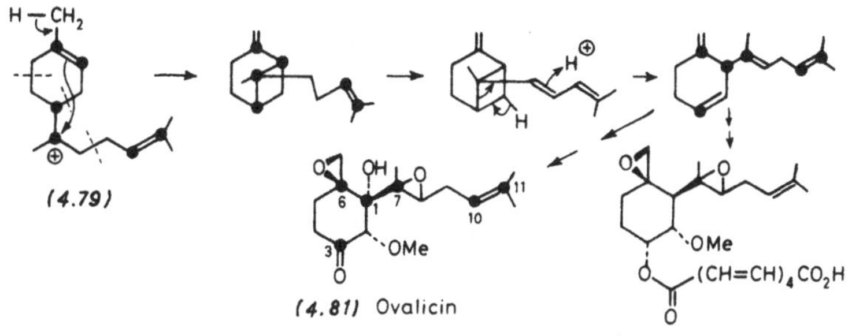

**(4.79)**

**(4.81)** Ovalicin

**(4.80)** Fumagillin

**Scheme 4.18**

tion of six intact acetate units and therefore no methyl migrations had occurred). The data are consistent with the pathway shown in Scheme 4.18 [86].

The $C_7$ side-chain of the *Penicillium* metabolite, mycophenolic acid **(4.84)**, has been shown to derive from mevalonate. The obviously poly-ketide nucleus has its origins in acetate, and the remaining carbons from methionine [marked ■ in **(4.84)**]. The phthalide **(4.82)** is a precursor. After adding it to the culture medium a new, farnesyl based, metabolite **(4.83)**, was isolated which was converted efficiently into **(4.84)** by the organism. The biosynthetic pathway follows as that shown [87].

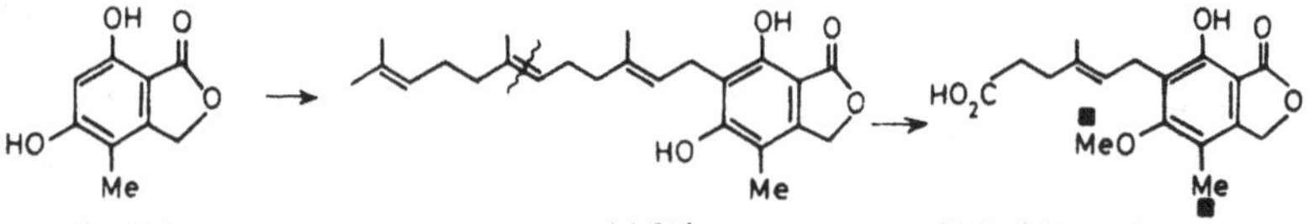

**(4.82)**    **(4.83)**    **(4.84)** Mycophenolic acid

Coriamyrtin **(4.85)**, tutin **(4.86)**, picrotoxinin **(4.87)** (part of the complex metabolite, picrotoxin) and the alkaloid, dendrobine **(4.88)**, are interesting sesquiterpenes which all embrace the structure **(4.93)**. Labelling patterns in **(4.85)** and **(4.86)** from [2-$^{14}$C]- and [4-$^{14}$C]-mevalonate and the specific incorporation of labelled copaborneol **(4.89;** ⌇ : bond broken) indicates the pathway to these compounds shown in Scheme

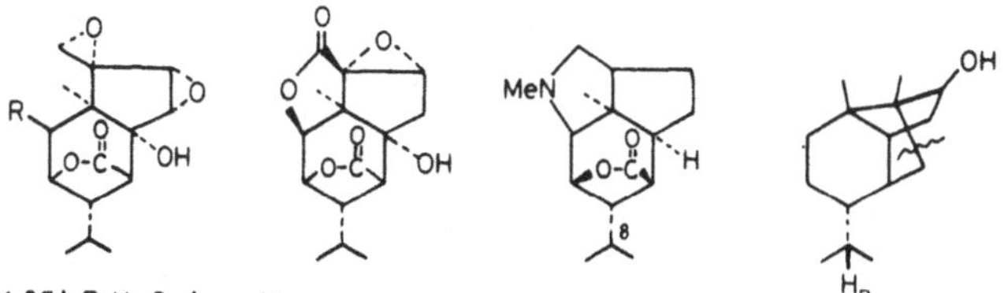

**(4.85)** R=H, Coriamyrtin
**(4.86)** R=OH, Tutin    **(4.87)** Picrotoxinin    **(4.88)** Dendrobine    **(4.89)**

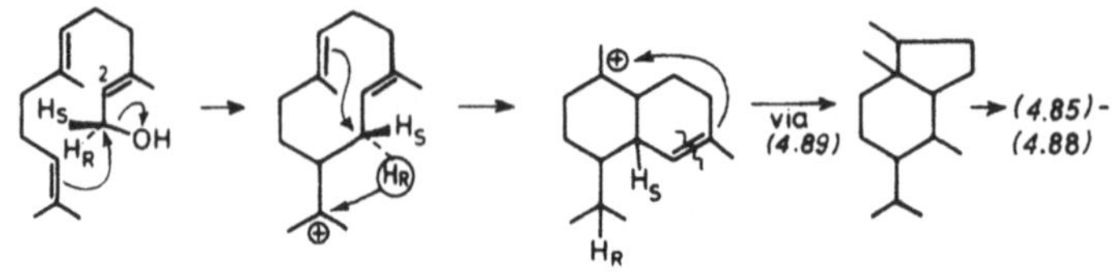


*(4.90)* *(4.91)* *(4.92)* *(4.93)*

**Scheme 4.19**

4.19. Additional information is available on dendrobine (*4.88*) biosynthesis and here it was shown that biosynthesis was accompanied by a 1,3-hydrogen shift, see (*4.91*) [$H_R$:a 5-$^3H_2$ mevalonate label located at C-8 in (*4.88*); five out of the six labels were retained]. A 1,2-shift of hydrogen was excluded in tutin biosynthesis, 2-*trans* rather than 2-*cis*-farnesol is a precursor for dendrobine (*4.88*) and further results show that it is the farnesol (1-*pro-R*)-hydrogen atom which migrates to C-8 in (*4.88*). The deduced pathway to dendrobine and related sesquiterpenes is summarized in Scheme 4.19 (cf Scheme 4.20) [88–91].

In the biosynthesis of sativene (*4.96*) all six of the C-5 mevalonate protons are retained and one of these, a (5-*pro-R*)-proton as in dendrobine formation, migrates to quench the carbonium ion on the isopropyl group in (*4.91*). The data are consistent with a pathway as shown in Scheme 4.20 which parallels that in Scheme 4.19. Interestingly, the isopropyl methyl groups retain their identity from mevalonate unlike tutin (*4.86*) which in this resembles iridoid terpenes. Loss or retention of identity in the methyl groups here may be interpreted as being the result of whether or not rotation of the isopropyl group occurs in (*4.91*) before quenching of the planar carbonium ion [92].

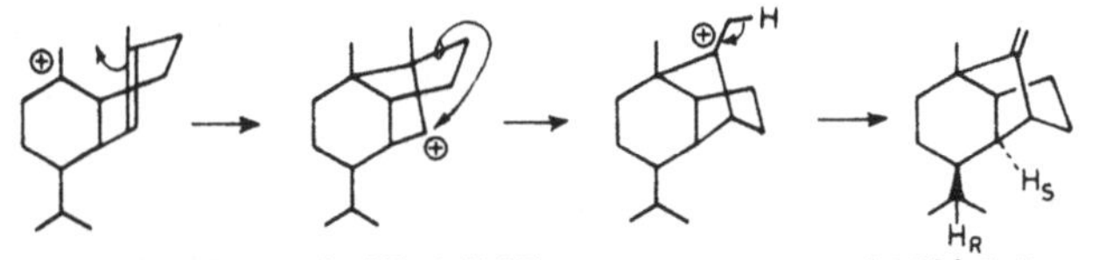

*(4.94)* ≡ *(4.92)* *(4.95)* cf *(4.89)* *(4.96)* Sativene

**Scheme 4.20**

(+)-Longifolene (*4.97*) can be seen to arise from (*4.90*), this time through an intermediate 11-membered, rather than a 10-membered [as (*4.91*)], ring. A 1,3-hydrogen shift is again observed and it is the (5-*pro-R*)-hydrogen which again migrates (Scheme 4.21). The formation of the fungal metabolite, avocettin (*4.99*), may *a priori* be thought of as being via (*4.91*) like other sesquiterpenes discussed above (Schemes 4.19

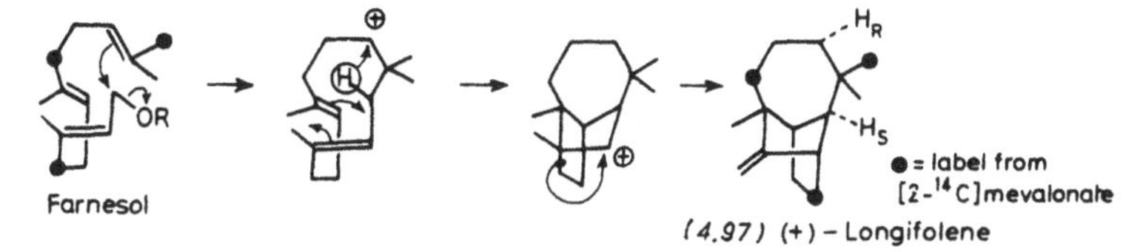

Farnesol    (4.97) (+) – Longifolene

● = label from [2-$^{14}$C]mevalonate

**Scheme 4.21**

and 4.20). However, it is the mevalonoid (5-*pro-S*)-proton which migrates in this case. This is an important difference and the formation of avocettin (*4.99*), sativene (*4.96*) coriamyrtin (*4.85*) through dendrobine (*4.88*), and longifolene (*4.97*) has been very carefully correlated with the conformation adopted by (*4.90*), which has specifically either the 2-*cis* or 2-*trans* configuration, during cyclization [as (*4.100*)]. One result is the particular hydrogen shifts observed. The other is the eventual stereochemistry of the sesquiterpene formed. (For a detailed discussion the reader is referred to [92].) In the biosynthesis of ( − )-longifolene it is the mevalonoid (5-*pro-S*)-hydrogen which migrates [92, 93]. A similar migration is observed in culmorin (*4.101*) biosynthesis [94].

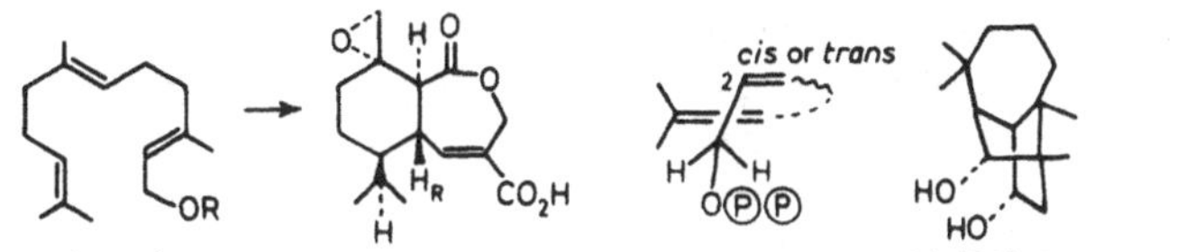

(4.98)        (4.99) Avocettin        (4.100)        (4.101) Culmorin

[1,2-$^{13}$C$_2$]Acetate gave a labelling pattern in 5-dihydrocoriolin C (*4.103*) which could be clearly interpreted in terms of a pathway based on the humulene skeleton (*4.102*) = (*4.65*) (Scheme 4.22) (cf Scheme 4.15). A similar route to related compounds, hirsutic and complicatic acids, and the illudins, is likely [95–98].

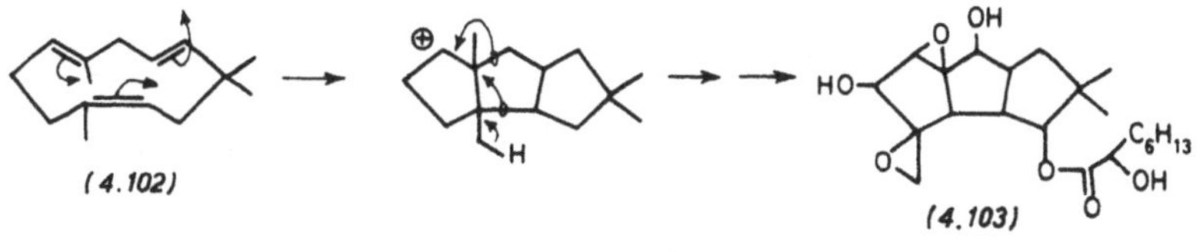

(4.102)        (4.103)

**Scheme 4.22**

Gossypol (*4.104*) is clearly formed by phenol oxidative coupling (section 1.3.1) of two molecules of (*4.105*). Careful examination of the biosynthesis of gossypol gave results which show that each unit of (*4.105*) is constructed from three units of mevalonate via farnesyl

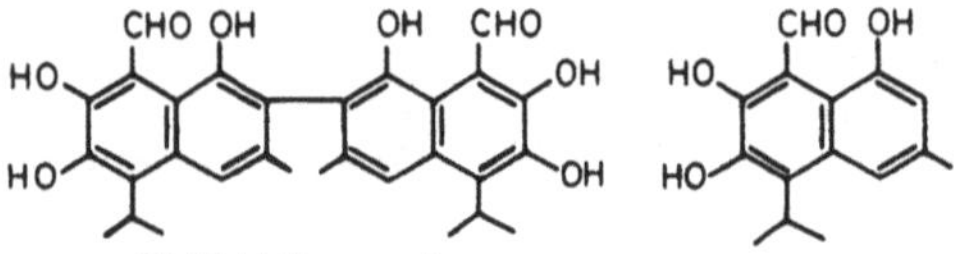

(4.104) Gossypol        (4.105)

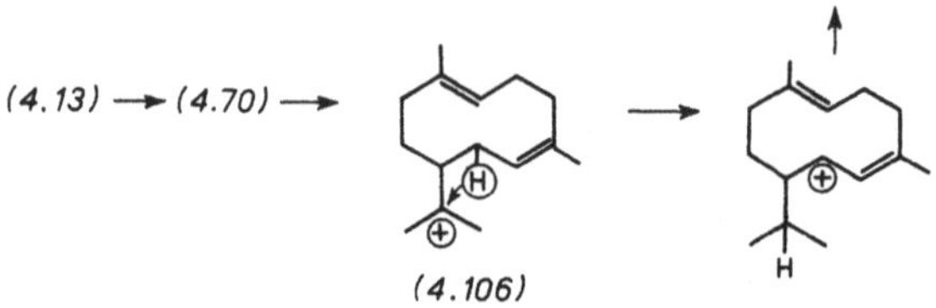

$(4.13) \longrightarrow (4.70) \longrightarrow$

(4.106)

pyrophosphate which is folded and cyclized to give (4.106) [i.e. similar to germacrene (4.71) above]; subsequent conversion into (4.105) involves a 1,3-hydrogen shift as shown [99].

### 4.7 DITERPENES

The diterpenes are elaborations of geranylgeranyl pyrophosphate (4.107). Cyclization may occur to give diterpenes, in a manner similar to that for sesquiterpenes discussed above, by attack of the distal double bond as shown in (4.107). An example is found in the biosynthesis of fusicoccin (4.108). The labelling pattern of (4.108), derived from [3-$^{13}$C]- and (4$R$)-[4-$^3$H]-mevalonic acid, is consistent with the route shown in Scheme 4.23. Mevalonic acid labelled with $^{13}$C and $^2$H has been used ingeniously to confirm the 1,2-shifts of protons in fusicoccin biosynthesis (by $^{13}$C n.m.r. analysis) [100, 101].

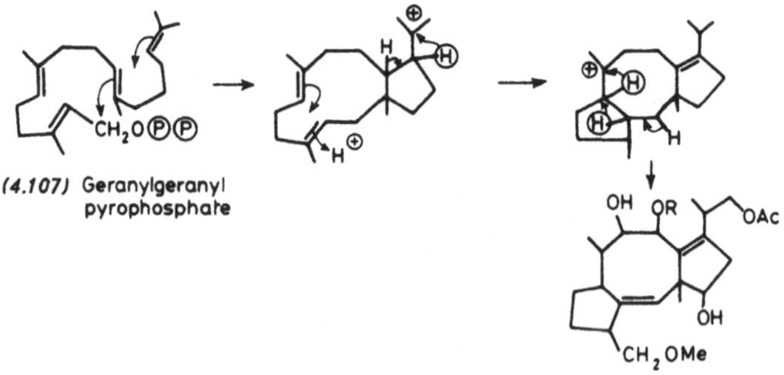

(4.107) Geranylgeranyl
pyrophosphate

(4.108) Fusicoccin

**Scheme 4.23**

An alternative path to diterpenes has some similarity to that which gives the triterpenes and steroids, as (4.107) to (4.109) in Scheme 4.24. No epoxide

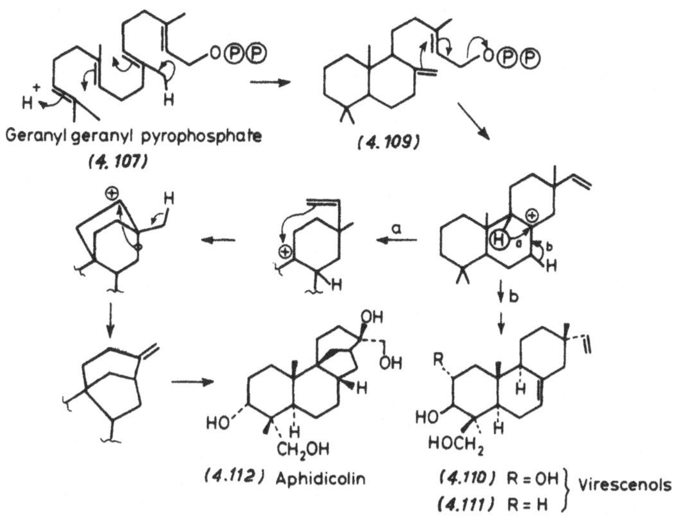

Scheme 4.24

is involved here though, nor is the constant absolute stereochemistry of the triterpenoid-steroid group maintained in the diterpenes.

Simple derivatives of (*4.109*), formed following a second cyclization, are virescenols A (*4.110*) and B (*4.111*) and aphidicolin (*4.112*). Use, in particular, of [1,2-$^{13}$C$_2$]acetate as precursor has established the biosynthetic pathway shown in Scheme 4.24. Allylic displacement of pyrophosphate in the cyclization of (*4.109*) which affords the virescenols occurs with overall *anti* stereochemistry and this is also found in the related biosynthesis of rosenonolactone (below) [102, 103].

Detailed study has been carried out on the biosynthesis of the fungal metabolite, rosenonolactone (*4.113*). [$^{14}$C]Acetate and [$^{14}$C]mevalonate labelling is consistent with the path shown in Scheme 4.25, in which a

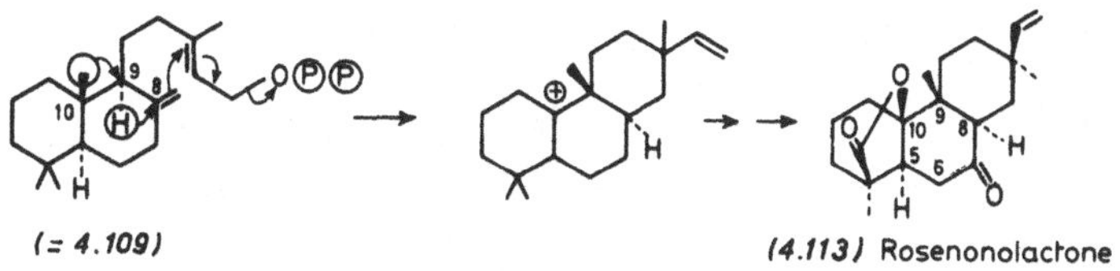

Scheme 4.25

methyl group migrates from C-10 to C-9. There is also a proposed proton shift from C-9 to C-8. The correctness of this proposal was demonstrated when a mevalonoid (4-*pro-R*)-tritium atom was found at C-8

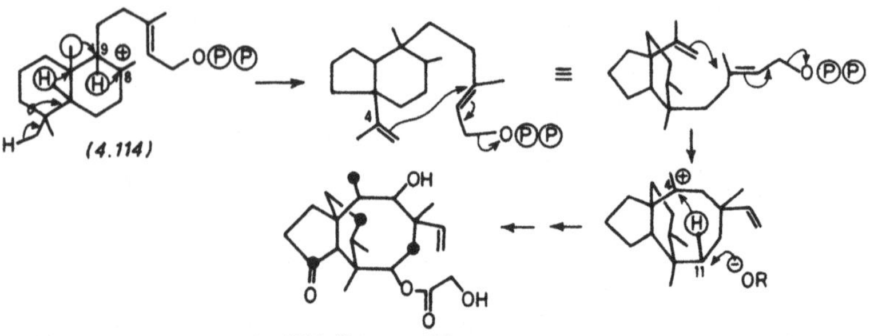

(4.115) Pleuromutilin

**Scheme 4.26**

(incorporation of other [³H]mevalonates also excluded $\Delta^{1(10)}$, $\Delta^{5(10)}$, and $\Delta^{5(6)}$ intermediates) [104]. More extensive migration and also bond fracture within (4.114) [the carbonium ion preceding (4.109)] is thought to occur in the biosynthesis of pleuromutilin (4.115). Again it is the mevalonoid (4-*pro-R*)-hydrogen which migrates from C-9 to C-8, and in a subsequent step (see Scheme 4.26) it is a C-5 (*pro-S*)-proton which moves from C-11 to C-4 (the labels from [2-¹⁴C]mevalonate = ●) [105].

The gibberellins, e.g. gibberellic acid (4.119), first isolated from the fungus, *Gibberella fujikuroi*, are important plant-growth hormones. Labelling by [2-¹⁴C]mevalonate and [1-¹⁴C]acetate established (4.119) to be a modified diterpene. A number of kauranoid diterpenes, including ent-kaurene (4.116), were found with gibberellic acid (4.119) in *G. fujikuroi*. This finding provided an important clue to the course of gibberellic acid (4.119) biosynthesis and (4.116) was then confirmed as a precursor for (4.119) (Scheme 4.27). A soluble enzyme system has also been obtained from fungal and plant sources which will effect the conversion (4.107) → (4.116). Biosynthesis beyond (4.116) involves stepwise oxidation of one of the

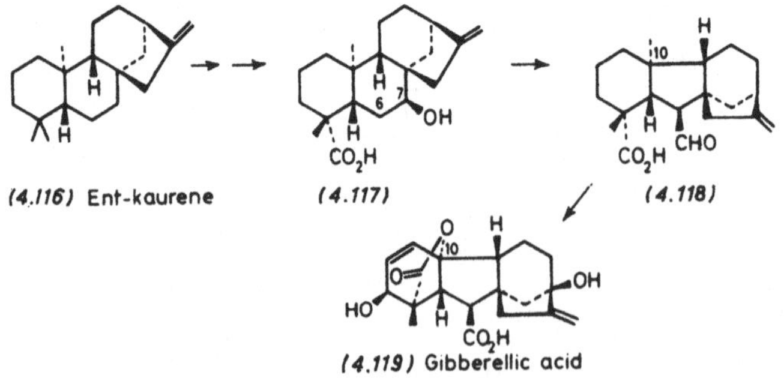

(4.116) Ent-kaurene    (4.117)    (4.118)

(4.119) Gibberellic acid

**Scheme 4.27**

gem-dimethyl groups, and then hydroxylation at C-7 to give (*4.117*). Ring contraction then occurs with proton loss from C-6.

Gibberellin $A_{12}$ aldehyde (*4.118*) is the first detectable gibbane inter-mediate in *G. fujikuroi*. From it a web of oxidation reactions leads through to various gibberellins including gibberellic acid (*4.119*), which also involves loss of the C-10 methyl group, probably via a carboxy-function [106–109].

## 4.8 SESTERPENES [110]

The ophiobolins are examples of this relatively small class of terpenes. Ophiobolin F (*4.121*) has been shown to derive from geranylfarnesyl pyrophosphate (*4.120*). During biosynthesis a C-2 (*pro-R*)-mevalonoid hydrogen undergoes a stereospecific 1,5-shift from C-8 to C-15 (Scheme 4.28).

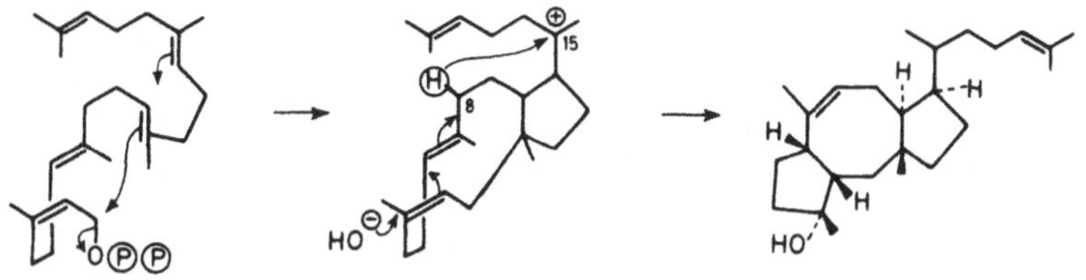

(*4.120*) Geranylfarnesyl pyrophosphate

(*4.121*) Ophiobolin F

**Scheme 4.28**

## 4.9 CAROTENOIDS AND VITAMIN A [3, 5, 111]

Most of the natural carotenoid pigments are tetraterpenes, being formed from two $C_{20}$ geranylgeranyl pyrophosphate units through phytoene (*4.124*).

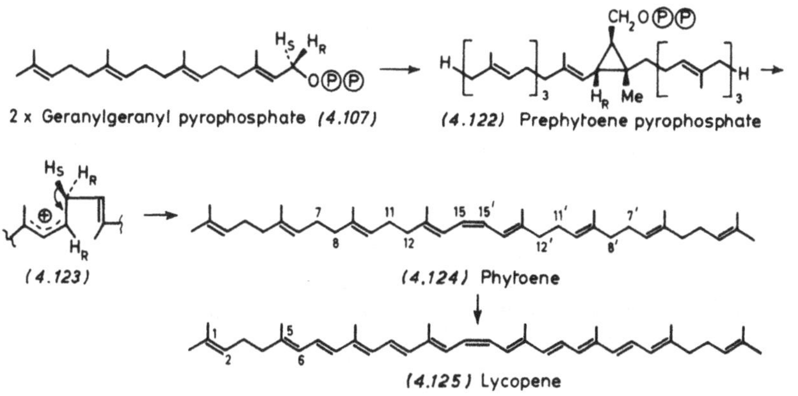

2 x Geranylgeranyl pyrophosphate (*4.107*)

(*4.122*) Prephytoene pyrophosphate

(*4.123*)

(*4.124*) Phytoene

(*4.125*) Lycopene

The mechanism of phytoene formation has much in common with squalene biosynthesis. In particular, head-on fusion of the two $C_{20}$ units occurs via pre-phytoene pyrophosphate (*4.122*), analogous to presqualene pyrophosphate (*4.48*) formation from two $C_{15}$ units (Scheme 4.12). Formation and collapse of (*4.48*) and (*4.122*) appear to be similar, the only apparent difference being in the last step: the carbonium ion (*4.123*) is quenched by proton loss to give phytoene (*4.124*), whereas in squalene formation the corresponding ion is neutralized by hydride donation from NADPH [63].

Most of the phytoene produced *in vivo* appears to be the 15,15'-*cis*-isomer (*4.124*). Formation of (*4.124*) involves loss of the (1-*pro-S*)-proton from each of the two molecules of (*4.107*). In some bacteria the 15,15'-*trans* compound is produced by loss of the (1-*pro-S*)-proton from one molecule of (*4.107*) and the (1-*pro-R*)-proton from the other [at the stage of (*4.123*)].

Desaturation of phytoene (*4.124*) occurs in a stepwise manner, each step extending the conjugation of the system, and leading finally to lycopene (*4.125*) (the desaturation sequence is $\Delta^{11,12}$, then $\Delta^{7,8}$ or $\Delta^{11',12'}$, then respectively $\Delta^{11',12'}$ or $\Delta^{7,8}$, then $\Delta^{7',8'}$). The stereochemistry of hydrogen elimination has been shown to be *trans* in experiments with mevalonate (*4.13*) stereospecifically tritiated at C-2 and C-5 [corresponds, respectively, to C-8 and C-7 in (*4.124*), for example] [112, 113].

Reactions at the C-1,2 double bond of lycopene (*4.125*) lead to a series of acyclic carotenoids characteristic of photosynthetic bacteria, or by cyclization to the mono- and bi-cyclic carotenoids typical of plants. Cyclization is believed to be initiated by protonation at C-2 and to proceed as shown in Scheme 4.29; the three ring types, $\beta$, $\epsilon$ and $\gamma$, are not interconverted. The bio-

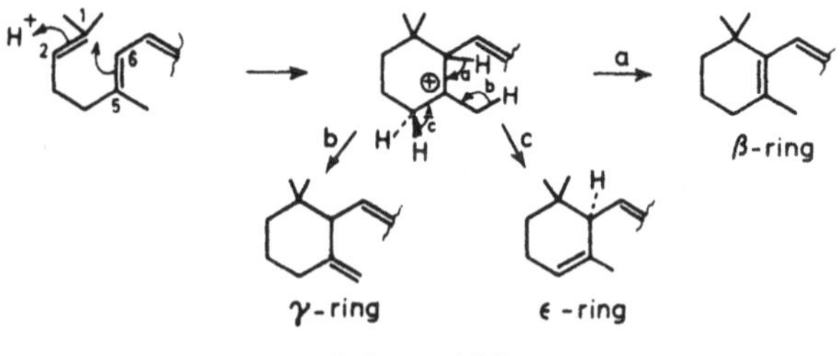

Scheme 4.29

synthesis of $\beta$-carotene (*4.126*) proceeds by two successive reactions to give two $\beta$-rings (Scheme 4.29). $\alpha$-Carotene, on the other hand, is formed by cyclization to give a $\beta$-type ring and then one of the $\epsilon$-type (Scheme 4.29). In certain non-photosynthetic bacteria cyclization may be initiated by an electrophilic species resulting in $C_{45}$ or $C_{50}$ carotenoids, e.g. (*4.128*).

In the cyclization of phytoene (*4.124*) a particular stereochemistry is expected for proton and carbon addition to the C-1 double bond. [2-$^{13}$C]Mevalonate labels the (*E*)-methyl group at C-1 in lycopene

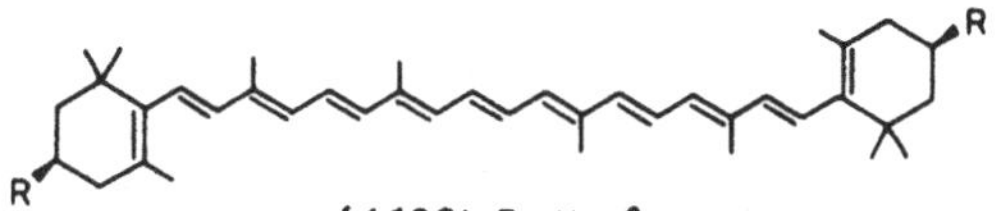

(4.126) R = H, β-Carotene
(4.127) R = OH, Zeaxanthin

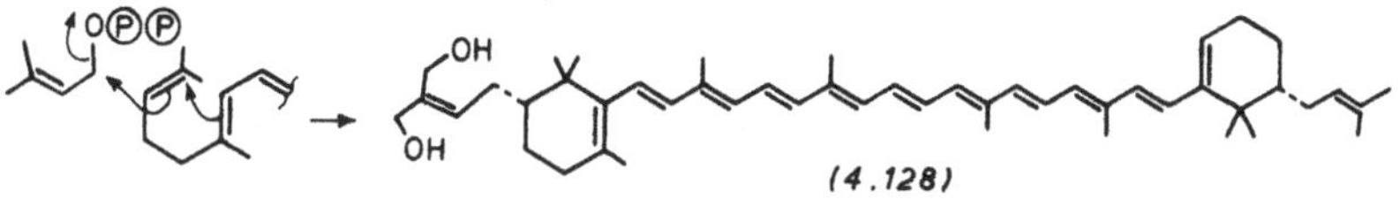

(4.128)

[(4.129) = (4.125); ● = label] and the 1α-methyl group in zeaxanthin [(4.130) = (4.127)] (see Scheme 4.30). Zeaxanthin formed in bacterial cells suspended in deuterium oxide had a 2β-deuteron. It follows that cyclization occurs as shown in Scheme 4.30 [114]. The stereochemistry in the $C_{50}$ carotenoids, e.g. (4.128), implies a different manner of addition to the C-1 double bond.

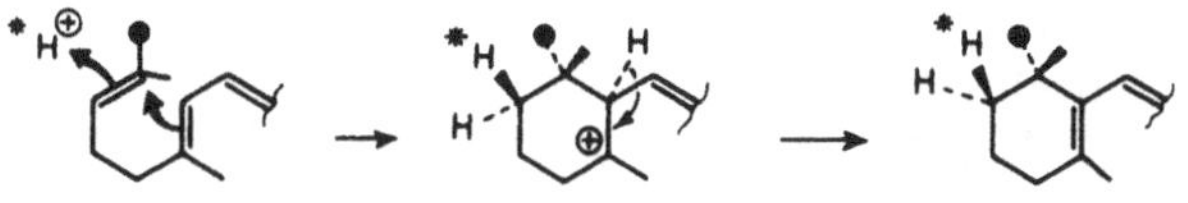

(4.129)                    (4.130)

**Scheme 4.30**

Carotenoids found in photosynthetic bacteria may derive by water or hydrogen addition across the C-1 double bond. Since their formation is inhibited by the same substances as inhibit carotenoid cyclization, it may be concluded that these additions and cyclization are similar: hydrogen and water addition initiated by protonation at C-2.

Modification of the basic carotenoid skeletons occurs to give a variety of metabolites. The most important reaction is C-3 hydroxylation in cyclic carotenoids [as in (4.127)]. In plants and bacteria, it has been shown that zeaxanthin (4.127) is formed by hydroxylation of β-carotene (4.126). The reaction is typical of a mixed-function oxidase: the hydroxyl oxygen comes from molecular oxygen and the hydroxylation occurs with retention of stereochemistry (section 1.2.3).

The most important carotenoid metabolites in animals are the vitamins A: vitamin $A_1$ (4.133), its 3,4-didehydro-derivative (vitamin $A_2$), retinaldehyde (4.132) and its 3,4-didehydro-derivative. These vitamins are the basis of rhodopsin and other visual pigments. Retinaldehyde is formed by oxidative cleavage of β-carotene via the peroxide (4.131) and the reaction is catalysed by an enzyme in the intestinal mucosa.

β-Carotene ⟶ (4.126)

(4.131)

(4.132) R = CHO, Retinaldehyde
(4.133) R = CH₂OH, Vitamin A₁

## REFERENCES

Further reading: [1]–[6].

1. Porter, J.W. and Spurgeon, S.L. (eds) (1981) Vol. 1, (1983) Vol. 2, *Biosynthesis of Isoprenoid Compounds*, Wiley, New York.
2. Goodwin, T.W. (1971) In *Rodd's Chemistry of Carbon Compounds*, 2nd edn (ed. S. Coffey), vol. IIE, pp. 54–137.
3. Goodwin, T.W. (1974) In *Rodd's Chemistry of Carbon Compounds*, vol. II Supplement, pp. 237–89.
4. Hanson, J.R. (1979) In *Comprehensive Organic Chemistry* (ed. D.H.R. Barton and W.D. Ollis), Pergamon, Oxford, vol. 5, pp. 989–1023.
5. Britton, G. (1979) In *Comprehensive Organic Chemistry* (ed. D.H.R. Barton and W.D. Ollis), Pergamon, Oxford, vol. 5, pp. 1025–42.
6. Harrison, D.M. (1985) *Nat. Prod. Rep.* **2**, 525–55; and refs cited.
7. Ruzicka, L. (1959) *Proc. Chem. Soc.*, 341–60.
8. Tavormina, P.A. and Gibbs, M.H. (1956) *J. Amer. Chem. Soc.*, **78**, 6210.
9. Rétey, J., von Stetten, E., Coy, U. and Lynen, F. (1970) *Eur. J. Biochem.*, **15**, 72–6.
10. Popják, G. and Cornforth, J.W. (1966) *Biochem. J.*, **101**, 553–68.
11. Clayton, R.B. (1965) *Quart, Rev. Chem. Soc.*, **19**, 168–230.
12. Mulheim, L.J. and Ramm, P.J. (1972) *Chem. Soc. Rev.*, **1**, 259–91.
13. Goad, L.J. (1970) In *Natural Substances Formed Biologically from Mevalonic Acid* (ed. T.W. Goodwin), Academic Press, New York, pp. 45–77.
14. Woodward, R.B. and Bloch, K. (1953) *J. Amer. Chem. Soc.*, **75**, 2023–4.
15. Cornforth, J.W., Gore, I. and Popják, G. (1957) *Biochem. J.*, **65**, 94–109.
16. Popják, G., Edmond, J., Anet, F.A.L. and Easton, N.R. jun., (1977) *J. Amer. Chem. Soc.*, **99**, 931–5.
17. Schechter, I. and Bloch, K. (1971) *J. Biol. Chem.*, **246**, 7690–6.
18. *Methods in Enzymology* (1969) (ed. R.B. Clayton), Academic Press, New York, vol. 15.
19. van Tamelen, E.E., Willet, J.D., Clayton, R.B. and Lord, K.E. (1966) *J. Amer. Chem. Soc.*, **88**, 4752–4.
20. Corey, E.J., Russey, W.E., Ortiz de Montellano, P.R. (1966) *J. Amer. Chem. Soc.*, **88**, 4750–1.

21. van Tamelen, E.E. and Hopla, R.E. (1979) *J. Amer. Chem. Soc.*, **101**, 6112–4.
22. Ono, T., Takahashi, K., Odani, S. *et al.* (1980) *Biochem. Biophys. Res. Comm.*, **96**, 522–8.
23. Barton, D.H.R., Mellows, G., Widdowson, D.A. and Wright, J.J. (1971) *J. Chem. Soc. (C)*, 1142–8.
24. Hérin, M., Sandra, P. and Krief, A. (1979) *Tetrahedron Lett.*, 3103–6.
25. Cornforth, J.W., Cornforth, R.H., Pelter, A. *et al.* (1959), *Tetrahedron*, **5**, 311–39.
26. Mandgal, R.K., Tchen, T.T. and Bloch, K. (1958) *J. Amer. Chem. Soc.*, **80**, 2589–90.
27. Clifford, K.H. and Phillips, G.T. (1976) *Eur. J. Biochem.*, **61**, 271–86.
28. Miller, W.L. and Gaylor, J.L. (1970) *J. Biol. Chem.*, **245**, 5375–81.
29. Hornby, G.M. and Boyd, G.S. (1970) *Biochem. Biophys. Res. Comm.*, **40**, 1452–4.
30. Caspi, E., Ramm, P.J. and Gain, R.E. (1969) *J. Amer. Chem. Soc.*, **91**, 4012–3.
31. Ramm, P.J. and Caspi, E. (1969) *J. Biol. Chem.*, **244**, 6064–73.
32. Akhtar, M., Alexander, K., Boar, R.B. *et al.* (1978) *Biochem. J.*, **169**, 449–63.
33. Gibbons, G.F., Pullinger, C.R. and Mitropoulos, K.A. (1979) *Biochem. J.*, **183**, 309–15.
34. Aberhart, D.J. and Caspi, E. (1971) *J. Biol. Chem.*, **246**, 1387–92.
35. Wilton, D.C. and Akhtar, M. (1970) *Biochem. J.*, **116**, 337–9.
36. Sih, C.J. and Whitlock, H.W. jun., (1968) *Ann. Rev. Biochem.*, **37**, 661–94.
37. Georghiu, P.E. (1977) *Chem. Soc. Rev.*, **6**, 83–107.
38. de Luca, H.F. and Schnoes, H.K. (1976) *Ann. Rev. Biochem.*, **45**, 631–66.
39. Rees, H.H. and Goodwin, T.W. (1974) *Biochem. Soc. Trans.*, **2**, 1027–32.
40. Barton, D.H.R., Corrie, J.E.T., Marshall, P.J. and Widdowson, D.A. (1973) *Bio-org. Chem.*, **2**, 363–73.
41. Altman, L.J., Han, C.Y., Bertolino, A. *et al.* (1978) *J. Amer. Chem. Soc.*, **100**, 3235–7.
42. Goad, L.J. and Goodwin, T.W. (1972) *Progr. Phytochem.*, **3**, 113–98.
43. Knights, B.A. (1973) *Chem. Brit.*, **9**, 106–11.
44. Seo, S., Uomori, A., Yoshimura, Y. and Takeda, K. (1983) *J. Amer. Chem. Soc.*, **105**, 6343–4.
45. Bimpson, T., Goad, L.J. and Goodwin, T.W. (1969) *Chem. Comm.*, 297–8.
46. Caspi, E. and Hornby, G.M. (1968) *Phytochemistry*, **7**, 423–7.
47. Tschesche, R., Hulpke, H. and Fritz, R. (1968) *Phytochemistry*, **7**, 2021–6.
48. Tschesche, R. and Spindler, M. (1978) *Phytochemistry*, **17**, 251–5.
49. Seo, S., Tomita, Y. and Tori, K. (1975) *J. Chem. Soc. Chem. Comm.*, 270–1.
50. Rees, H.H., Britton, G. and Goodwin, T.W. (1968) *Biochem. J.*, **106**, 659–65.
51. Barton, D.H.R., Jarman, T.R., Watson, K.G. *et al.* (1974) *J. Chem. Soc. Chem. Comm.*, 861–2.
52. Delprino, L., Belliano, G., Cattel, L. *et al.* (1983) *J. Chem. Soc. Chem. Comm.*, 381–2.
53. Aberhart, D.J. and Caspi, E. (1979) *J. Amer. Chem. Soc.*, **101**, 1013–9.
54. Arigoni, D. (1958) *Experientia*, **14**, 153–5.
55. Ruzicka, L. (1963) *Pure Appl. Chem.*, **6**, 493–523.
56. Poulter, C.D., Wiggins, P.L. and Le, T.L. (1981) *J. Amer. Chem. Soc.*, **103**, 3926–7; and following paper.

57. Clifford, K., Cornforth, J.W., Mallaby, R. and Phillips, G.T. (1971) *J. Chem. Soc. Chem. Comm.*, 1599–600.
58. Cornforth, J.W. (1969) *Quart. Rev. Chem. Soc.*, **23**, 125–40.
59. Beytia, E., Qureshi, A.A. and Porter, J.W. (1973) *J. Biol. Chem.*, **248**, 1856–67.
60. Muscio, F., Carlson, J.P., Kuehl, L. and Rilling, H.C. (1974) *J. Biol. Chem.*, **249**, 3746–9.
61. Popják, G., Ngan, H.-L. and Agnew, W. (1975) *Bio-org. Chem.*, **4**, 279–89.
62. Ortiz de Montellano, P.R., Castillo, R., Vinson, W. and Wei, J.S. (1976) *J. Amer. Chem. Soc.*, **98**, 3020–1.
63. Altman, L.J., Kowerski, R.C. and Laungani, D.P. (1978) *J. Amer. Chem. Soc.*, **100**, 6174–82.
64. Sandifer, R.M., Thompson, M.D., Gaughan, R.G. and Poulter, C.D. (1982) *J. Amer. Chem. Soc.*, **104**, 7376–8.
65. Poulter, C.D. and Rilling, H.C. (1981) in ref. 1, Vol. 1, pp. 413–41.
66. Banthorpe, D.V., Charlwood, B.V. and Francis, M.J.O. (1972) *Chem. Rev.*, **72**, 115–49.
67. Croteau, R., Felton, N.M. and Wheeler, C.J. (1985) *J. Biol. Chem.*, **260**, 5956–62.
68. Cane, D.E. (1985) *Accounts Chem. Research*, **18**, 220–6.
69. Banthorpe, D.V. and Ekundayo, O. (1976) *Phytochemistry*, **15**, 109–12.
70. Banthorpe, D.V., Doonan, S. and Gutowski, J.A. (1977) *Phytochemistry*, **16**, 85–92.
71. Pattenden, G., Popplestone, C.R. and Storer, R. (1975) *J. Chem. Soc. Chem. Comm.*, 290–1.
72. Yeowell, D.A. and Schmid, H. (1964) *Experientia*, **20**, 250–2.
73. Cordell, G.A. (1976) *Chem. Rev.*, **76**, 425–60.
74. Cane, D.E., Ha, H.-J., Pargellis, C. *et al.* (1985) *Bioorg. Chem.*, **13**, 246–65.
75. Milborrow, B.V. (1983) in ref. 1, Vol. 2, pp. 413–6.
76. Horgan, R., Neill, S.J., Walton, D.C. and Griffin, D. (1983) *Biochem. Soc. Trans.*, **11**, 553–7.
77. Morikawa, K., Hirose, Y. and Nozoe, S. (1971) *Tetrahedron Lett.*, 1131–2.
78. Ballesia, F., Grandi, R., Marchesini, A. *et al.* (1975) *Phytochemistry*, **14**, 1737–40.
79. Croteau, R. and Loomis, W.D. (1972) *Phytochemistry*, **11**, 1055–66.
80. Souček, M. (1962) *Coll. Czech. Chem. Comm.*, **27**, 2929–33.
81. Baker, F.C. and Brooks, C.J.W. (1976) *Phytochemistry*, **15**, 689–94.
82. Hoyano, Y., Stoessl, A. and Stothers, J.B. (1980) *Can. J. Chem.*, **58**, 1894–6.
83. Overton, K.H. and Picken, D.J. (1976) *J. Chem. Soc. Chem. Comm.*, 105–6.
84. Birnbaum, G.I., Huber, C.P., Post, M.L. *et al.* (1976) *J. Chem. Soc. Chem. Comm.*, 330–1.
85. Stoessl, A. and Stothers, J.B. (1983) *Can. J. Chem.*, **61**, 1766–70.
86. Cane, D.E. and Levin, R.H. (1976) *J. Amer. Chem. Soc.*, **98**, 1183–8.
87. Colombo, L., Gennari, C. and Scolastico, C. (1978) *J. Chem. Soc. Chem. Comm.*, 434.
88. Corbella, S., Gariboldi, P., Jommi, G. and Sisti, M. (1975) *J. Chem. Soc. Chem. Comm.*, 288–9.
89. Biollaz, M. and Arigoni, D. (1969) *J. Chem. Soc. Chem. Comm.*, 633–4.

90. Corbella, A., Gariboldi, P., Jommi, G. and Scolastico, C. (1969) *J. Chem. Soc. Chem. Comm.*, 634-5.
91. Edwards, O.E., Douglas, J.L. and Mootoo, B. (1970) *Can. J. Chem.*, **48**, 2517-24.
92. Arigoni, D. (1975) *Pure Appl. Chem.*, **41**, 219-45.
93. Dorn, F., Bernasconi, P. and Arigoni, D. (1975) *Chimia (Switz.)*, **29**, 24-5.
94. Hanson, J.R. and Nyfeler, R. (1975) *J. Chem. Soc. Chem. Comm.*, 824-5.
95. Tanabe, M., Suzuki, K.T. and Jankowski, W.C. (1974) *Tetrahedron Lett.*, 2271-4.
96. Hanson, J.R., Marten, T. and Nyfeler, R. (1976) *J. Chem. Soc. Perkin I*, 876-80.
97. Mellows, G., Mantle, P.G., Feline, T.C. and Williams, D.J. (1973) *Phytochemistry*, **12**, 2717-20.
98. Hikino, H., Miyase, T. and Takemoto, T. (1976) *Phytochemistry*, **15**, 121-3.
99. Masciadri, R., Angst, W. and Arigoni, D. (1985) *J. Chem. Soc. Chem. Comm.*, 1573-4.
100. Banerji, A., Jones, R.B., Mellows, G. *et al.* (1976) *J. Chem. Soc. Perkin I*, 2221-8.
101. Banerji, A., Hunter, R., Mellows, G. *et al.* (1978) *J. Chem. Soc. Chem. Comm.*, 843-5.
102. Cane, D.E., Hasler, H., Materna, J. *et al.* (1981) *J. Chem. Soc. Chem. Comm.*, 280-2.
103. Ackland, M.J., Hanson, J.R., Yeoh, B.L. and Ratcliffe, A.H. (1985) *J. Chem. Soc. Perkin I*, 2705-7.
104. Achilladelis, B. and Hanson, J.R. (1969) *J. Chem. Soc. (C)*, 2010-4.
105. Arigoni, D. (1968) *Pure Appl. Chem.*, **17**, 331-48.
106. Cross, B.E. (1968) *Progress in Phytochem.*, **1**, 195-222.
107. Graebe, J.E., Hedden, P. and MacMillan, J. (1975) *J. Chem. Soc. Chem. Comm.*, 161-3.
108. Dawson, R.M., Jeffries, P.R. and Knox, J.R. (1975) *Phytochemistry*, **14**, 2593-7.
109. Evans, R. and Hanson, J.R. (1975) *J. Chem. Soc. Perkin I*, 633-6.
110. Cordell, G.A. (1974) *Phytochemistry*, **13**, 2343-64.
111. *Pure Appl. Chem.* (1985) **57**, 639-821.
112. Goodwin, T.W. (1972) *Biochem. Soc. Symp.*, **35**, 233-44.
113. McDermott, J.C.B., Britton, G. and Goodwin, T.W. (1973) *Biochem. J.*, **134**, 1115-7.
114. Britton, G., Lockley, W.J.S., Patel, N.J. *et al.* (1977) *J. Chem. Soc. Chem. Comm.*, 655-6.

# 5 The shikimic acid pathway

## 5.1 INTRODUCTION

A compound of unsuspected importance was isolated in 1885 from the fruit of *Illicium religiosum*. To this compound was given the name shikimic acid, a name derived from *shikimi-no-ki* which is the Japanese name for the plant. Shikimic acid (*5.7*), it transpired from the very elegant studies of much later investigators, is a key intermediate in the biosynthesis of the aromatic amino acids, L-phenylalanine, L-tyrosine and L-tryptophan, in plants and micro-organisms (animals cannot carry out *de novo* synthesis using this pathway).

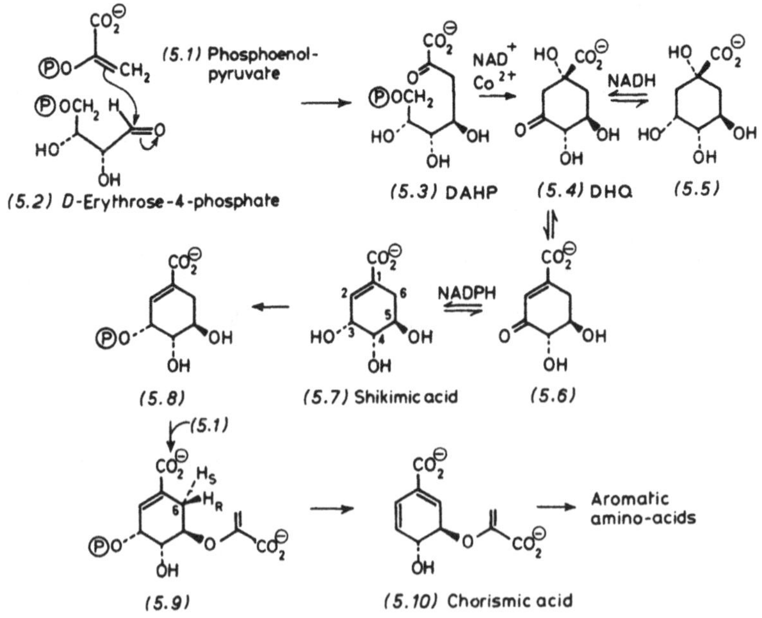

Scheme 5.1

These three aromatic amino acids are individually important precursors for numerous secondary metabolites, and so to some extent are earlier biosynthetic intermediates related to shikimic acid, as the ensuing discussion in this chapter and in Chapters 6 and 7 will show.

The biosynthetic pathway through shikimic acid (*5.7*) to aromatic amino acids, outlined in Scheme 5.1 (acids are shown as anions) is called the shikimic acid or shikimate pathway [1, 2, 5, 6]. It has its origins in carbohydrate metabolism and shows several interesting features, much of it known from detailed examination of the steps involved. The first step is a stereospecific aldol-type condensation between phosphoenolpyruvate (*5.1*) and D-erythrose-4-phosphate (*5.2*) to give 3-deoxy-D-arabinoheptulosonic acid 7-phosphate (*5.3*; DAHP), in which addition occurs to the *si*-face of the double bond in (*5.1*) and the *re*-face of the carbonyl group in (*5.2*) and which has been rationalized in terms of the mechanism shown in Scheme 5.2 [7].

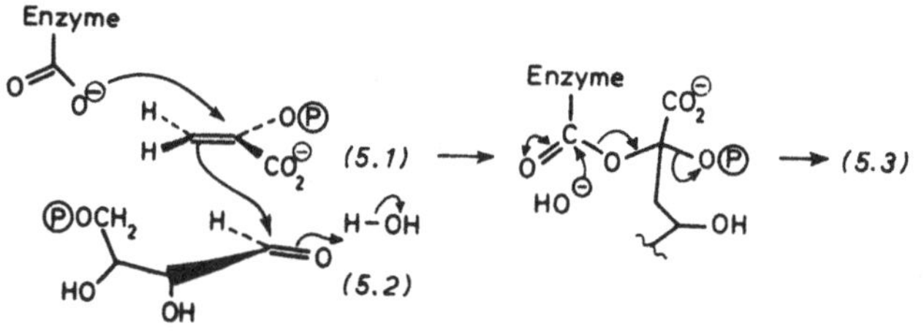

**Scheme 5.2**

Ring closure in DAHP (*5.3*) affords dehydroquinic acid (DHQ) (*5.4*). The mechanism which has been deduced for this reaction is illustrated in Scheme 5.3, in which, apart from reduction and oxidation, *syn*-elimination of inorganic phosphate occurs followed in the last step by an intramolecular aldol reaction, through a chair-like transition state (*5.11*) [8]. DHQ undergoes reversible dehydration to give dehydroshikimic acid (*5.6*). The addition and elimination of water takes place in an unusual *cis* sense and has been postulated to occur in a two-step sequence involving an enamine (*5.12*) formed through the keto-group in (*5.4*) [9].

Two unexceptional steps lead through shikimic acid (*5.7*) to its 3-phosphate (*5.8*). Condensation with phosphoenolpyruvate (*5.1*) gives (*5.9*). This type of biological reaction is almost unique, only one other example being known. The mechanism suggested is illustrated in Scheme 5.4. It has been found from the results of ingenious labelling experiments that the addition reaction has the opposite stereochemistry to the later elimination step [10, 11].

1,4-Conjugate elimination of phosphoric acid converts (*5.9*) into chorismic acid (*5.10*), a reaction which it was shown involves *trans* elimination of the two groups lost [loss of the (6-*pro-R*)-hydrogen] [7].

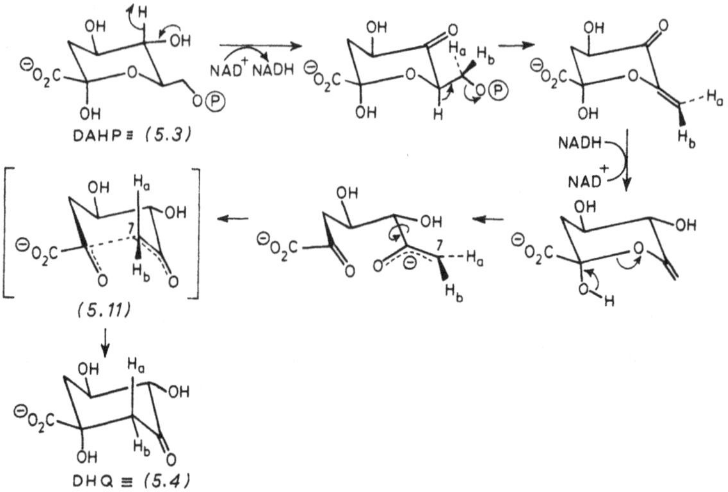

**Scheme 5.3**

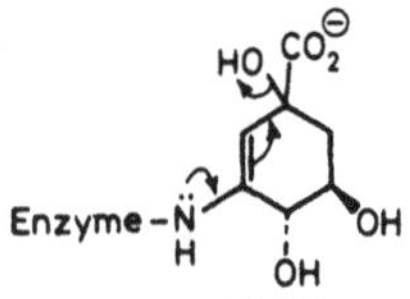

**Scheme 5.4**

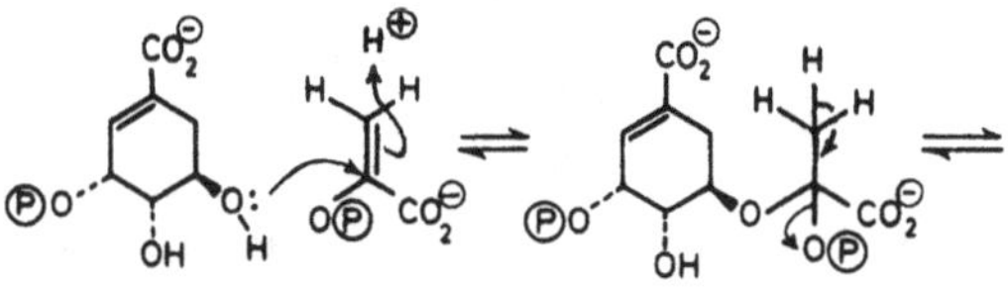

Perhaps the shikimate pathway should be called the chorismate pathway because it is at chorismic acid (*5.10*) that the single line of biosynthesis fragments into different lines whose termini are the vital aromatic amino acids and diverse other compounds. It appears, however, that the biosynthesis of some microbial metabolites (section 7.6.1) may divert from the main line at a stage close to dehydroquinic acid (*5.4*).

Amination of chorismic acid (*5.10*) leads through anthranilic acid (*5.13*) to tryptophan (*5.14*). The formation of phenylalanine (*5.17*) and tyrosine (*5.18*), on the other hand, proceeds via prephenic acid (*5.16*), whose formation from chorismic acid (*5.10*) = (*5.15*) involves what is formally a Claisen

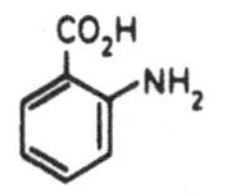

(*5.13*) 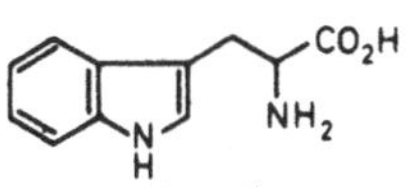  (*5.14*)

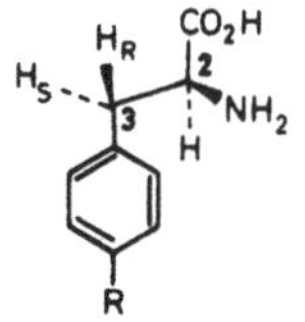

(*5.17*) R=H, Phenylalanine
(*5.18*) R=OH, Tyrosine

rearrangement—a unique biological example. The same reaction can also be achieved by heating aqueous solutions of chorismic acid (*5.10*), but calculation shows that the enzyme (chorismate mutase) from *Aerobacter aerogenes* enhances the rate of reaction (at pH 7.5 and 37°C) by $1.9 \times 10^6$ compared to the purely thermal one. The enzymic reaction appears not to be a concerted reaction but to involve the stepwise rearrangement illustrated in Scheme 5.5 [12].

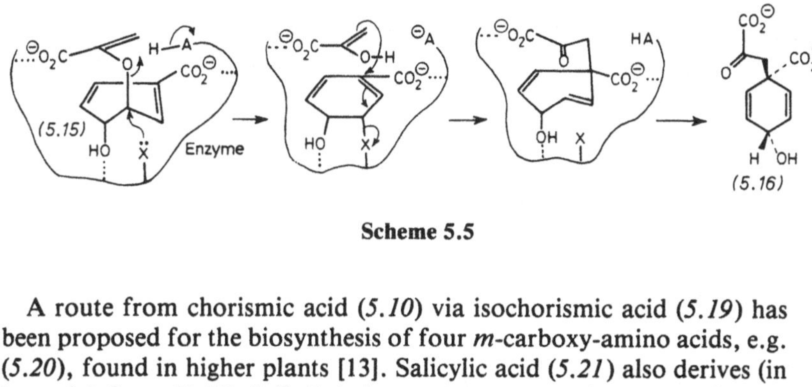

**Scheme 5.5**

A route from chorismic acid (*5.10*) via isochorismic acid (*5.19*) has been proposed for the biosynthesis of four *m*-carboxy-amino acids, e.g. (*5.20*), found in higher plants [13]. Salicylic acid (*5.21*) also derives (in bacteria) from (*5.19*) [14]. It is interesting to note that aromatization may occur, in micro-organisms, of intermediates of the shikimic acid

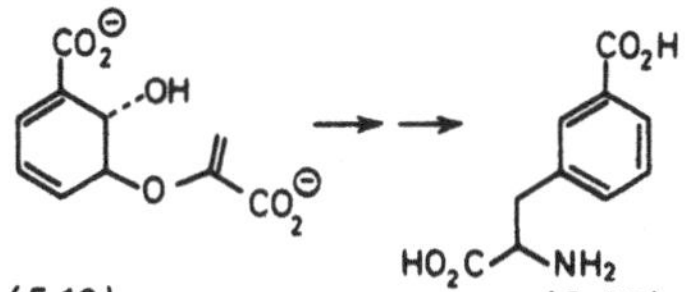

(*5.19*)  (*5.20*) 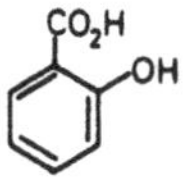  (*5.21*) 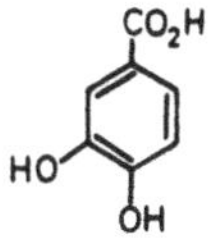  (*5.22*)

pathway just after [as for (5.21)] or before chorismic acid (5.10). This leads, for example, from dehydroshikimic acid (5.6) to protocatechuic acid (5.22) [15, 16]. Aromatization in this way provides access to phenolic compounds by a route different from those via polyketides (Chapter 3) and, in plants, via aromatic amino acids.

In higher plants the polymer, lignin, and various aromatic secondary metabolites, notably many alkaloids (Chapter 6) and flavonoids (section 5.4) are formed from the aromatic amino acids, L-phenylalanine (5.17) and/or L-tyrosine (5.18). [For some alkaloids as well as some microbial metabolites, tryptophan (5.14) is the source of their particular aromatic rings.] There are for these metabolites common pathways leading from phenylalanine, and in some plants tyrosine, to phenylpropanoid ($C_6$-$C_3$) intermediates. The first step from phenylalanine involves the enzyme L-phenylalanine ammonia lyase (known, perhaps affectionately, as PAL), an enzyme widely distributed and well-characterized. Elimination of ammonia occurs to give cinnamic acid (5.23). It involves loss of the (3-*pro-S*)-proton of L-phenylalanine (5.17), and thus occurs in the *anti*-sense [L-tyrosine ammonia lyase functions to remove also the (3-*pro-S*)-proton in tyrosine] [17, 18].

*p*-Coumaric acid (5.24) is the product derived by loss of ammonia from tyrosine (5.18). Much more commonly this acid derives by hydroxylation of cinnamic acid (5.23) formed by ammonia loss from phenylalanine; hydroxylation is accompanied by the usual NIH shift (section 1.3.2) for this *para*-hydroxylation (also *ortho*-hydroxylation [19]). Up to two further phenolic hydroxy-groups may be introduced on (5.24), Scheme 5.6 [20, 21].

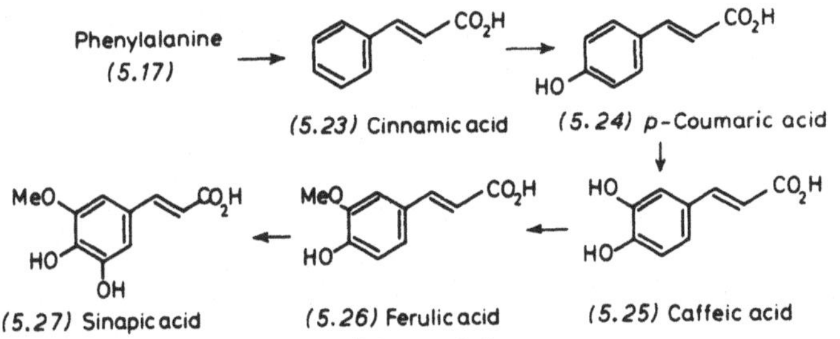

Scheme 5.6

It is interesting to note that in *Mentha* species rosmarinic acid (5.28) has different origins for its two $C_6$-$C_3$ units: independently phenylalanine (5.17)

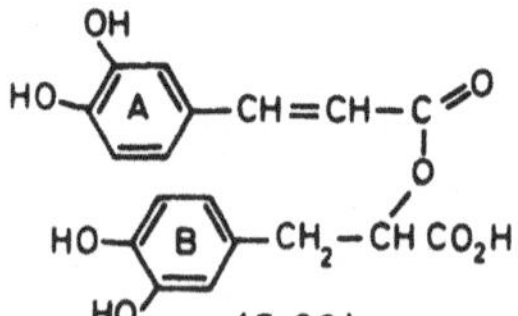

for A and tyrosine (*5.18*) for B [22]. [Commonly in higher plants (*5.17*) is not converted into (*5.18*).]

From the cinnamic acids, (*5.24*)–(*5.27*), β-oxidation and truncation of the side-chains yields a series of benzoic acids [23]. (For gallic acid biosynthesis see [24].) The plant lignins are formed essentially by oxidative polymerization of various cinnamyl alcohols corresponding to the cinnamic acids, (*5.24*), (*5.26*) and (*5.27*) [25].

Podophyllotoxin (*5.33*) is a representative of one of two groups of tumour-inhibitory lignans [26]. This group is distinguished by the presence of three methoxy groups on ring D; the other group, e.g. 4′-demethylpodophyllotoxin (*5.35*), has a 4′-hydroxy group rather than a methoxy group. The two groups have a partially independent biogenesis. Podophyllotoxin (*5.33*) is formed by hydroxylation of (*5.32*) whilst (*5.35*) derives from (*5.34*) (Scheme 5.7).

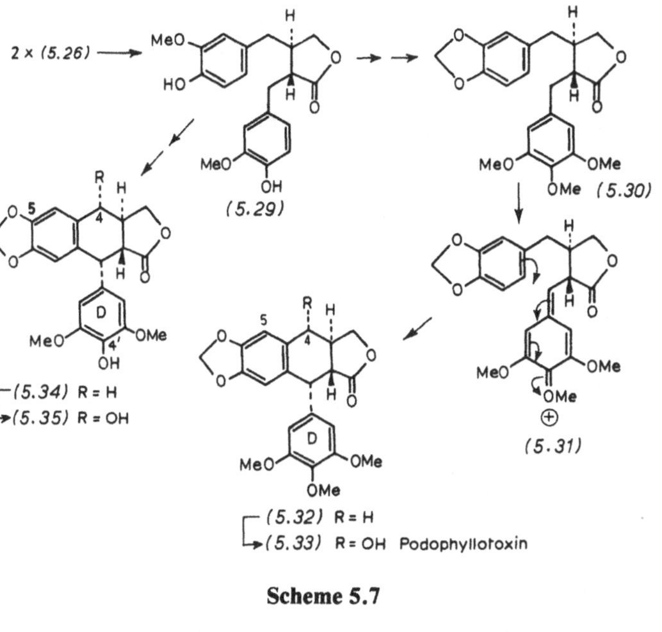

**Scheme 5.7**

Phenylalanine (*5.17*), cinnamic acid (*5.23*) and ferulic acid (*5.26*) were found to be good precursors for (*5.33*) and (*5.35*). Degradation of (*5.33*) which had been derived from [*methyl*-$^{14}$C]ferulic acid (*5.26*) showed that this acid is used to construct both halves of the lignan and is used equally for both. Thus it is likely that both of the phenylpropane units have the same substitution pattern at the coupling stage. This pattern was deduced to be that seen in ferulic acid (4-hydroxy-3-methoxy).

The results of further experiments have allowed identification of yatein (*5.30*) as an important intermediate in the biosynthesis of (*5.33*). It is reasonable to conclude (see above) that (5.30) is formed from matairesinol (5.29). This compound could then also afford the 4'-demethyl series exemplified by (5.35). Other lignans may be formed by C-5 hydroxylation of (5.33) or (5.35) also by diversion from the proposed intermediate (5.31) without further ring closure [27, 28].

The cyanogenic glycosides, e.g. prunasin (5.36) [29], and the glucosinolates, e.g. (5.37) [30], are biosynthesized from amino acids [31], in this case phenylalanine (5.17). A generalized pathway to cyanogenic glycosides is shown in Scheme 5.8.

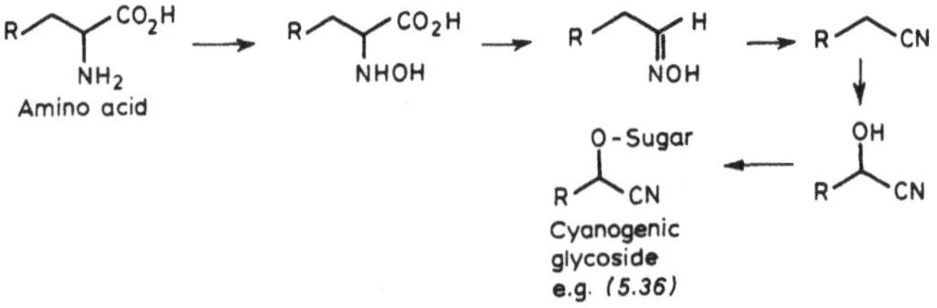

**Scheme 5.8**

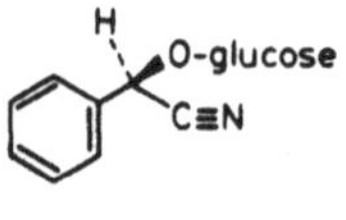

(5.36) 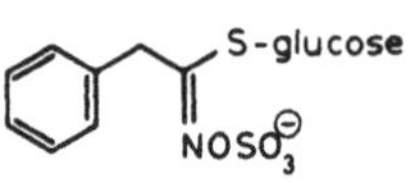          (5.37)

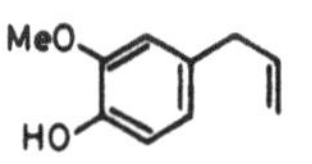

(5.38) 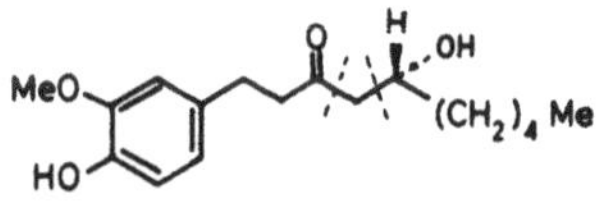          (5.39) [6]- Gingerol

The allylphenols, e.g. eugenol (5.38) [32] also derive from phenylalanine. Ferulic acid (5.26) is a precursor for both (5.38) and [6]-gingerol (5.39). The remaining carbon atoms in (5.39) derive from a single acetate, with loss of its carboxy-group, and an intact molecule of hexanoic acid [33].

Psilotin (5.40) is biosynthesized in part from phenylalanine (5.17) by way of p-coumaric acid (5.24) and with the incorporation of one acetate/malonate unit; (5.41) is a likely intermediate [34]. Phenylalanine (5.17) by way of benzoic acid ($C_6$-$C_1$ unit) serves as a part source for mucidin (5.42). The remainder of the molecule derives from acetate as

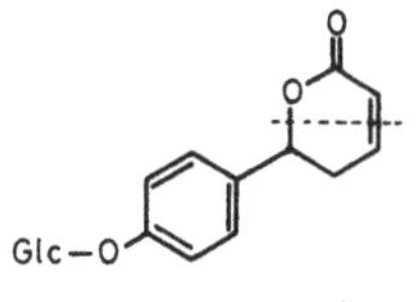

(5.40) Psilotin 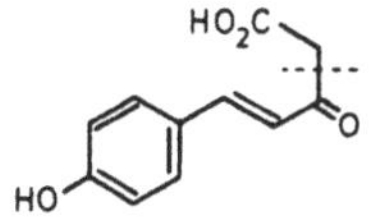  (5.41) 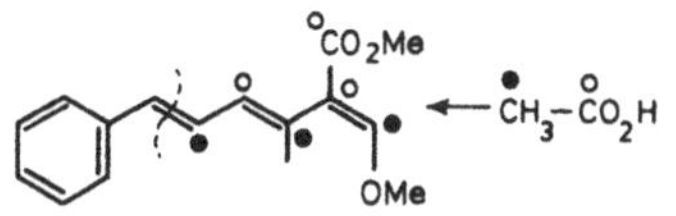  (5.42) Mucidin

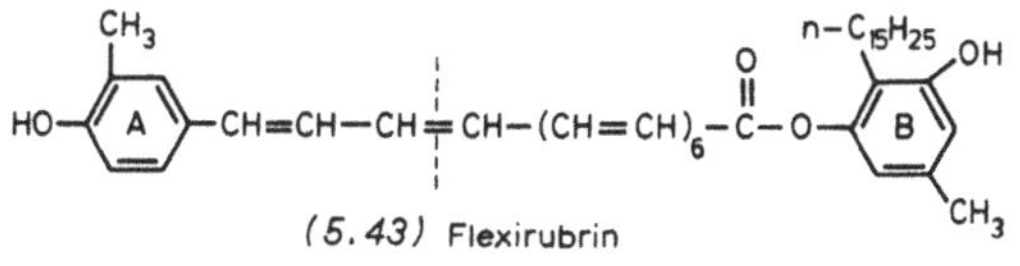

(5.43) Flexirubrin

shown and the *C*-methyl and (*O*-methyl) groups arise from the methyl group of methionine [35].

Flexirubin (*5.43*) also derives in part from acetate. Tyrosine (*5.18*) and methionine (ring A and attached *C*-methyl group) are also involved. Ring B is formed via a symmetrical intermediate and is built up independently of the polyene chain from acetate [36].

Stilbenes, e.g. resveratrol (*5.44*) are also elaborated from phenylalanine [by way of *p*-coumaric acid (*5.24*)] and three molecules of acetate/malonate reasonably via (*5.45*) [37]. The biosynthesis of lunularic acid (*5.46*) is very similar [38] as is the biosynthesis of phenanthrenes, e.g. isobatatasin I (*5.47*). A slightly different pathway, which involves *m*-coumaric acid (*5.48*), and its dihydro derivative, instead of (*5.24*), is apparent for the dihydrophenanthrene hircinol (*5.49*) [39].

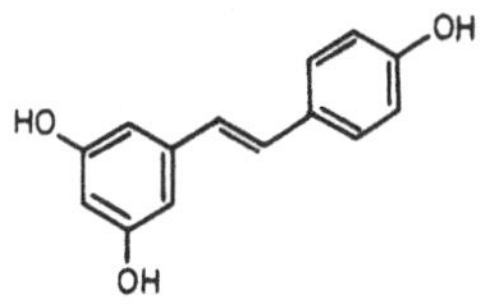

(5.44) 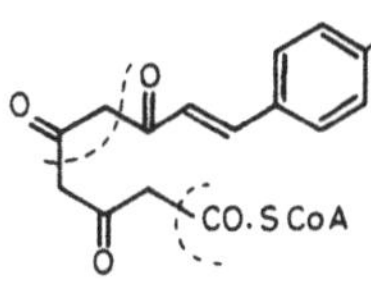  (5.45)

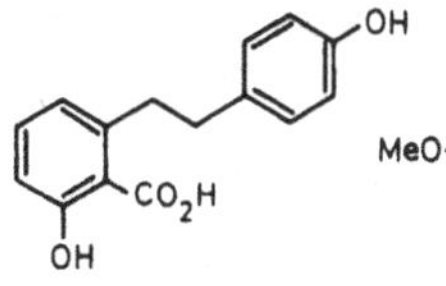

(5.46) Lunaric acid 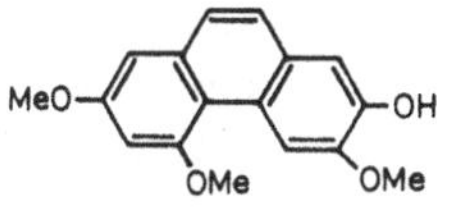  (5.47) 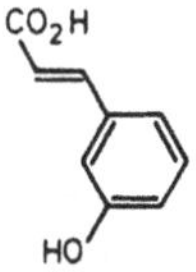  (5.48) 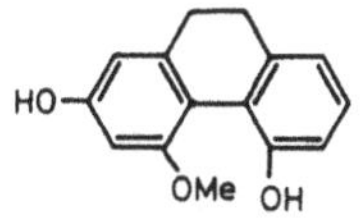  (5.49) Hircinol

There are apparent similarities in the assembly of the building blocks for these metabolites with the assembly of the units for flavonoids (section 5.4) [cf (*5.45*) and (*5.101*)].

The xanthone magniferin (*5.50*) again is elaborated from *p*-coumaric

*(5.50)* Magniferin

acid *(5.24)* and acetate/malonate, by way of a benzophenone intermediate [40].

Further mixed biosynthetic origins have been found for bakuchiol *(5.51)*: mevalonate and phenylalanine [by way of *p*-coumaric acid *(5.24)*, with loss of the carboxyl group] [41].

*(5.51)* Bakuchiol

pro–R–S edge   pro–S–R edge

ÓH
*(5.52)*

*(5.53)* Ketomycin

Finally in this section the *Streptomyces* metabolite ketomycin *(5.53)* takes us back to pre-aromatic intermediates in the shikimate pathway for it is elaborated from shikimic acid via chorismic acid and prephenic acid *(5.52)* (labelling shown is that present at C-1 and C-6 of shikimate); phenylalanine *(5.17)* was a poor precursor. The side chain of *(5.53)* appears to arise by shortening of the pyruvyl moiety in *(5.52)*. It is interesting to note the conversion of *(5.52)* into *(5.53)* is stereospecific in the sense that the two enantiotopic edges of the ring in *(5.52)* are distinguished [42].

## 5.2 QUINONES [43, 44]

The widely distributed natural quinones are formed by a diversity of routes: several are known for each of the benzoquinones, naphthoquinones and anthraquinones. This can be attributed to the importance which the quinone moiety of these compounds has in the economy of living systems.

The isoprenoid quinones, e.g. the ubiquinones *(5.59)*, are essential metabolites, being involved in electron transport in living systems. In the ubiquinones a particular chain length is favoured from $n = 6$ in certain micro-organisms to $n = 10$ in most mammals. Mevalonic acid *(5.54)* is well established as the source of polyprenyl side-chains in these metabolites. It is probable that the side-chain is assembled as a polyprenyl pyrophosphate which then couples with the aromatic fragment. The evidence is that polyprenyl pyrophosphate synthetases and 4-hydroxybenzoate:polyprenyltransferases have been isolated from living systems and ubiquinones co-occur with polyprenyl alcohols [as *(5.55)*] [45, 46].

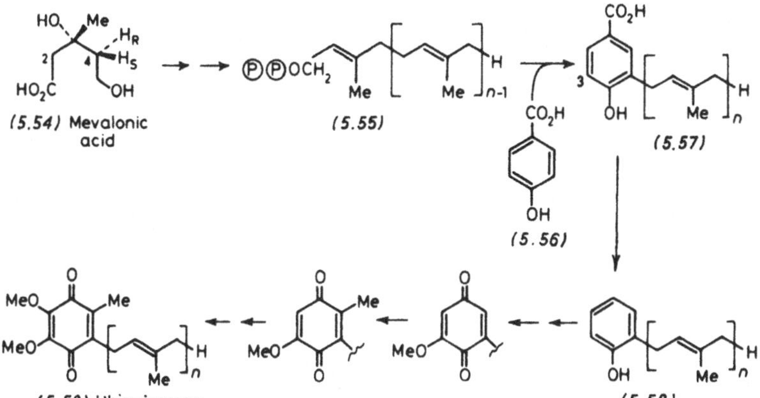


(5.54) Mevalonic acid (5.55) (5.56) (5.57) (5.58) (5.59) Ubiquinones

**Scheme 5.9**

4-Hydroxybenzoic acid (*5.56*) has been shown to stand as the key intermediate in ubiquinone biosynthesis in living systems from micro-organisms to mammals. In animals, phenylalanine and tyrosine serve as precursors, but in bacteria chorismic acid (*5.10*) is the precursor [47, 48]. Interlocking evidence obtained from bacteria in experiments with mutants (and genetic analysis), cell-free preparations, and isolation and identification of intermediates allows clear delineation [43, 48] of the sequence of biosynthesis as that shown in part in Scheme 5.9; beyond (*5.57*) there are different steps leading to (*5.59*) in prokaryotes and eukaryotes. In the latter, hydroxylation at C-3 and *O*-methylation precede decarboxylation but in prokaryotes decarboxylation to give (*5.58*) is the first step [49].

The plastoquinones are isoprenoid benzoquinones, (*5.63*), found in plants and algae which function in photosynthetic electron transport. They include methyl substituents on the quinone ring instead of the methoxy-groups seen in the ubiquinones. Tyrosine (*5.18*) is the source of the quinone ring [via (*5.61*) and homogentisic acid (*5.62*)] but unlike the ubiquinones the side-chain is not lost completely: C-3 is retained as a *C*-methyl group; the other methyl group derives from methionine (with retention of all three methyl protons). The tyrosine labelling site was deduced by means of a clever degradation sequence [carried out on material biosynthesized from [1,6-$^{14}C_2$]shikimic acid (*5.60*)]; see Scheme 5.10. The chemical introduction of the third methyl group and Kuhn-Roth (chromic acid-sulphuric acid) oxidation of (*5.63*) and (*5.64*) gave a pair of results: ~25% of the total radioactivity in each compound appeared in the acetic acid. This is in accord with the pattern shown (●, ◓); the alternative (■, ⬒) would have given acetic acid containing 25% of the activity of (*5.63*) and 50% of the activity of (*5.64*) [50, 51].

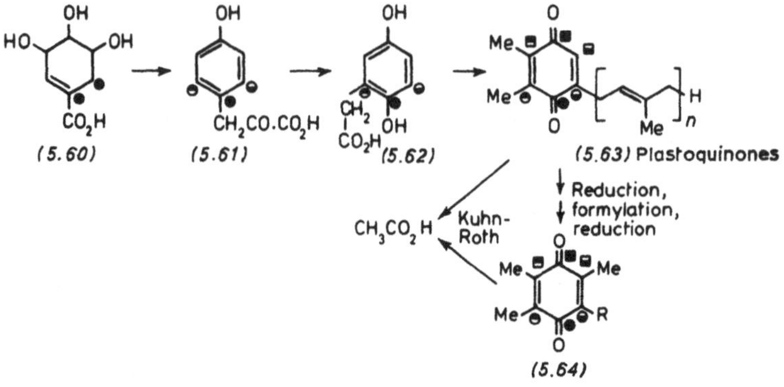

**Scheme 5.10**

The normal catabolic pathway of tyrosine (*5.18*) is via 4'-hydroxy-phenylpyruvic acid [as (*5.61*)] and homogentisic acid [as (*5.62*)], and is, as we have seen (Scheme 5.10), followed in plastoquinone formation. A similar course of biosynthesis has been deduced for $\alpha$- and $\gamma$-tocopherol (*5.65*) and (*5.66*), respectively (* = methionine derived methyl groups) [50, 51].

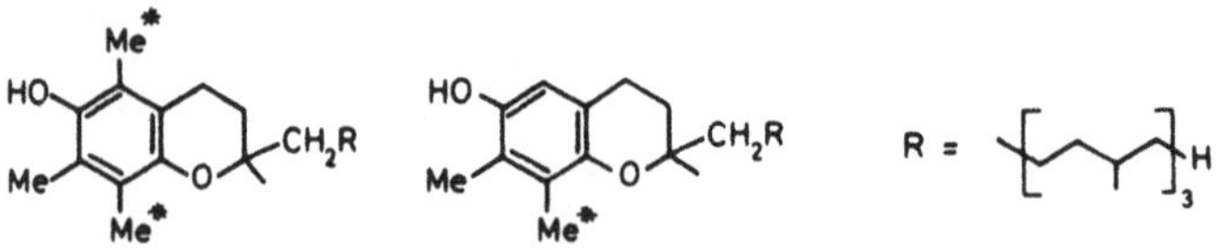

(*5.65*) $\alpha$-Tocopherol    (*5.66*) $\gamma$-Tocopherol

In the conversion of homogentisic acid (*5.62*) into (*5.63*) one of the reactions which occurs is decarboxylation with the generation of a methyl group. The stereochemistry of this reaction has been examined for the biosynthesis plastoquinone-9 (*5.63*; $n = 9$) and $\alpha$-tocopherol (*5.65*) with chirally deuteriated samples of homogentisic acid (*5.67*) (obtained by enzymic conversion from chirally deuteriated tyrosine) [53]. The chirality of the methyl group generated (section 1.2.2) established that the conversion proceeds with retention of configuration; the tentative mechanism for the process is illustrated in Scheme 5.11, in

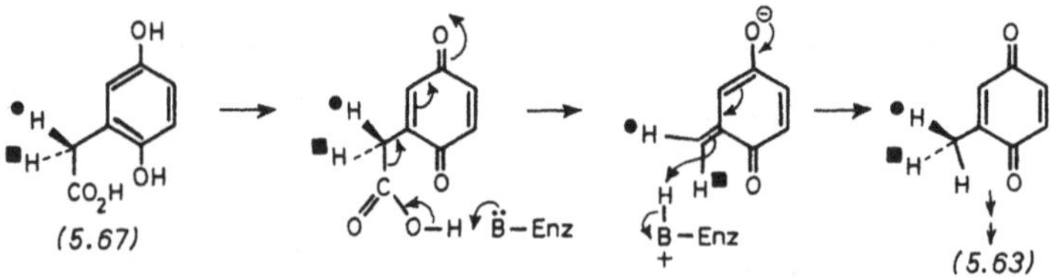

**Scheme 5.11**

which a proton is added to the same side of the molecule as that from which the carboxyl is lost.

Helicobasidin (*5.69*) has been shown to have entirely mevalonoid origins: [2-$^{13}$C]mevalonate labelled the sites indicated [C-8 and C-10 have become equivalent; only one site will be labelled in the precursor isoprenoid: see (*5.68*)]. Farnesyl pyrophosphate (*5.68*) is the appropriate (C$_{15}$) mevalonoid intermediate and in the course of biosynthesis a hydrogen shift (from C-6) occurs, see Scheme 5.12 (cf section 4.6) [54, 55].

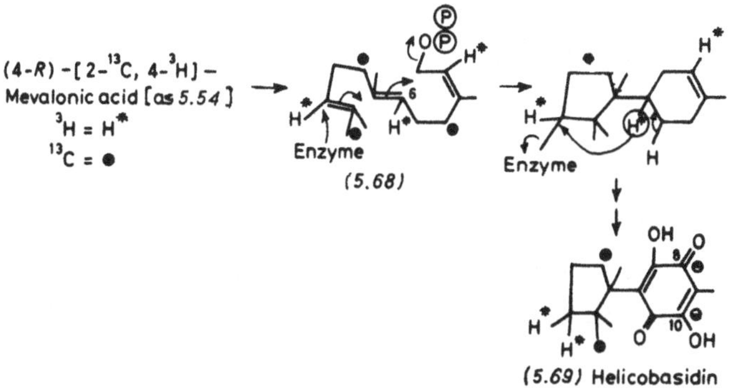

(*5.69*) Helicobasidin

**Scheme 5.12**

The naphthoquinones, exemplified in higher plants by the phylloquinones, e.g. vitamin K$_1$ (*5.73*), and in bacteria by the menaquinones, e.g. vitamin K$_2$ (*5.74*), resemble the benzoquinones (above) in structure and function. The polyprenyl side-chains derive from mevalonic acid and the nuclear *C*-methyl groups from methionine. The origin of the remaining atoms is more fascinating.

Experiments with [1,6-$^{14}$C$_2$]- and [3-$^3$H]-shikimic acid establish that the bacterial menaquinones and lawsone (*5.75*), from the plant *Impatiens balsamina*, derive from shikimate (*5.7*) with C-1 and C-2 appearing at the naphthoquinone ring junction (Scheme 5.13). The carboxy-group of shikimic acid is retained (at *) on naphthoquinone formation, and (*5.7*) thus accounts for seven of the ten nuclear carbon atoms in e.g. (*5.73*). [The (6-*pro-S*)-hydrogen is also retained, but the (6-*pro-R*)-proton is lost [56, 57].] The remaining C$_3$ unit, importantly, was identified as having its origins in glutamic acid or its transamination product, 2-ketoglutaric acid (*5.70*). Incorporation was of C-2, C-3, and C-4, with loss of C-1 and C-5 [58].

The very efficient incorporation of 2-succinylbenzoate (*5.71*) into plant and bacterial naphthoquinones identifies (*5.71*) as a later intermediate derived from shikimic acid and (*5.70*). Further corroboration comes from the

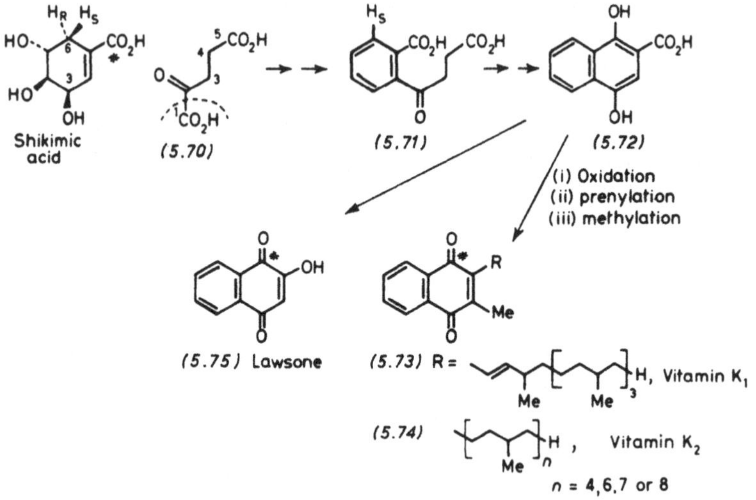

**Scheme 5.13**

identification of radioactive (*5.71*) in *I. balsamina* following the feeding of [U-$^{14}$C]glutamic acid [59, 60].

It is now apparent that diversion from the shikimate pathway into naphthoquinone biosynthesis occurs at chorismic acid (*5.10*) which is converted into isochorismic acid (*5.76*) and thence into (*5.71*). Enzyme preparations from *Escherichia coli* have been found to catalyse the formation of 2-succinylbenzoic acid (*5.71*) from 2-ketoglutaric acid (*5.70*) and isochorismic acid (*5.76*) [not (*5.10*)] in the presence of thiamine pyrophosphate. The diene (*5.77*) has been identified as an intermediate in the overall conversion and a likely pathway is illustrated in Scheme 5.14 [44, 61, 62]. Formation of the second aromatic ring which yields (*5.72*) occurs via the activated thioester (*5.78*) [44, 63].

In the steps which lie beyond (*5.72*) at no point are there any symmetrical intermediates, e.g. 1,4-naphthoquinone (*5.79*) (see labelling in Scheme 5.13) although the apparently related quinone, juglone (*5.80*), does derive from this compound [57, 64].

> Other routes have been discovered for naphthoquinone elaboration. The plant quinone shikonin (*5.82*) derives from mevalonic acid (*5.54*) and the benzoquinone precursor 4-hydroxybenzoic acid (*5.56*) (derived in turn from phenylalanine and cinnamic acid), by way of (*5.81*) [65, 66]. Homogentisic acid (*5.62*), the plastoquinone precursor, is a precursor for chimaphilin (*5.84*), possibly via (*5.83*). The biosynthesis of naphthoquinones and benzoquinones from polyketides is found widely in bacteria and two examples have been identified in plants, plumbagin

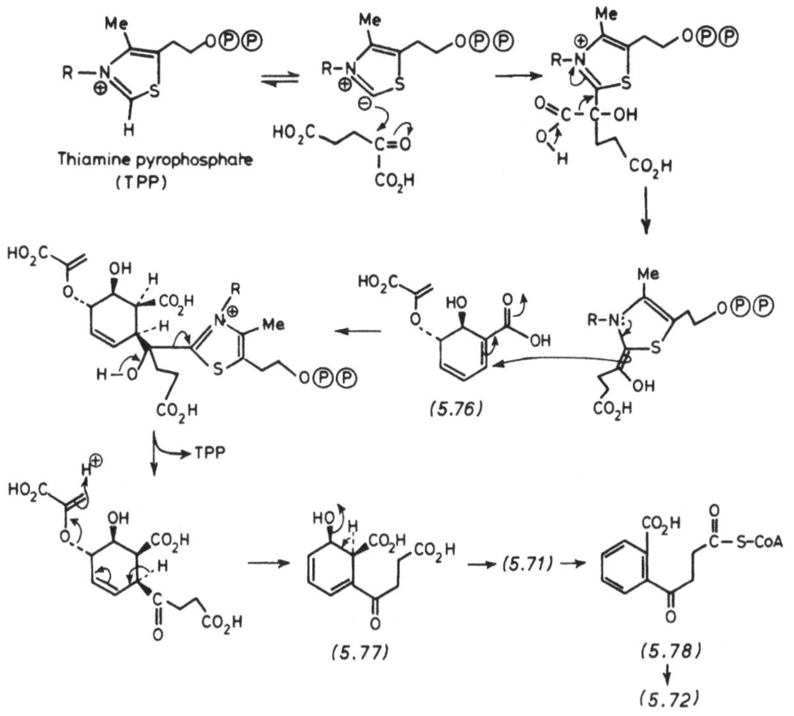

**Scheme 5.14**

(5.85) and 7-methyljuglone (5.86) [67]. Anthraquinones may also have polyketide origins [43, 68, 69]. In some cases plant anthraquinones may derive from 2-succinylbenzoic acid (5.88) e.g. (5.87) and lucidin

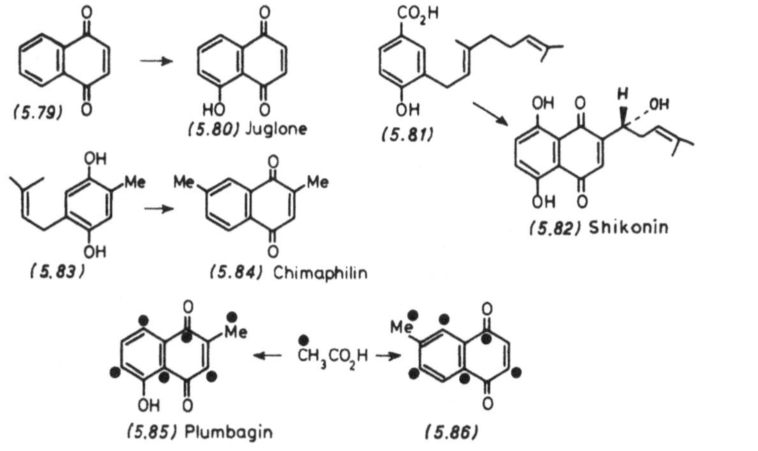

(5.87)

(5.75)

(5.88)

(5.89) R = primoverosyl

(5.90) α-Dunnione

(5.91) Molugin

3-*O*-primerveroside (*5.89*). These anthraquinones also incorporate mevalonic acid (*5.54*) as a prenyl unit but, curiously, in different ways. In the case of (*5.87*) prenylation occurs at C-2 of a derivative of (*5.88*) [trace the label in (*5.88*) and see also the dotted lines in (*5.87*)]. Lawsone (*5.75*) is an intermediate after (*5.88*) in the biosynthesis of e.g. α-dunnione (*5.90*), which is elaborated by the same species (*Streptocarpus dunnii*). *O*-Prenylation of (*5.75*) followed by Claisen rearrangement and further ring closure yields (*5.90*) [70].

In the case of the anthraquinone (*5.89*), in *Galium mollugo*, prenylation occurs at the position corresponding to C-3 of (*5.88*) [again trace the labels and see the dotted lines in (*5.89*)]. Mollugin (*5.91*) which is also produced by this species derives from an intermediate in the biosynthesis of (*5.89*) [71].

## 5.3 COUMARINS [72, 73]

Coumarin (*5.93*) the simplest of compounds of this type is probably an artefact arising by enzymatic cleavage of the glucoside (*5.92*). The biosynthesis of (*5.92*)/(*5.93*) depends crucially on unusual *ortho*-hydroxylation of cinnamic acid (*5.23*), and the appropriate enzyme has been detected [74]. The hydroxycoumarin, umbelliferone (*5.94*), arises through *ortho*-hydroxylation of *p*-coumaric acid (*5.24*). Daphnin (*5.95*) and cichorin (*5.96*) are deduced to derive via umbelliferone (*5.94*) [75]. Furanocoumarins, e.g. (*5.98*), derive from isoprenyl coumarins [as (*5.97*)]. The biosynthesis of (*5.98*) involves (*5.94*) and 7-demethyl-suberosin (*5.97*) [76, 77] (cf furoquinoline alkaloid biosynthesis, section 6.5).

Novobiocin (*5.99*) is produced by a micro-organism, *Streptomyces niveus*. The origin of the coumarin residue is in tyrosine (and is thus different from the origins of plant coumarins), and the ring oxygen derives from the carboxyl oxygen of tyrosine [78].

*(5.92)*      *(5.93)* Coumarin    *(5.94)* Umbelliferone

*(5.95)* R¹=H, R²=OH, Daphnin
*(5.96)* R¹=OH R²=H, Cichorin

*(5.97)*      *(5.98)* Marmesin

*(5.99)* Novobiocin

## 5.4 FLAVONOIDS [4]

The flavonoids are found universally in plants as the largest single group of oxygen ring compounds. They account for a variety of colours found in plant tissues and some, the rotenoids, are insecticidal. The basic skeleton is found in the flavones, e.g. *(5.104)*; most of the variations on this pattern are those arising from hydroxylation (and *O*-methylation and glucoside formation), and oxidation of ring B, as found in the anthocyanidins [as *(5.111)*]. Some further variation is associated with rearrangement in ring B, found in the iso-flavones [as *(5.117)*], and rotenones [as *(5.118)*].

The oxygenation pattern shown in *(5.104)* is characteristic and may be traced back to the origins of the system, namely three acetate units joined head-to-tail (ring A) and a *p*-coumaryl fragment [as *(5.100)*; rings C and B]. The key stage in the biosynthesis of all flavonoids is reached with the formation of an intermediate chalcone [as *(5.102)*] ⇌ flavanone [as *(5.103)*]. Labelling and enzyme evidence points to the formation of chalcones occurring by the condensation of *p*-coumaryl-coenzyme A with three malonyl-CoA (≃ acetyl-CoA, section 3.2.1) units; chalcones are the first identifiable intermediates (Scheme 5.15) [79–82]. The chalcone-flavanone interconversion has been shown to be stereospecific as illustrated in Scheme 5.15 [83]. The chalcone, echinatin *(5.113)*, shows an unusual hydroxylation pattern because the unsaturated ketone function in a normal chalcone has been transposed; transposition occurs on the chalcone *(5.112)* [84].

The hydroxylation pattern of the chalcone is discriminating: those with hydroxy-groups as in *(5.102)* are converted into 5,7-dihydroxyflavonoids [as

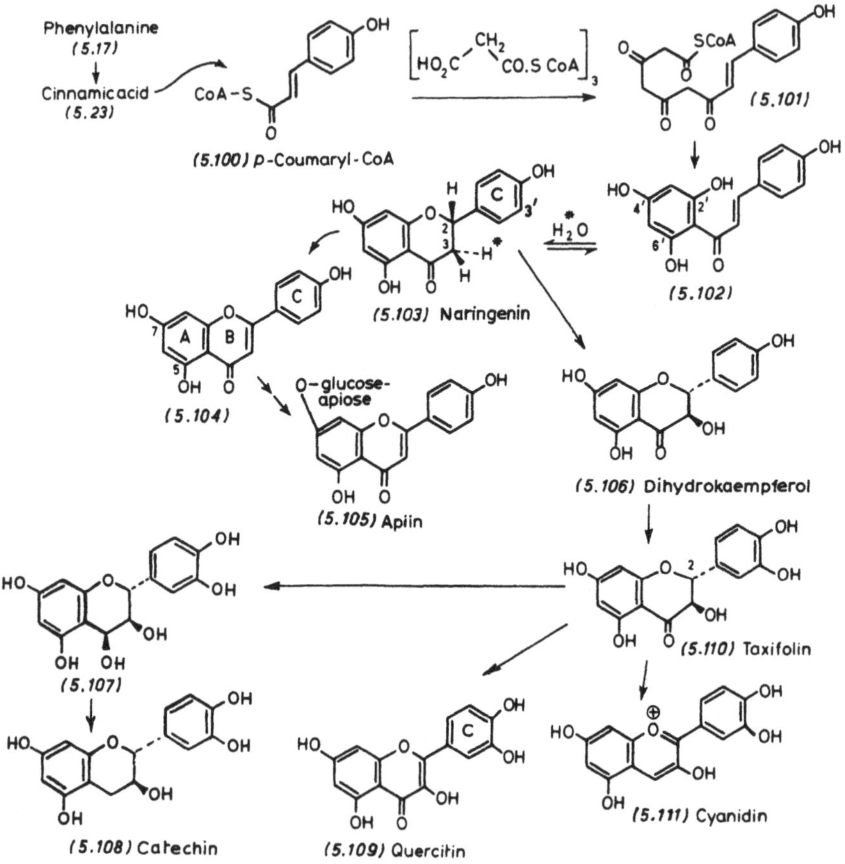


**Scheme 5.15**

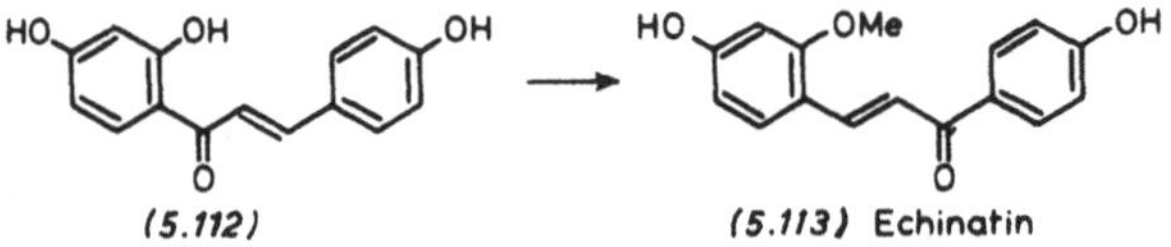


(5.112) (5.113) Echinatin

(*5.104*)], those with hydroxy-groups at C-2′ and C-4′ only [i.e. one oxygen lost from C-6′ during chalcone formation, see (*5.102*)] afford selectively 7-hydroxy-flavonoids [see (*5.104*)]. The intact incorporation of chalcones into flavonoids has been proved using doubly labelled precursors [85] (for further discussion of this technique see section 2.2.1).

The weight of evidence from precursor feeding experiments indicates that the extra hydroxy-groups found in ring C, of e.g. quercitin (*5.109*) are introduced after the chalcone stage of biosynthesis [86]. Work with isolated enzymes, however, indicates that some of the chalcone synthases accept both the CoA esters of *p*-coumaric acid (*5.24*) and caffeic acid (*5.25*) as substrates and for some substrate acceptability is pH dependent [87–89]. Cyclization of the chalcone [as (*5.102*)] to give the flavanone [as (*5.103*)] is catalysed by chalcone isomerase [90]. Hydroxylation at C-3 of flavanones, e.g. naringenin (*5.103*) which yields dihydrokaempferol (*5.106*) stereospecifically, is catalysed by a specific oxygenase; substrate with an additional hydroxy group, at C-3′ [see (*5.103*)], was also acceptable [91].

Studies with cell suspension cultures of parsley have given information about the appearance of enzymes involved in the biosynthesis of apiin (*5.105*) from phenylalanine (*5.17*). The deduced pathway which proceeds via the flavanone (*5.103*) is illustrated in scheme 5.15 [92].

Dihydroflavonols [as (*5.106*)] are important intermediates in the biosynthesis of other flavonoids. This is illustrated for the biosynthesis of cyanidin (*5.111*) and quercitin (*5.109*) in Scheme 5.15; dihydrokaempferol (*5.106*) was an excellent precursor and its role as an intermediate was confirmed in experiments with a cell culture [93, 94]. Catechin (*5.108*) and related metabolites are biosynthesized as shown in Scheme 5.15 [95].

Chlorflavonin (*5.115*) is a novel flavonoid produced by the fungus *Aspergillus candidus*. Its biosynthesis is markedly different to plant flavonoids. Benzoic acid appears to act as a starter unit to which four acetate units are added [see (*5.114*)] [96].

(*5.114*)                    (*5.115*)

The formation of isoflavonoids, e.g. genistein (*5.117*), involves a rearrangement of the flavonoid skeleton seen in e.g. naringenin (*5.103*). This is proved [97] to occur by an intramolecular 1,2-shift of ring C [see (*5.103*)]. Isoflavone synthase activity capable of effecting this profound biosynthetic step e.g. of (*5.103*) into (*5.117*), has been obtained. The enzyme is a monooxygenase requiring molecular oxygen and NADPH as cofactors. The rearrangement is proposed to occur as shown in Scheme 5.16; (*5.116*) has been identified tentatively as an intermediate, its dehydration affording genistein (*5.117*) [98, 99].

Rotenone (*5.118*) and amorphigenin (*5.119*) derive by the general flavanone-isoflavone pathway (Schemes 5.15 and 5.16) with specialized steps beyond the simple isoflavone intermediate. The labelling of rotenone [(*5.118*); ●, ▲ = $^{14}$C] by phenylalanine (by way of cinnamic and *p*-coumaric acids) is illustrated. Careful work has allowed detailed delineation of the biosynthetic pathway to the rotenoids e.g. (*5.118*). The extra carbon

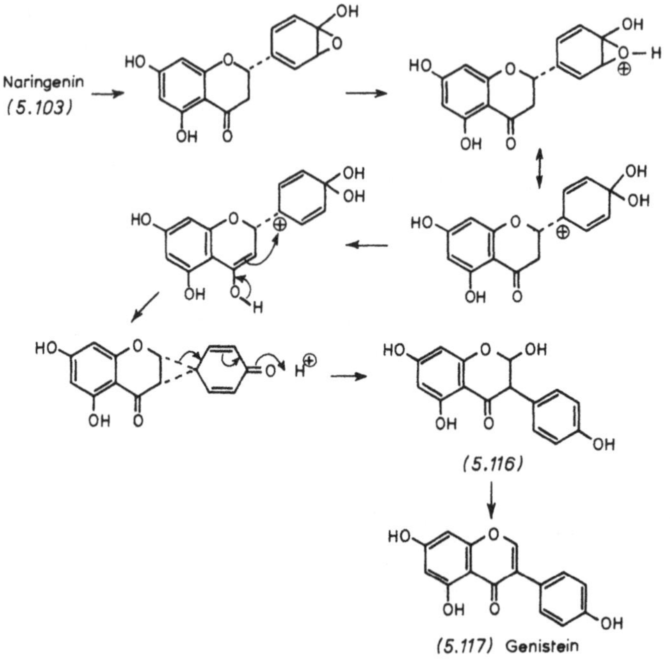

**Scheme 5.16**

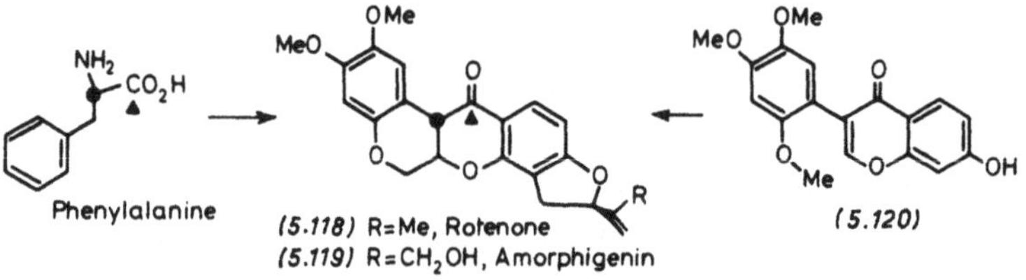

atom derives oxidatively from a methoxy group, i.e. (5.120) is an intermediate. Rotenone (5.118) is a precursor for amorphigenin (5.119) [99–101].

The pterocarpin phytoalexins, e.g. (+)-maackiain (5.125), are elaborated through and beyond the isoflavone pathway (Scheme 5.17). The biosynthetic pathway to maackiain begins with the chalcone (5.121) and proceeds by way of the isoflavones formonetin (5.122) and (5.123). Reduction of the double bond first yields (5.124) and then reduction of the carbonyl group leads to maackiain (5.125). This metabolite is an intermediate via its 6a-hydroxy derivative in the biosynthesis of (+)-pisatin (5.126). Related pathways which also begin with (5.121) lead

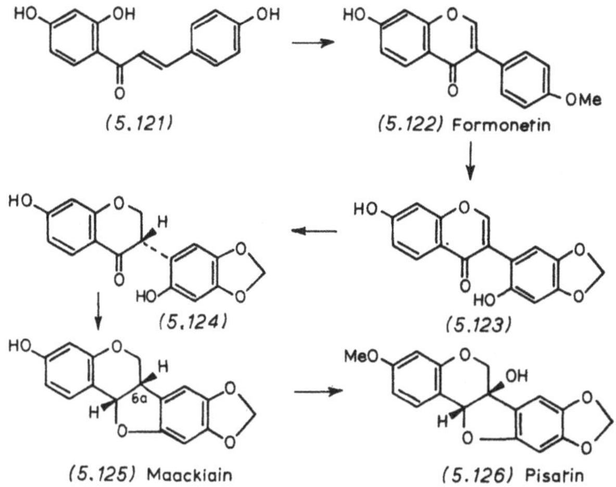

(5.121)     (5.122) Formonetin

(5.124)     (5.123)

(5.125) Maackiain     (5.126) Pisatin

**Scheme 5.17**

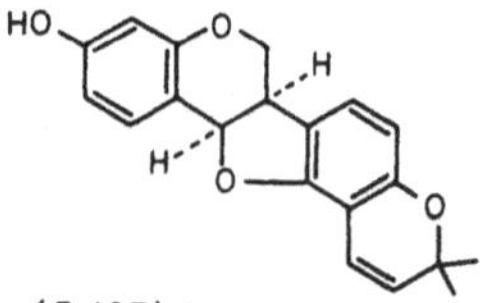

(5.127) Phaseollin

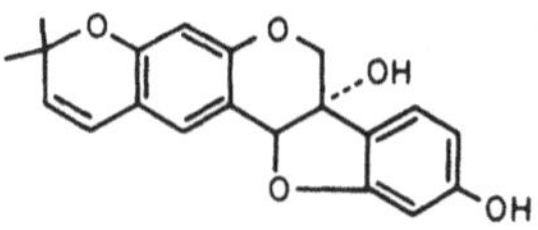

(5.128)

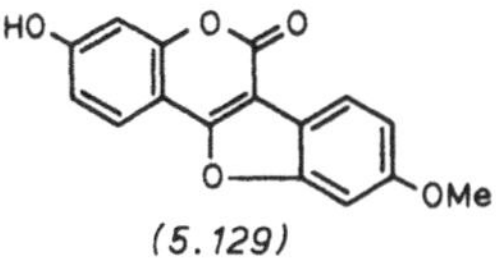

(5.129)

to phaseollin (5.127) [102, 103], glyceollins, e.g. (5.128) [104], and coumestans, e.g. (5.129) [105].

Calophyllolide (5.130) is one of a group of neoflavonoids which can be thought of as arising from the same basic units as other flavonoids but in a different manner (Scheme 5.18). In accord with this, label from

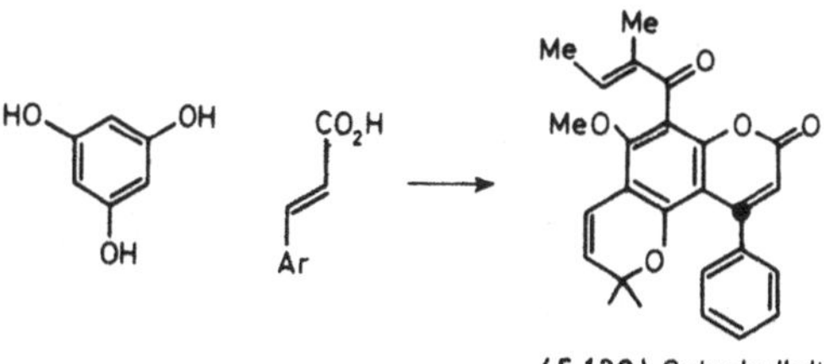

(5.130) Calophyllolide

**Scheme 5.18**

[3-$^{14}$C]phenylalanine [as (*5.17*)] was found to be located as shown in (*5.130*) (●: $^{14}$C label) [106].

## REFERENCES

Further reading: [1]–[5].

1. Haslam, E. (1979) In *Comprehensive Organic Chemistry* (eds D.H.R. Barton and W.D. Ollis), Pergamon, Oxford, vol. 5, pp. 1167–205.
2. Haslam, E. (1974) *The Shikimate Pathway*, Butterworths, London.
3. *Natural Product Reports*.
4. Hahlbrock, K. and Grisebach, H. (1975) In *The Flavonoids* (eds J.B. Harborne, T.J. Mabry and H. Mabry), Chapman and Hall, London, pp. 866–915.
5. Weiss, U. and Edwards, J.M. (1980) *The Biosynthesis of Aromatic Compounds*, Wiley, New York.
6. Gibson, F. and Pittard, J. (1968) *Bact. Rev.*, **32**, 465–92.
7. Floss, H.G., Onderka, D.K. and Carroll, M. (1972) *J. Biol. Chem.*, **247**, 736–44.
8. Widlanski, T.S., Bender, S.L. and Knowles, J.R. (1987) *J. Amer. Chem. Soc.*, **109**, 1873–5.
9. Turner, M.J., Smith, B.W. and Haslam, E. (1975) *J. Chem. Soc. Perkin I*, 52–5.
10. Grimshaw, C.E., Sogo, S.G., Copley, S.D. and Knowles, J.R. (1984) *J. Amer. Chem. Soc.*, **106**, 2699–700.
11. Lee, J.J., Asano, Y., Shieh, T.-L. *et al.* (1984) *J. Amer. Chem. Soc.*, **106**, 3367–8.
12. Guilford, W.J., Copley, S.D. and Knowles, J.R. (1987) *J. Amer. Chem. Soc.*, **109**, 5013–9.
13. Larsen, P.O. and Wieczorkowska, E. (1975) *Biochim. Biophys. Acta*, **381**, 409–15.
14. Marshall, B.J. and Ratledge, C. (1972) *Biochim. Biophys. Acta*, **264**, 106–16.
15. Scharf, K.H., Zenk, M.H., Onderka, D.K. *et al.* (1971) *J. Chem. Soc. Chem. Comm.*, 765–6.
16. Haslam, E. (1986) In *The Shikimic Acid Pathway* (ed. E.E. Conn) (Recent Advances in Phytochemistry, Vol. 20), Plenum, New York, pp. 163–200.
17. Camm, E.L. and Towers, G.H.N. (1973) *Phytochemistry*, **12**, 961–73.
18. Strange, P.G., Staunton, J., Wiltshire, H.R. *et al.* (1972) *J. Chem. Soc. Perkin I*, 2364–72.
19. Ellis, B.E. and Amrhein, N. (1971) *Phytochemistry*, **10**, 3069–72.
20. Amrhein, N. and Zenk, M.H. (1969) *Phytochemistry*, **8**, 107–13.
21. Neish, A.C. (1964) In *Biochemistry of Phenolic Compounds* (ed. J.B. Harborne), Academic Press, New York, pp. 294–359.
22. Ellis, B.E. and Towers, G.H.N. (1970) *Biochem. J.*, **118**, 291–7.
23. Zenk, M.H. (1966) In *Biosynthesis of Aromatic Compunds* (ed. G. Billek), Pergamon, pp. 45–60.
24. Dewick, P.M. and Haslam, E. (1969) *Biochem. J.*, **113**, 537–42.
25. Higuchi, T. (1971) *Adv. Enzymol.*, **34**, 207–83.
26. Pelter, A.W., in ref. 16, pp. 201–41.

27. Kamil, W.M. and Dewick, P.M. (1986) *Phytochemistry*, **25**, 2093-102.
28. Kamil, W.M. and Dewick, P.M. (1986) *Phytochemistry*, **25**, 2089-92.
29. Tapper, B.A., Zilg, H. and Conn, E.E. (1972) *Phytochemistry*, **11**, 1047-53.
30. Matsuo, M., Kirkland, D.F. and Underhill, E.W. (1972) *Phytochemistry*, **11**, 697-701.
31. Dewick, P.M. (1984) in ref. 3, Vol. 1, pp. 545-9.
32. Klischies, M., Stöckigt, J. and Zenk, M.H. (1975) *J. Chem. Soc. Chem. Comm.*, 879-80.
33. Denniff, P., MacLeod, I. and Whiting, D.A. (1980) *J. Chem. Soc., Perkin I*, 2637-44.
34. Leete, E., Muir, A. and Towers, G.H.N. (1982) *Tetrahedron Lett.*, **23**, 2635-8.
35. Nerud, F., Sedmera, P. Souchová, Z. *et al.* (1982) *Coll. Czech. Chem. Comm.*, **47**, 1020-5.
36. Achenbach, H., Böttger-Vetter, A., Hunkler, D. *et al.* (1983) *Tetrahedron*, **39**, 175-85.
37. Rupprich, N. and Kindl, H. (1978) *Hoppe-Seylers Z. Physiol. Chem.*, **359**, 165-72.
38. Abe, S. and Ohta, Y. (1984) *Phytochemistry*, **23**, 1379-81.
39. Fritzemeier, K.-H., Kindl, H. and Schlösser, E. (1984) *Z. Naturforsch.*, **39c**, 217-21.
40. Fugita, M. and Inoue, T. (1981) *Phytochemistry*, **20**, 2183-5.
41. Banerji, A. and Chintalwar, G.J. (1984) *Phytochemistry*, **23**, 1605-6.
42. Takeda, Y., Mak V., Chang, C.-C. *et al.* (1984) *J. Antibiot.*, **37**, 868-75.
43. Bentley, R. (1975) In *Biosynthesis (Specialist Periodical Reports)* (ed. T.A. Geissman), The Chemical Society, London, vol. 3, pp. 181-246.
44. Leistner, E., in ref. 16, pp. 243-61.
45. Allen, C.M., Alworth, W., MacRae, A. and Bloch, K. (1967) *J. Biol. Chem.*, **242**, 1895-902.
46. Hemming, F.W., Morton, R.A. and Pennock, J.F. (1963) *Proc. Roy. Soc.*, **158B**, 291-310.
47. Olson, R.E. (1966) *Vitam. Horm.*, **24**, 551-74.
48. Young, I.G., Stroobant, P., MacDonald, C.G. and Gibson, F. (1973) *J. Bacteriol.*, **114**, 42-52.
49. Sippel, C.J., Goewert, R.R., Slachmann, F.N. and Olson, R.E. (1983) *J. Biol. Chem.*, **258**, 1057-61.
50. Threlfall, D.R. (1972) *Biochim. Biophys. Acta*, **280**, 472-80.
51. Whistance, G.R. and Threlfall, D.R. (1971) *Phytochemistry*, **10**, 1533-8.
52. Marshall, P.S., Morris, S.R. and Threlfall, D.R. (1985) *Phytochemistry*, **24**, 1705-11.
53. Krügel, R., Grumbach, K.-H., Lichenthaler, H. and Rétey, J. (1985) *Bioorg. Chem.*, **13**, 187-96.
54. Adams, P.M. and Hanson, J.R. (1972) *J. Chem. Soc. Perkin I*, 586-8.
55. Tanabe, M., Suzuki, K.T. and Jankowski, W.C. (1973) *Tetrahedron Lett.*, 4723-6.
56. Baldwin, R.M., Snyder, C.D. and Rapoport, H. (1973) *J. Amer. Chem. Soc.*, **95**, 276-8.
57. Scharf, K.-H., Zenk, M.H., Onderka, D.K. *et al.* (1971) *J. Chem. Soc. Chem. Comm.*, 576-7.
58. Campbell, I.M., Robins, D.J., Kelsey, M. and Bentley, R. (1971) *Biochemistry*, **10**, 3069-78.

59. Grotzinger, E. and Campbell, I.M. (1974) *Phytochemistry*, **13**, 923–6.
60. Dansette, P. and Azerad, R. (1970) *Biochem. Biophys. Res. Comm.*, **40**, 1090–5.
61. Weische, A. and Leistner, E. (1985) *Tetrahedron Lett.*, **26**, 1487–90.
62. Emmons, G.T., Campbell, I.M. and Bentley, R. (1985) *Biochem. Biophys. Res. Comm.*, **131**, 956–60.
63. Kolkmann, R. and Leistner, E. (1985) *Tetrahedron Lett.*, **26**, 1703–4.
64. Müller, W.-U. and Leistner, E. (1978) *Phytochemistry*, **17**, 1739–42.
65. Yazaki, K., Fukui, H. and Tabata, M. (1986) *Chem. Pharm. Bull.*, **34**, 2290–3.
66. Yazaki, K., Fukui, H. and Tabata, M. (1986) *Phytochemistry*, **25**, 1629–32.
67. Durand, R. and Zenk, M.H. (1971) *Tetrahedron Lett.*, 3009–12.
68. Steglich, W., Arnold, R., Lösel, W. and Reiniger, W. (1972) *J. Chem. Soc. Chem. Comm.*, 102–3.
69. Curtis, R.F., Hassall, C.H. and Parry, D.R. (1971) *J. Chem. Soc. Chem. Comm.*, 410.
70. Inoue, K., Ueda, S., Nayeshiro, H. *et al.* (1984) *Phytochemistry*, **23**, 313–8.
71. Inoue, K., Shiobara, Y., Nayeshiro, H. *et al.* (1984) *Phytochemistry*, **23**, 307–11.
72. Brown, S.A. (1966) In *Biosynthesis of Aromatic Compounds* (ed. G. Billek), Pergamon, pp. 15–24.
73. Brown, S.A., in ref. 16, pp. 287–316.
74. Gestetner, B. and Conn, E.E. (1974) *Arch. Biochem. Biophys.*, **163**, 617–24.
75. Satô, M. and Hasegawa, M. (1972) *Phytochemistry*, **11**, 657–62.
76. Austin, D.J. and Brown, S.A. (1973) *Phytochemistry*, **12**, 1657–67.
77. Brown, S.A. and Steck, W. (1973) *Phytochemistry*, **12**, 1315–24.
78. Bunton, C.A., Kenner, G.W., Robinson, M.J.T. and Webster, B.R. (1963) *Tetrahedron*, **19**, 1001–10.
79. Grisebach, H. and Barz, W. (1969) *Naturwiss.*, **56**, 538–44.
80. Grisebach, H. (1968) In *Recent Advances in Phytochemistry*, **1**, 379–406.
81. Hrazdina, G., Kreuzaler, F., Hahlbrock, K. and Grisebach, H. (1976) *Arch. Biochem. Biophys.*, **175**, 392–9.
82. Light, R.J. and Hahlbrock, K. (1980) *Z. Naturforsch.*, **35c**, 717–21.
83. Hahlbrock, K., Zilg, H. and Grisebach, H. (1970) *Eur. J. Biochem.*, **15**, 13–18.
84. Ayabe, A. and Furuya, T. (1982) *J. Chem. Soc., Perkin I*, 2725–34.
85. Patschke, L., Grisebach, H. and Barz, W. (1964) *Z. Naturforsch.*, **19b**, 1110–3.
86. Patschke, L. and Grisebach, H. (1965) *Z. Naturforsch.*, **20b**, 1039–42.
87. Beerhues, L. and Wiermann R. (1985) *Z. Naturforsch.*, **40c**, 160–5.
88. Hinderer, W. and Seitz, H.U. (1985) *Arch. Biochem. Biophys.*, **240**, 265–72.
89. Stotz, G., Spribille, R. and Forkmann, G. (1984) *J. Plant Physiol.*, **116**, 173–83.
90. Mol, J.N.M., Robbins, M.P., Dixon, R.A. and Veltkamp, E. (1985) *Phytochemistry*, **24**, 2267–9.
91. Britsch, L. and Grisebach, H. (1986) *Eur. J. Biochem.*, **156**, 569–77.
92. Hahlbrock, K., Ebel, J., Ortmann, R. *et al.* (1971) *Biochim. Biophys. Acta*, **244**, 7–15.
93. Fritsch, H., Hahlbrock, K. and Grisebach, H. (1971) *Z. Naturforsch.*, **26b**, 581–5.
94. Fritsch, H. and Grisebach, H. (1975) *Phytochemistry*, **14**, 2437–41.
95. Stafford, H.A. and Lester, H.H. (1985) *Plant Physiol.*, **78**, 791–4.

96. Burns, M.K., Coffin, J.M., Kurobane, I. *et al.* (1981) *J. Chem. Soc., Perkin I*, 1411-6.
97. Al-Ani, H.A.M. and Dewick, P.M. (1984) *J. Chem. Soc., Perkin I*, 2831-8.
98. Kochs, G. and Grisebach, H. (1986) *Eur. J. Biochem.*, **155**, 311-8.
99. Crombie, L., Holden, I., Van Bruggen, N. and Whiting, D.A. (1986) *J. Chem. Soc., Chem. Comm.*, 1063-5.
100. Crombie, L. (1984) *Nat. Prod. Rep.*, **1**, 3-19.
101. Begley, M.J., Crombie, L., Rossiter, J. *et al.* (1986) *J. Chem. Soc., Chem. Comm.*, 353-6.
102. Sweigard, J.A., Matthews, D.E. and VanEtten, H.D. (1986) *Plant Physiol.*, **80**, 277-9.
103. Banks, S.W. and Dewick, P.M. (1983) *Z. Naturforsch.*, **38c**, 185-8.
104. Banks, S.W. and Dewick, P.M. (1983) *Phytochemistry*, **22**, 2729-33.
105. Martin, M. and Dewick, P.M. (1980) *Phytochemistry*, **19**, 2341-6.
106. Kunesch, G. and Polonsky, J. (1969) *Phytochemistry*, **8**, 1221-6.

# 6 Alkaloids

## 6.1 INTRODUCTION

Basic nitrogenous metabolites isolated from plants are called alkaloids. There are very many of them, and they have diverse structures [1, 2]. Yet an examination of these structures indicates that alkaloids can be classified within a few groups, as will be apparent from the discussion below. This is because they are formed very largely from a handful of $\alpha$-amino acids, lysine, ornithine, phenylalanine, tyrosine and tryptophan, and the skeletons of these amino acids are retained largely intact in the derived alkaloids. Mevalonate and acetate are further important starting points from primary metabolism.

Economy in the use of a few primary metabolites has been of importance in allowing the development of useful ideas [3] about alkaloid biosynthesis, e.g. tropine (6.3) and hygrine (6.2) could be thought of simply as deriving from ornithine and acetate (see Scheme 6.1). Two further things have been impor-

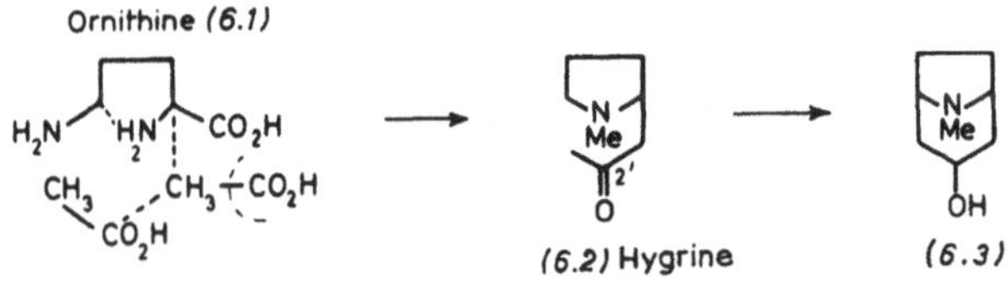

Scheme 6.1

tant in useful speculation about alkaloid biosynthesis: first, structures of similar alkaloids often suggest plausible biosynthetic relationships, e.g. hygrine (6.2), formed in the same species as esters of (6.3), might well be an intermediate in the formation of tropine (6.3); secondly alkaloid formation involves simple, almost repetitive, reactions, and, most widely, those of the

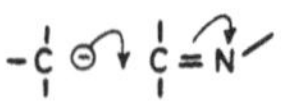

Scheme 6.2

type summarized in Scheme 6.2. Such a reaction is proposed for the formation of hygrine (*6.2*) and is one that can be carried out in the laboratory (see Scheme 6.3 and section 6.2.2).

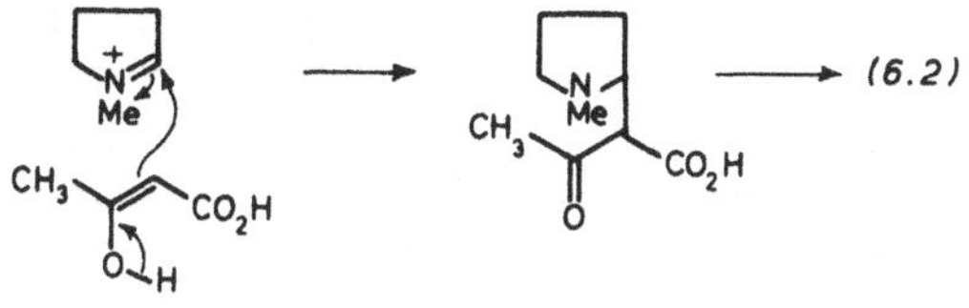

**Scheme 6.3**

Another, crucially important, reaction type, encountered in the biosynthesis of aromatic alkaloids, is the joining together of benzene rings which is thought to occur by what is called phenol oxidative coupling (for further discussion see section 1.3.1).

In the living plant, the roles (and there is surely not just one role) that alkaloids have to play are largely obscure, but it is to be noted that plants tend to synthesize alkaloids of similar structure sometimes by different pathways (see e.g. section 6.2.1). Clearly, particular structural types are (or have been) important to the survival of at least some plants.

## 6.2 PIPERIDINE AND PYRROLIDINE ALKALOIDS

### 6.2.1 Piperidine alkaloids

Simple examples of piperidine alkaloids are *N*-methylpelletierine (*6.19*) and the hemlock alkaloid, coniine (*6.14*). In these bases the structural relationship is manifestly close. This relationship is similarly apparent between anabasine (*6.20*) and anatabine (*6.7*) which are, moreover, found in the same

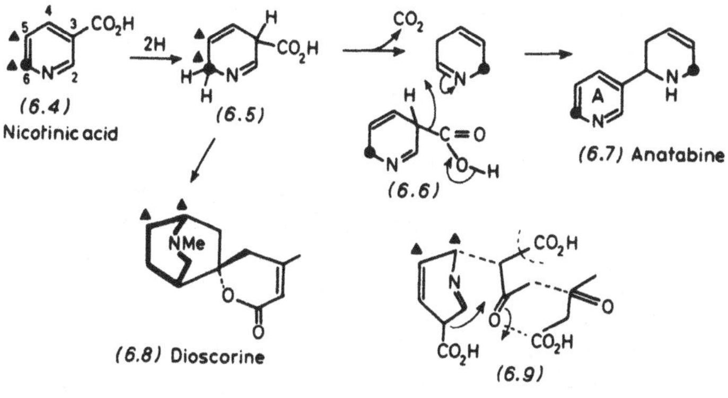

**Scheme 6.4**

plant. It is however, clear, from biosynthetic experiments, that whilst the piperidine rings of *N*-methylpelletierine (*6.19*) and anabasine (*6.20*) derive from the amino acid lysine (*6.17*) those of coniine (*6.14*) and, most surprisingly, anatabine (*6.7*) have quite different origins. It is proved that anatabine (*6.7*) is formed from two molecules of nicotinic acid (*6.4*) [4] (the labelling results, and a suggested pathway, are illustrated in Scheme 6.4). Only the pyridine ring of anabasine derives from nicotinic acid (*6.4*). For both alkaloids, however, the point of attachment of ring A is C-3, the site of the carboxy-group in (*6.5*). Presumably this group assists in the condensation between the two heterocycles in the course of the biosynthesis of each alkaloid (see Scheme 6.4, cf Scheme 6.17). For evidence on the likely intermediacy of (*6.5*) in the biosynthesis of the analogous base, nicotine, see section 6.2.2.

Dioscorine (*6.8*) is, like anatabine, exceptional in that the piperidine ring (heavy bonding) also derives from nicotinic acid (*6.4*); the remaining atoms derive from acetate (acetoacetate) [see (*6.9*)] [5].

The entire skeleton of coniine (*6.14*) derives from acetate, with labelling of C-2′, C-2, C-4, and C-6 by [1-$^{14}$C]acetate. A polyketide pathway immediately seems likely and 5-oxo-octanoic acid (*6.11*) and the corresponding aldehyde (*6.12*), with the necessary functionality for inclusion of the nitrogen atom, were shown, by tracer and enzymic evidence, to be implicated in coniine biosynthesis, as was γ-coniceine (*6.13*) which is also a hemlock alkaloid. These results [6, 7] lead in a straightforward way to the pathway shown in Scheme 6.5. Most curiously, however, it has been found that octanoic acid

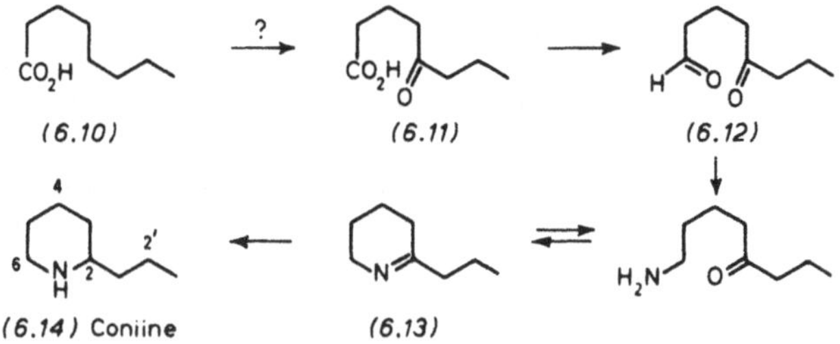

Scheme 6.5

(*6.10*) is a significant and specific precursor for coniine (*6.14*). It follows from this that coniine may be formed oxidatively from a C-8 fatty acid, (*6.10*), rather than reductively from a C-8 polyketide. If this is so, then this alkaloid is almost unique in the field of secondary metabolism since very few other acetate-derived metabolites so far investigated are known to arise through fatty acids.

Pinidine (*6.15*) [8], with a structure similar to that of coniine (*6.14*), also derives by simple linear combination of acetate units: C-2, C-4, C-6, and C-9 were labelled by [1-$^{14}$C]acetate. One of the acetate carboxy groups is lost

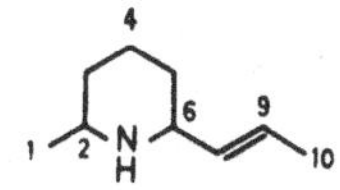

*(6.15)* Pinidine 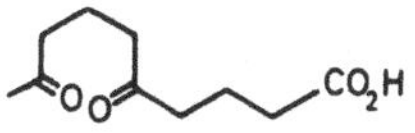          *(6.16)*

during biosynthesis (from C-1 or C-10). Non-radioactive sodium acetate fed at the same time as diethyl [1-$^{14}$C]malonate diluted radioactivity from C-2 which must therefore be part of the 'starter' acetate unit (see Chapter 3), i.e. the carboxy group is lost from C-10. This, in the light of coniine biosynthesis, leads to *(6.16)*, and possibly decanoic acid, as intermediates in pinidine formation but so far negative results have been obtained in feeding experiments with these compounds.

The pathways deduced for coniine *(6.14)* and pinidine *(6.15)* on the one hand, and anatabine *(6.7)* and dioscorine *(6.8)* on the other, are exceptional. That deduced for *N*-methylpelletierine *(6.19)* can be considered much more typical of piperidine alkaloids because the piperidine ring originates from lysine *(6.17)*. Two other alkaloids, anabasine *(6.20)* and sedamine *(6.21)* have similar origins and much of the evidence for the three alkaloids is interlocking so they are best discussed together.

The results of extensive experiments with variously labelled samples of lysine closely define the way in which this amino acid is assimilated into the alkaloids. Thus C-2 and C-6 of the precursor become C-2 and C-6, respectively, in *(6.19)* [9], *(6.21)* [10, 11], and *(6.20)* [12] (Scheme 6.6). Although cadaverine *(6.26)* is also an alkaloid precursor it cannot be an intermediate *following* lysine because any label at C-2 or C-6 of the amino acid would become spread over both C-1 and C-5 of the symmetrical diamine *(6.26)*; a single lysine label would thus be spread over both C-2 and C-6 of the alkaloids.

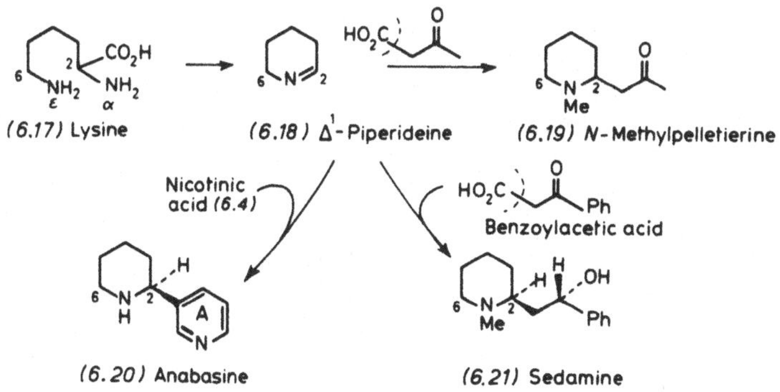

Scheme 6.6

Tritium at C-2 and C-6 of lysine is retained on formation of sedamine *(6.21)* [10, 11]. These results, supported by those with [$^{15}$N]-labelled samples of lysine, indicate that alkaloid formation involves retention of the C-6 amino-group and loss of the one at C-2. Loss of the C-6 amino group would require oxidation of C-6 and tritium loss, but removal of the other amino-group is not expected, necessarily, to result in tritium loss from C-2. The reaction may be one of deaminative decarboxylation (see below) leading to $\Delta^1$-piperideine (6.18); [6-$^{14}$C]-labelled (6.18) was specifically incorporated into anabasine (6.20) with labelling of C-6 [13].

The origins of the skeletal fragments of (6.19), (6.20), and (6.21) not accounted for by lysine (and $\Delta^1$-piperideine) are as follows. The $C_3$ side-chain in *N*-methylpelletierine (6.19) has its origins in acetate plausibly via acetoacetate [9]. The side-chain of sedamine (6.21), on the other hand, derives from phenylalanine, probably by way of its deamination product, cinnamic acid [10, 11]. Benzoylacetic acid, a normal *in vivo* transformation product of cinnamic acid, may also reasonably be included in the pathway to this alkaloid, as it is in, e.g., the biosynthesis of lobeline (6.34) and of phenanthroindolizidine alkaloids (see Section 6.2.2). The pyridine ring of anabasine (6.20) arises from nicotinic acid [14] by way presumably of (6.5) (cf nicotine, Section 6.2.2). The acid precursors (see Scheme 6.6) for (6.19), (6.20) and (6.21), have in common an electron-donating functionality (Scheme 6.7) which may react with $\Delta^1$-piperideine (6.18), possibly with concomitant decarboxylation, to give the alkaloids (cf fatty acid biosynthesis, section 1.1.2).

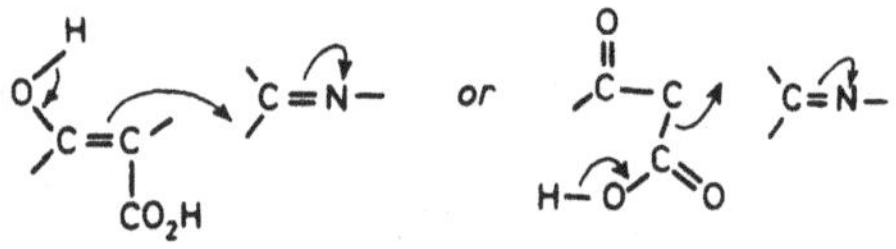

**Scheme 6.7**

In the case of sedamine (6.21) and *N*-methylpelletierine (6.19) a further step of methylation is required and it was at one time thought that the incorporation of lysine with distinction between C-2 and C-6, as well as the incorporation of cadaverine (6.26), could be accommodated in terms of $\epsilon$-*N*-methyl-lysine [see (6.17)], *N*-methylcadaverine (6.24) and *N*-methyl-$\Delta^1$-piperideine (6.22). However, it has been shown that (6.22) is not implicated in anabasine (6.20) biosynthesis but does afford *in vivo* the unnatural base (6.23) [15], and that *N*-methylcadaverine (6.24) is not present in plants producing (6.21) (*Sedum* species) and (6.20). Most importantly

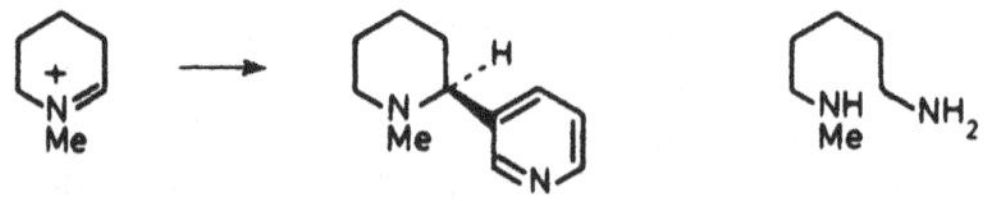

(6.22)          (6.23)          (6.24)

it has been shown [16] that although ε-*N*-methyl-lysine [see (*6.17*)] is a natural constituent of *Sedum acre*, it was formed from [6-³H]lysine and [*Me*-¹⁴C]methionine (source of the *N*-methyl groups) with an isotope ratio which differed from that of sedamine (*6.21*) formed in the same experiment, clearly excluding a precursor–product relationship between the two compounds.

The piperidine alkaloids discussed above, as we have seen, are formed from L-lysine (*6.17*) with distinction maintained between C-2 and C-6. This means that a symmetrical compound such as cadaverine (*6.26*) cannot be involved as an *intermediate*, and yet cadaverine is an alkaloid precursor. A solution [16] to this problem involving common, unsymmetrical, pyridoxal-linked intermediates (*6.25*) and (*6.27*) has been proposed but it needs modification in the light of other evidence. First, cadaverine is utilized in alkaloid biosynthesis with loss of the 1-*pro-S* proton [see (*6.27*)]. Second, L-lysine is decarboxylated with retention of configuration (see Scheme 6.8); protonation occurs on the same face as that from which the carboxy group is lost [17]. This means that the proton present on C-2 of L-lysine would become the 1-*pro-S* proton in (*6.27*) which is lost on oxidation to give (*6.25*) and thence (*6.18*); and it is known to be retained. Therefore (*6.27*), a pyridoxal-linked intermediate which may be formed from cadaverine (*6.26*), cannot be interconvertable with the pyridoxal-linked intermediate (*6.25*) formed from lysine. One may then conclude that when administered to the plant the lysine undergoes decarboxylation to give (*6.25*) which cyclizes directly (Scheme 6.8) to give (*6.18*) and is not first protonated to give (*6.27*).

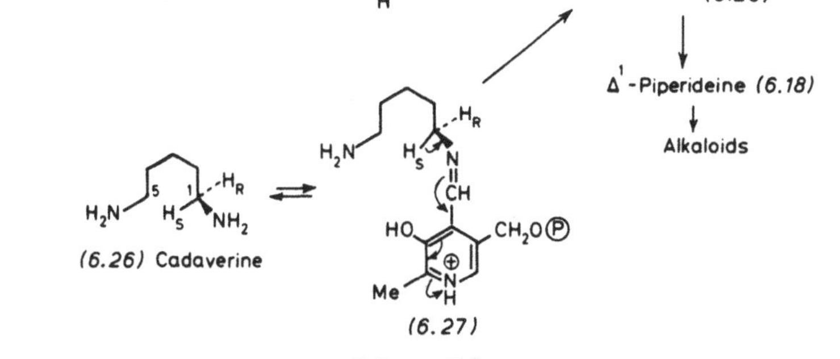

**Scheme 6.8**

Administered cadaverine on the other hand undergoes oxidation (by diamine oxidase) to give (*6.18*) in a quite separate sequence possibly via (*6.27*) and (*6.25*) (Scheme 6.8).

In examples to be discussed below, we shall see that some alkaloids are formed from lysine through a symmetrical intermediate. One concludes for these alkaloids that biosynthesis from lysine always proceeds by decarboxylation to give free cadaverine (*6.26*), which then undergoes further reaction.

In contrast to alkaloid formation which is preferentially from L-lysine, the genesis of pipecolic acid (*6.28*), which is found widely in plants, animals and micro-organisms, is preferentially from D-lysine [18, 19]. Significantly the biosynthetic pathway to pipecolic acid, illustrated in Scheme 6.9, is different from that which leads to piperidine alkaloids (Scheme 6.8) [11].

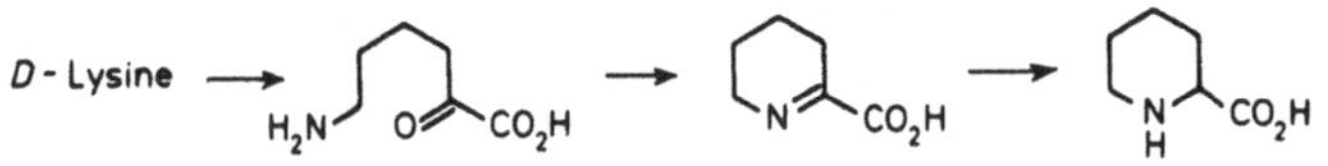


(*6.28*) Pipecolic acid

**Scheme 6.9**

The unit (*6.29*) found in sedamine (*6.21*) and *N*-methylpelletierine (*6.19*) is one which occurs widely in piperidine alkaloids, a number of which have had their biosynthesis investigated, and proved to be largely as expected: anaferine (*6.31*) [9] and pseudopelletierine (*6.32*) [16] (for a full discussion on related pyrrolidine alkaloids see section 6.2.2), (*6.33*) [20], lobeline (*6.34*) [21], (*6.35*) [21] and also lobinaline (*6.36*) which is formed from two molecules of 2-phenacylpiperidine (*6.30*) (see dotted line) [22]. Securinine (*6.37*) [23, 24] is manifestly not composed from the (*6.29*) unit. Carbons 6–13

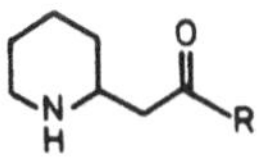

(*6.29*)
(*6.30*) R = Ph

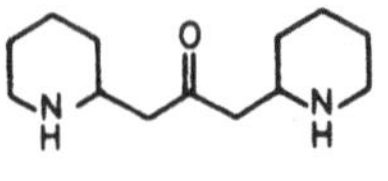

(*6.31*)

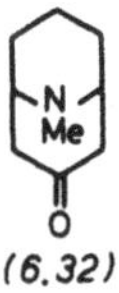

(*6.32*)

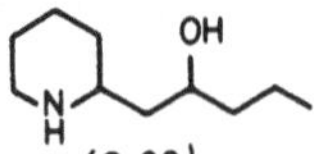

(*6.33*)

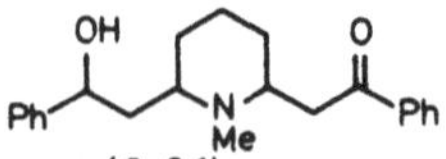

(*6.34*)

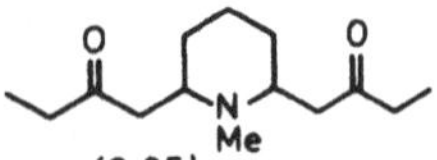

(*6.35*)

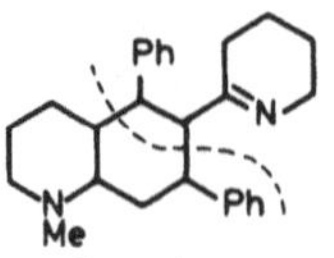

(*6.36*)

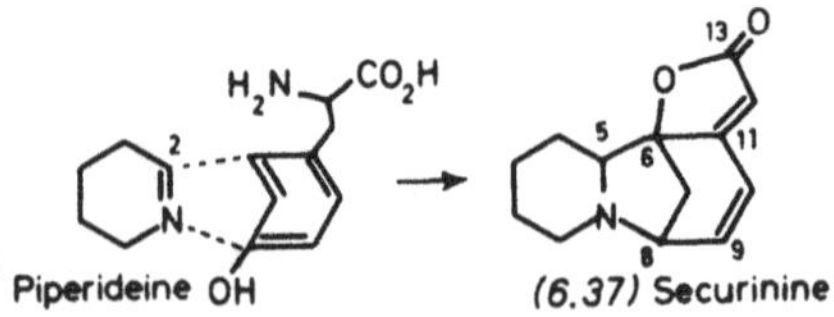


$\Delta^1$ Piperideine OH
Tyrosine

(*6.37*) Securinine

derive from the aromatic amino acid tyrosine, in a pathway for which no analogies can as yet be seen. It is clear, however, that at some stage condensation occurs with $\Delta^1$-piperideine, since both it, lysine, and cadaverine are specific precursors for the piperidine ring of securinine (6.37) in the expected manner; label from C-2 of the first two precursors appears at C-5 of the alkaloid, thus defining the orientation of the $\Delta^1$-piperideine double bond as that shown. It is interesting to note, from these results, that incorporation of lysine avoids a symmetrical intermediate and it is so far the only base with a nitrogen atom common to two rings which does so (see the discussion of Lythraceae and *Lycopodium* alkaloids below).

Experience with the biosynthetic pathways to anatabine (6.7) and anabasine (6.20) does not allow us to predict with security the likely origins of adenocarpine (6.39) and santiaguine (6.41). Indeed, here lysine is the progenitor for all four nitrogen heterocycles in santiaguine; incorporation is without intervention of a symmetrical intermediate (see Scheme 6.10). Biosynthesis has been shown [25] to proceed from lysine through $\Delta^1$-piperideine (6.18) and its dimer (6.38) to adenocarpine (6.39) and thence to santiaguine (6.41). This alkaloid arises then by a $2\pi + 2\pi$ dimerization of adenocarpine, although an alternative, if apparently less important, route involves combination of truxillic acid (6.40) with (6.38); (6.40) itself arises from phenylalanine via cinnamic acid. Concerted chemical $2\pi + 2\pi$ dimerization is photochemically mediated, but interestingly occurs *in vivo* whether the plants are in light or dark.

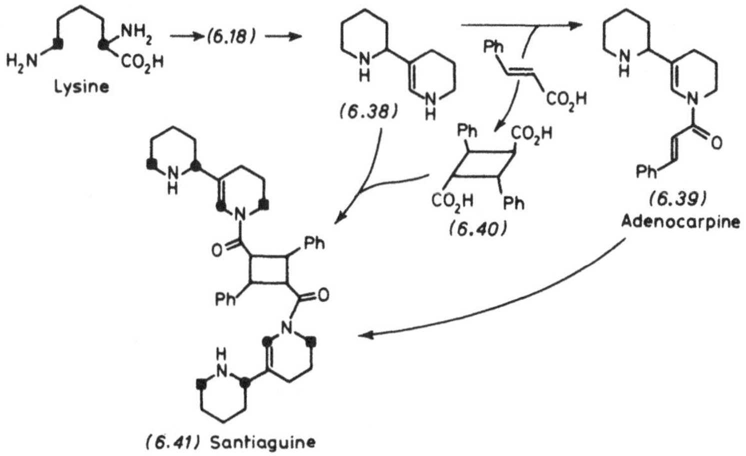

Scheme 6.10

The biosynthesis of the structurally complex alkaloids of the primitive club moss (*Lycopodium* species), e.g. lycopodine (6.43) and cernuine (6.44), has been analysed simply in terms of a pathway which involves the

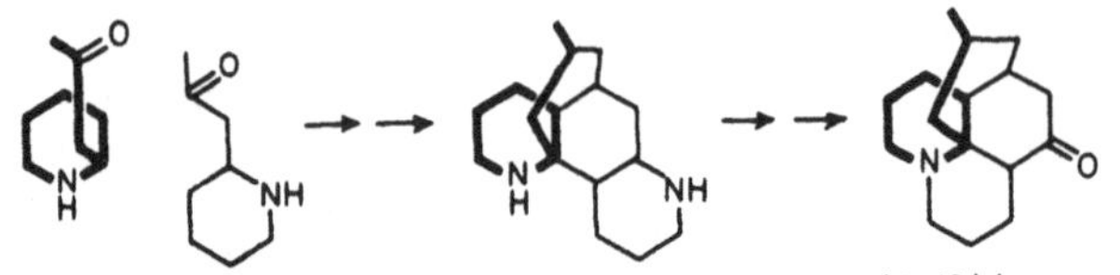

*(6.42)* Pelletierine                    *(6.43)* Lycopodine

*(6.44)* Cernuine

**Scheme 6.11**

dimerization of two pelletierine units (Scheme 6.11). Substance is given to this hypothesis by the appropriate incorporations of acetate, lysine (*6.17*), and $\Delta^1$-piperideine (*6.18*) into lycopodine (*6.43*) and cernuine (*6.44*). Significantly, these precursors were incorporated equally into both putative pelletierine ($C_8N$) units [26].

The crucial test for the hypothesis lay with the examination of pelletierine (*6.42*) as a precursor. The results were quite unexpected for, although pelletierine (*6.42*) was incorporated as an intact unit, it labelled only one of the $C_8N$ units (shown with heavy bonding) in lycopodine (*6.43*) and cernuine (*6.44*). It was subsequently shown that pelletierine (*6.42*) is a normal constituent of the plant used to study lycopodine biosynthesis and is formed from lycopodine precursors. Unlabelled pelletierine, when fed with radioactive cadaverine or $\Delta^1$-piperideine, very efficiently diluted activity in the one $C_8N$ unit (heavy bonding) of lycopodine (*6.43*) for which it is a precursor. It follows that pelletierine is a normal intermediate in the biosynthesis of lycopodine (and cernuine), but unlike the other precursors tested is involved in the biosynthesis of only one unit. The equal labelling of both $C_8N$ units by $\Delta^1$-piperideine (*6.18*) (and lysine) requires that the second $C_8N$ unit must be closely related to pelletierine (*6.42*): significant differences in structure would be associated with differing pathways and hence unequal incorporation of label into the two $C_8N$ units. The results argue thus for the pelletierine-dimer hypothesis but in modified form. It is not easy, however, to propose modifications to the pathway which will fit the results. Hypotheses surrounding (*6.45*) and (*6.46*) as intermediates have not been

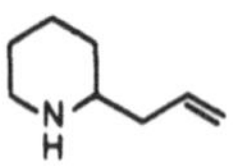

*(6.45)* 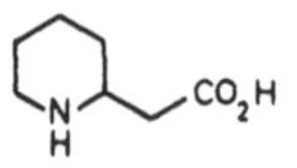            *(6.46)* 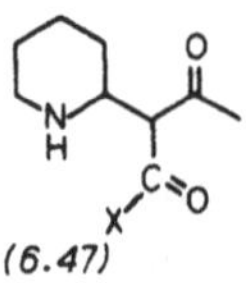            *(6.47)*

supported by subsequent experimentation. So the puzzle remains. One possibility not so far examined is (*6.47*), potentially an intermediate in the condensation of (*6.18*) with acetoacetate (cf Scheme 6.7).

Both the *Lycopodium* [26] and Lythraceae [27, 29] alkaloids, e.g. decodine (*6.49*) and cryogenine (*6.48*) derive from lysine in such a way that C-2 and C-6 become equivalent, logically via cadaverine (*6.26*) (cf Scheme 6.8). This contrasts with the other piperidine alkaloids discussed above.

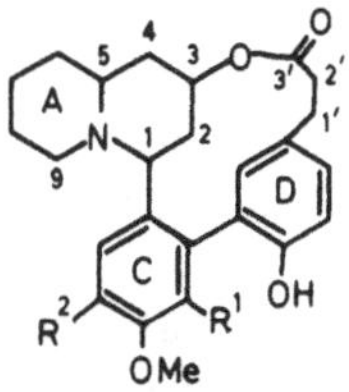

(*6.48*) $R^1 = H$  $R^2 = OMe$, $\Delta^{1'}$, Cryogenine
(*6.49*) $R^1 = OH$, $R^2 = H$, Decodine
(*6.50*) $R^1 = H$, $R^2 = OMe$, Decinine

L-Lysine, its decarboxylation product, cadaverine (*6.26*), and $\Delta^1$-piperideine (*6.18*) serve as specific precursors for ring A of decodine (*6.49*), labelling the alkaloid in complementary ways [27] [supporting, but less complete, evidence was obtained for decinine (*6.50*)]. Exactly similar results were obtained for the related bases lythrine and vertine using [4,5-$^{13}C_2$, 6-$^{14}C$]lysine as a precursor [29]. Ring D especially of (*6.48*) corresponds to a cinnamic acid (*6.52*) derivative and it has been shown that this $C_6$—$C_3$

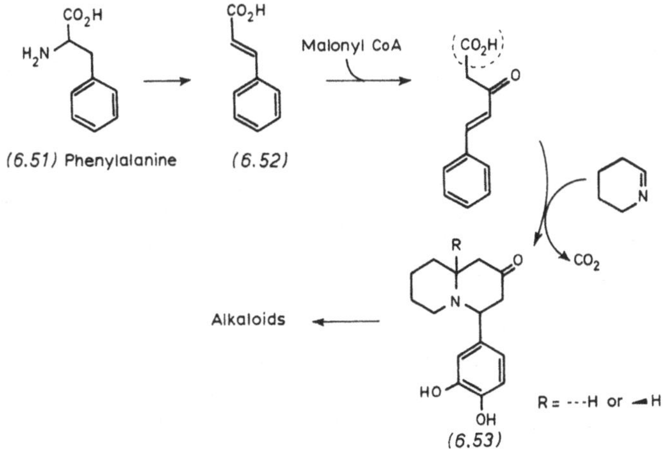

(*6.51*) Phenylalanine  (*6.52*)

Malonyl CoA

Alkaloids

$R = $ ---H or ◀H

(*6.53*)

**Scheme 6.12**

fragment in decodine (*6.49*) derives appropriately from phenylalanine (*6.51*), a known precursor for cinnamic acid. A second $C_6$—$C_3$ unit [as (*6.52*)] is present in (*6.48*): label in the side chain of phenylalanine appears appropriately at C-1, -2, and -3 of (*6.48*) [28]. Early evidence had pointed to pelletierine (*6.42*) as an intermediate in the biosynthesis of Lythraceae alkaloids, but (*6.42*) does not in fact act as a precursor which is in agreement with the labelling results obtained with phenylalanine [28]. From the combined results the pathway outlined in Scheme 6.12 may be proposed. The dihydroxyquinolizidinones (*6.53*) are implicated *en route* to lythrine and vertine [29] and they are involved rather than their mono-*O*-methyl ethers. This suggests that in later phenol oxidative coupling (section 1.3.1) in the biosynthesis of Lythraceae alkaloids one of the two aromatic rings which is involved necessarily bears two hydroxy groups rather than the usual one [as (*6.53*)] (cf section 6.3.5).

Lupinine (*6.54*) is one representative of a group of quinolizidine alkaloids which includes bases (*6.55*), (*6.58*)-(*6.60*) and (*6.63*)-(*6.65*). It is structurally one of the simplest. Slightly more complex is sparteine (*6.55*) which like lupinine is a major alkaloid of the bitter form of the yellow lupin. Specific incorporation of cadaverine in the manner shown (Scheme 6.13) indicates an origin for both exclusively in cadaverine with two and three molecules being involved. [2-$^{14}$C]Lysine labelled the same sites, so, as with other alkaloids having nitrogen common to two rings (except securinine), incorporation is through a symmetrical intermediate [30, 31].

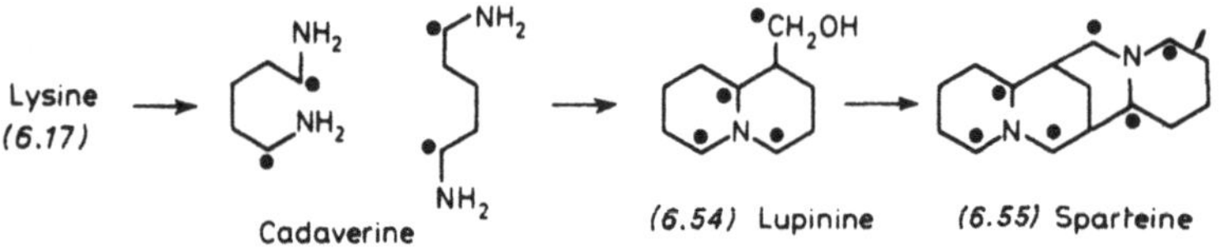

Scheme 6.13

The L-isomer of lysine is the alkaloid precursor, its decarboxylation to cadaverine proceeding with orthodox retention of configuration. Clear definition of the way in which cadaverine is utilized as a key precursor in the biosynthesis of quinolizidine alkaloids has come with the use of [1-$^{13}$C, 1-*amino*-$^{15}$N]cadaverine (*6.56*; labels as shown) and (1*R*)- and (1*S*)-[1-$^2$H]cadaverine [as (*6.62*)] [32, 33, 34].

The former precursor (*6.56*) was incorporated into lupinine (*6.57*), sparteine (*6.58*), lupanine (*6.59*) and angustifoline (*6.60*) and the alkaloids were analysed by $^{13}$C n.m.r. spectroscopy. In the case of lupinine (*6.57*), for example, C-4, C-6, C-10, and C-11 showed equally enriched $^{13}$C n.m.r. signals (cf Scheme 6.13) with $^{13}$C-$^{15}$N coupling visible for C-6. The presence of this coupling (to N-5) and no coupling between C-4 and N-5 excludes a symmetrical intermediate such as (*6.61*) in biosynthesis. (For a different

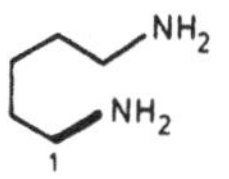

(6.56)

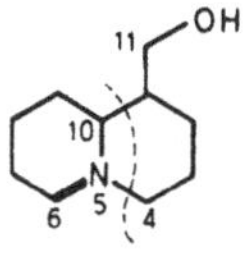

(6.57) Lupinine

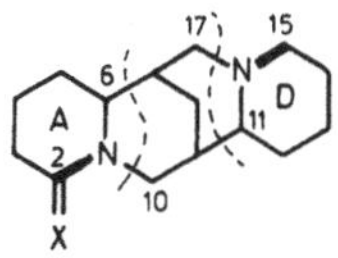

(6.58) X = H₂
(6.59) X = O Lupanine

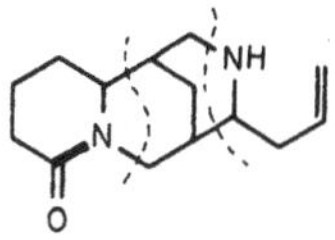

(6.60) Angustifoline

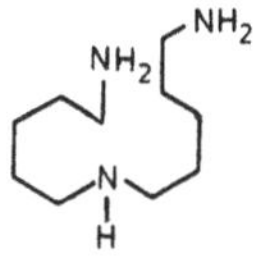

(6.61)

result in the biosynthesis of pyrrolizidine alkaloids, see below.) In support (*6.61*) was found to be neither a precursor for lupinine (*6.57*) nor could it be trapped in a feeding experiment with radioactive lysine.

Similar results were obtained for the other alkaloids: the dotted lines in the formulae (*6.58*)–(*6.60*) indicate the individual cadaverine units and the thickened bonds indicate the intact $^{13}C$-$^{15}N$ units carried over from (*6.56*).

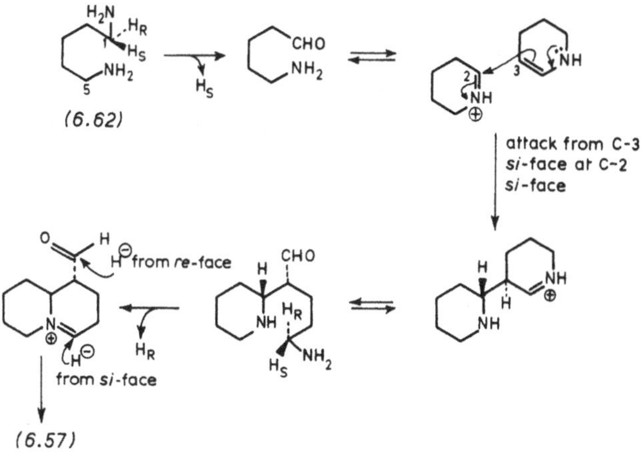

**Scheme 6.14**

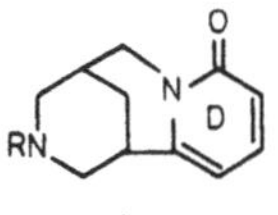

(6.63) R = Me
(6.64) R = H

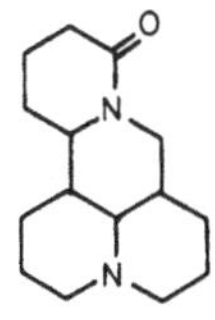

(6.65) Matrine

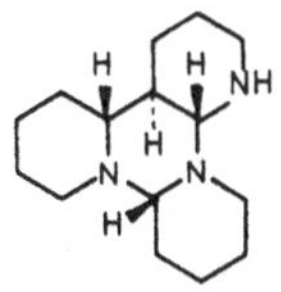

(6.66)

The cadaverine precursors (*6.62*) which were chirally deuteriated at C-1, on incorporation into lupinine (*6.57*), (+)- and (−)-sparteine, lupanine (*6.59*) and angustifoline (*6.60*) gave results which reveal the fate of the C-1 hydrogen atoms in (*6.62*) during the elaboration of the alkaloids. The course of biosynthesis for lupinine (*6.57*) is shown in Scheme 6.14. Interestingly and non-intuitively it appears from the results for *N*-methylcytisine (*6.63*) that it is ring A of an intermediate related to (*6.58*) which is cleaved and ring D which becomes the pyridone in (*6.63*) [cf angustifoline (*6.60*)] [35].

$\Delta^1$-Piperideine (*6.18*) is a normal intermediate in quinolizidine alkaloid biosynthesis after cadaverine. It has been shown to be a specific precursor for matrine (*6.65*), lupanine (*6.59*), and lupinine (*6.57*). For example, the labelling of (*6.59*) by (*6.18*) is as follows: label at C-6 in (*6.18*) appears at C-2, C-15, and, by inference, at C-10, whilst label at C-2 appears at C-17, C-11 and, by inference, at C-6. The results with $\Delta^1$-piperideine (*6.18*) provide support for an attractive hypothesis that the C-15 quinolizidine alkaloids such as lupanine, sparteine and matrine are elaborated from the $\Delta^1$-piperideine trimer (*6.66*) [34, 36].

The implied relationship between lupinine (*6.54*) and sparteine (*6.55*) is given substance by the observation that the bicyclic base is the precursor for the tetracyclic one [37]. Results of experiments with $^{14}CO_2$ indicate that sparteine (*6.55*) and lupanine (*6.59*) have a partially independent biogenesis. The latter serves as a precursor via its 5,6-dehydro-derivative for bases with a pyridone ring, e.g. thermopsine (*6.67*) [38, 39]. Since radioactivity from $^{14}CO_2$ appeared in tetracyclic bases before the tricyclic ones, e.g. (*6.68*) and (*6.64*), it follows that the tricyclic skeleton is formed by fragmentation of the tetracyclic one. It seems, moreover, that rhombifoline (*6.68*) plays a primary role in the formation of cytisine (*6.64*).

(*6.67*) Thermopsine

(*6.68*) Rhombifoline

In important work, crude enzyme preparations obtained from plant cell cultures have been found to catalyse the conversion of cadaverine into 17-oxosparteine [as (*6.58*)] in the presence of pyruvic acid as an amino-group receptor (i.e. the enzymic reaction is a transamination rather than an oxidation for the conversion of $CH_2NH_2$ into CHO). Alkaloid synthesis is further apparently associated with the chloroplasts and shows a light-dependent diurnal rhythm [40]. Results with deuteriated cadaverine and $\Delta^1$-piperideine in which C-17 in sparteine/lupinine is labelled excludes 17-oxosparteine as an intermediate on the way to these alkaloids; it seems likely that production of 17-oxosparteine is the result of further oxidation of sparteine which is the primary product.

## 6.2.2 Pyrrolidine alkaloids

Reactions which occur under so-called 'physiological conditions', in aqueous solution at pH 7 and room temperature, can provide a reliable guide to actual biological reactions. Thus the condensation of acetoacetic acid with *N*-methyl-$\Delta^1$-pyrroline (*6.73*) to give the naturally occurring alkaloid hygrine (*6.2*) [41], is also most probably the same reaction which occurs *in vivo* (see Scheme 6.3); a more complex example is given later in this section when discussing the biosynthesis of phenanthroindolizidine alkaloids and there are many more most notably in the area of terpenoid indole alkaloids (section 6.6.2).

On the other hand, the classical and spectacular synthesis of tropinone (*6.74*) by reaction of succindialdehyde, methylamine and acetonedicarboxylic acid under 'physiological conditions' is a quite misleading guide to the biosynthesis of the group of tropane alkaloids, exemplified by hyoscyamine (*6.76*) [42]. They are formed instead from the amino acid ornithine (*6.69*) (incorporation of [2-$^{14}$C]-labelled material) [43, 44] and putrescine (*6.72*) is also implicated (Scheme 6.15). Label from [1,4-$^{14}$C$_2$]putrescine was

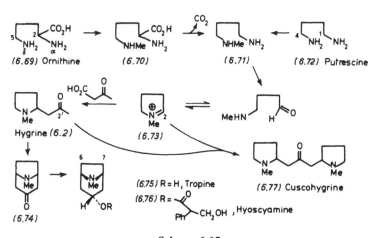

**Scheme 6.15**

confined to the bridgehead carbons of e.g. hyoscyamine (*6.76*) [45]. Equal labelling of the bridgehead carbons by putrescine is expected, but the single label from [2-$^{14}$C]ornithine, whilst it should be located at the bridgehead, might appear at one or both of the bridgehead atoms (cf the discussion of the analogous situation in the formation of piperidine alkaloids from lysine, section 6.2.1). An ingenious and elegant degradative sequence was developed which established the ornithine labelling pattern [43]. The labelled hyoscyamine (*6.78*) was converted into the racemate (*6.79*) (Scheme 6.16) which was resolved. Further degradation allowed isolation of the carbon atom to

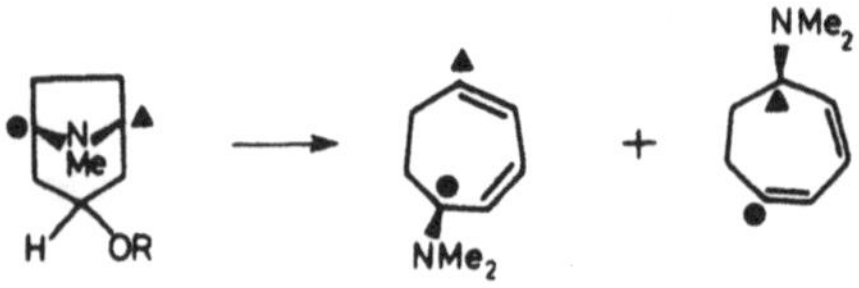

**Scheme 6.16**

which the amino-group, in each enantiomer, was attached. These carbon atoms correspond to the two bridgehead atoms in (*6.78*) and a distinction between them was thus achieved.

It was found that [2-$^{14}$C]ornithine labelled only one of the bridgehead positions. It follows from this that the incorporation of the amino acid is unsymmetrical and cannot therefore involve the symmetrical base putrescine as an intermediate. As a result of corroborating tracer experiments obtained for tropane alkaloids and cuscohygrine (*6.77*), it is currently believed [46] that unsymmetrical ornithine incorporation is achieved through methylation to *N*-methylornithine (*6.70*) (which has been isolated from a plant which produces tropane alkaloids [47] followed by decarboxylation to give *N*-methylputrescine (*6.71*); this latter unsymmetrical base may also be formed from putrescine (*6.72*). It is to be noted that this pathway differs from that deduced for piperidine alkaloids, where nonsymmetrical incorporation of the precursor amino acid (lysine) is more simply accounted for (section 6.2.1).

Confirmation of the role of *N*-methylputrescine in the biosynthesis of the tropane alkaloid, scopolamine (*6.81*) has been elegantly obtained with [1-$^{13}$C, $^{14}$C; *methylamino*-$^{15}$N]-*N*-methylputrescine (*6.80*). It gave (*6.81*) which was labelled appropriately and as shown [48].

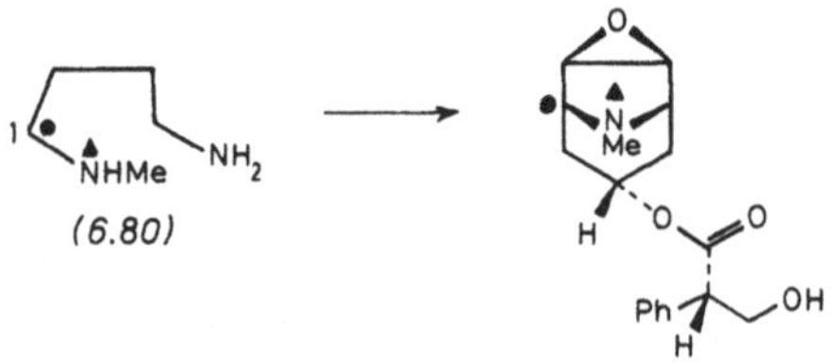

(*6.81*) Scopolamine

In the biosynthesis of nicotine (*6.83*) in *Nicotiana* species it has been found [48–51] that both ornithine (*6.69*) (the L-isomer rather than the D-isomer is involved [52]) and putrescine (*6.82*) are again involved in pyrrolidine-ring formation. For this alkaloid, however, incorporation of the amino acid is through at least one symmetrical intermediate, logically putrescine (*6.82*), because [2-$^{14}$C]ornithine gave nicotine (*6.83*) with label equally spread over C-2′ and C-5′; additional results were obtained with [$^{15}$N]ornithines.

The intact incorporations into nicotine (*6.83*) of *N*-methylputrescine (*6.71*) and *N*-methyl-Δ¹-pyrroline (*6.73*) further define the pathway to this alkaloid [significantly label from C-2 of the precursor (*6.73*) appeared at C-2' of the alkaloid]. *N*-Methyl-Δ¹-pyrroline (*6.73*) was isolated in radioactive form after feeding, for example, radioactive ornithine to *Nicotiana* species, thus establishing its role as an intermediate in nicotine biosynthesis. Additional persuasive evidence comes from the isolation from *Nicotiana* of an enzyme which catalyses the conversion of putrescine (*6.72*) into (*6.71*), and one which effects the transformation of (*6.71*) into (*6.73*) [53, 54].

(*R*)-[1-²H]Putrescine [as (*6.82*)] has been used to determine the stereochemistry of the oxidative conversion of (*6.71*) into (*6.73*). Deuterium label appeared in the derived nicotine at the positions corresponding to the 2'-proton and the 5'-*pro-R* proton. Therefore oxidation of (*6.71*) results in loss of the 1-*pro-S* hydrogen atom [see (*6.82*)] which is the stereochemistry expected of a reaction catalysed by a diamine oxidase. It is further apparent in results obtained with tritiated ornithine that decarboxylation of L-lysine to give (*6.82*) proceeds with orthodox retention of configuration [52].

The pyridine ring of nicotine has been proved to have its genesis in nicotinic acid (*6.4*) and the pyrrolidine ring becomes attached to the site from which the carboxy group is lost. The mechanism whereby the two rings join is suggested by the results obtained with deuteriated and tritiated nicotinic acid samples: hydrogen isotope appears to be lost from C-6, and C-6 only. This is not the result of hydroxylation at this site because 6-hydroxynicotinic acid is not a nicotine precursor. Instead a dihydronicotinic acid intermediate may be proposed which condenses with (*6.73*) leading to nicotine (*6.83*) (Scheme 6.17; H* corresponds to a C-6 tritium label) [55].

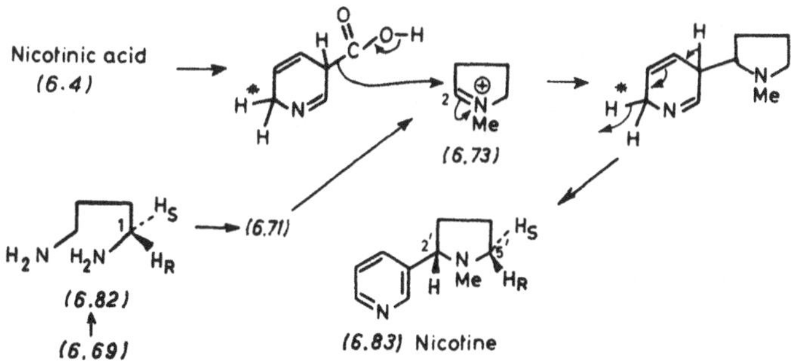

**Scheme 6.17**

The combined results for nicotine suggest the pathway: ornithine (*6.69*) → putrescine (*6.82*) → (*6.71*) → (*6.73*) which condenses with a dihydronicotinic acid to give nicotine (*6.83*) (Scheme 6.17).

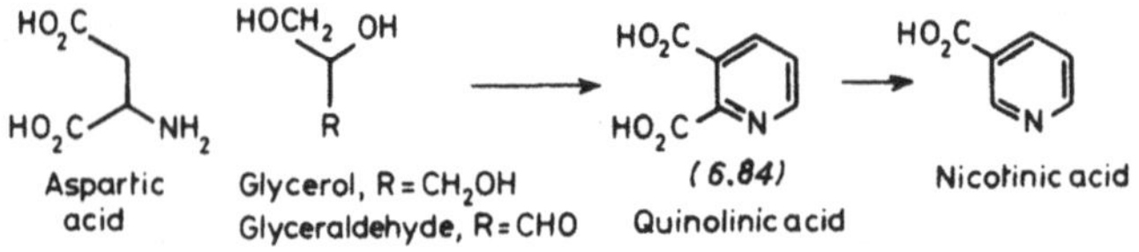

HO₂C  HOCH₂ OH          HO₂C          HO₂C

HO₂C⁀NH₂    R        HO₂C⁀N              N

Aspartic    Glycerol, R = CH₂OH      *(6.84)*        Nicotinic acid
acid        Glyceraldehyde, R = CHO   Quinolinic acid

**Scheme 6.18**

The biosynthesis of nicotinic acid in animals and some micro-organisms is well established to be by degradation of tryptophan, whereas in plants nicotinic acid has its origins in aspartic acid and glycerol (glyceraldehyde) via quinolinic acid *(6.84)* (Scheme 6.18). Such are the origins of the pyridine ring in nicotine *(6.83)* [and anabasine *(6.20)*, section 6.2.1] as well as several other pyridine alkaloids [56].

The observed occurrence of hygrine *(6.2)* and tropane alkaloids, e.g. *(6.76)*, in plants of the same species together with a consideration of their structures and the biosynthesis of nicotine *(6.83)* and piperidine alkaloids already discussed, indicates the sequence of biosynthesis for e.g. hyoscyamine *(6.76)*, shown in Scheme 6.15. This is supported by the observed incorporations of hygrine *(6.2)* and tropine *(6.75)* into hyoscyamine *(6.76)*. Hygrine is also a precursor for cuscohygrine *(6.77)*. Interestingly, and correctly, [*N-methyl*, 2'-$^{14}$C₂]hygrine [as *(6.2)*] gave *(6.77)* where the *specific* radioactivity in the *N*-methyl groups of the alkaloid was one half of that required to maintain the *N*-methyl:C-2' ratio of radioactivity from the hygrine [57]. Thus the second pyrrolidine ring of cuscohygrine arises not by degradation of hygrine but by condensation of the side-chain methyl group of hygrine with a further molecule of *(6.73)*.

Tropine *(6.75)* is a precursor for the more extensively oxygenated alkaloids, scopolamine *(6.81)* and meteloidine *(6.86)*, presumably by way of *(6.85)* (Scheme 6.19) [58, 59]. The two oxygenation sequences involve loss of the C-6 and C-7 β-protons from *(6.75)* [60].

The tropane alkaloids commonly occur as esters of tropic acid [as *(6.81)*] and tiglic acid [as *(6.86)*]. The former acid arises preferentially from phenylalanine possibly via cinnamic acid and the entire carbon skeleton is involved. The results of $^{14}$C-labelling studies show that, in the rearrangement which

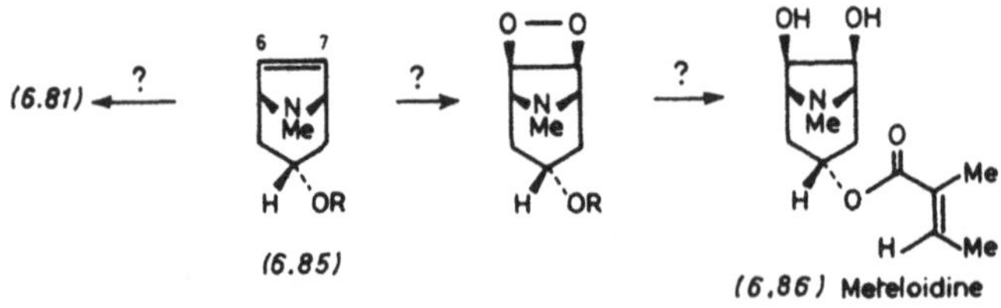

*(6.81)* ←―?―    6  7       ―?→        O—O        ―?→      OH  OH

         N            N                  N
         Me           Me                 Me
                                                         O
       H  OR        H  OR              H  O        Me

*(6.85)*                                        H   Me
                                        *(6.86)* Meteloidine

**Scheme 6.19**

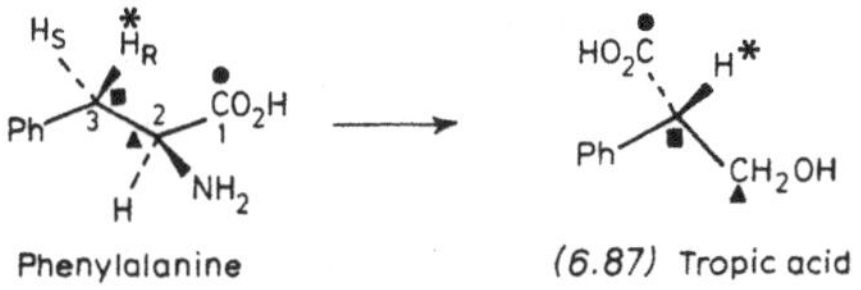

Phenylalanine                    (6.87) Tropic acid

**Scheme 6.20**

occurs during biosynthesis, it is the carboxy group which migrates (see Scheme 6.20). Proof of an intramolecular shift for this group was obtained by showing that $[1,3-^{13}C_2]$phenylalanine (majority of molecules doubly labelled) afforded *doubly* labelled hyoscyamine (6.76) and scopolamine (6.81) with labels confined to the expected sites. (Intermolecular migration would have led to two singly labelled species due to dilution with unlabelled materials in the plant.) Further evidence shows that during the rearrangement there is a stereospecific 1,2-shift of the 3-*pro-S* proton in phenylalanine in addition to the migration of the carboxy group, which is with retention of configuration at what was C-3 of phenylalanine (Scheme 6.20) [61]. It is interesting to note an apparently similar rearrangement of a phenylalanine residue in the biosynthesis of the microbial metabolite tenellin (section 7.2).

The biosynthesis of the tigloyl [as (6.86)] residues of, e.g. meteloidine (6.86), has been found to be similar to that leading to tiglic acid in animals. It derives from the amino acid L-isoleucine [62-64].

Hygrine (6.2) which is formed as an intermediate in the biosynthesis of tropane alkaloids must be formed unsymmetrically from ornithine, but this is not the only route. In *Nicandra physaloides* hygrine is formed via the symmetrical diamine putrescine and in *Erythroxylon coca* cuscohygrine (6.77) (which is elaborated from hygrine) is also formed via putrescine from ornithine [57, 65].

Cocaine has recently gained world-wide notoriety as a drug of abuse. Its structure (6.89) is similar to those of the sometimes clinically useful tropane alkaloids, e.g. (6.76). After twenty-five years of frustration, a change in feeding technique (painting the precursor onto the leaves of the plant, *Erythroxylon coca*) has at last resulted in an incorporation of ornithine

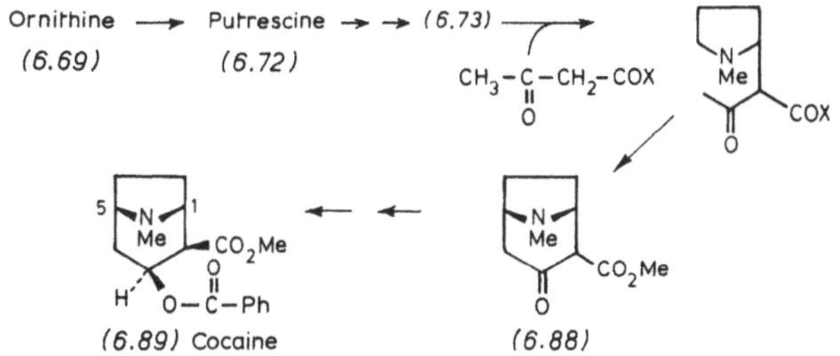

Ornithine ⟶ Putrescine → → (6.73) ⟶
(6.69)            (6.72)

$CH_3-\overset{\underset{||}{O}}{C}-CH_2-COX$

(6.89) Cocaine          (6.88)

**Scheme 6.21**

*(6.69)* into cocaine *(6.89)*. Label from [5-$^{14}$C]ornithine [as *(6.69)*] was equally incorporated into C-1 and C-5 of *(6.89)*. So, like the cuscohygrine *(6.77)* in this plant, cocaine is elaborated from ornithine via a symmetrical intermediate which is probably putrescine. The tropinone *(6.88)* is further an intact precursor for *(6.89)* (i.e. no ester hydrolysis). The likely pathway is outlined in Scheme 6.21 (cf Scheme 6.15). Appropriate incorporations of acetate were recorded and the benzoyl fragment derives by metabolism of phenylalanine [65, 66].

From the outcome of an excellent set of rigorous experiments, chiefly using precursors labelled with stable isotopes, a detailed picture of the biosynthesis of the basic portion of the pyrrolizidine alkaloids e.g. retrorsine *(6.90)* has emerged. The basic portion in this alkaloid, rotronecine *(6.94)* is formed from two molecules of L-ornithine *(6.69)* by way of putrescine *(6.92)*. In one

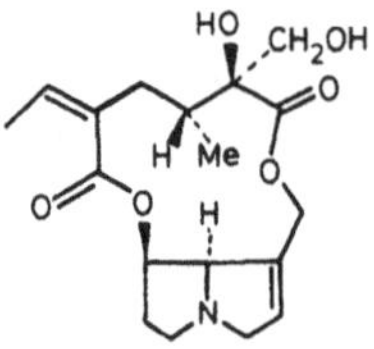

*(6.90)* Retrorsine 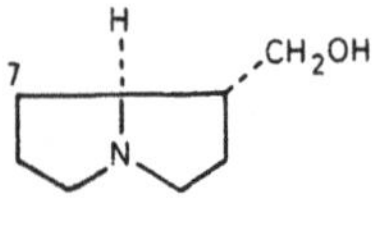          *(6.91)*

set of experiments [1-*amino*-$^{15}$N; 1-$^{13}$C]putrescine [as *(6.92)*] was incorporated into retronecine *(6.94)*. The $^{13}$C n.m.r. spectrum of this material showed enriched signals for C-3 and C-5 which were of approximately equal intensity. The signals were each made up of a doublet, arising from $^{13}$C–$^{15}$N

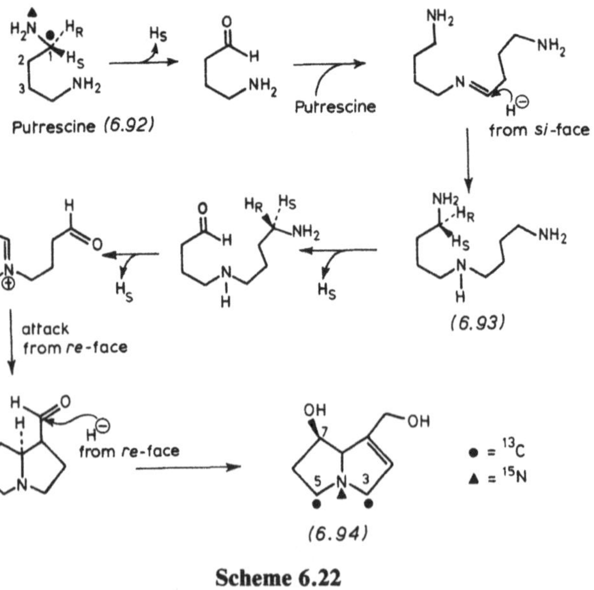

**Scheme 6.22**

coupling, and a singlet ($^{13}C$ next to $^{14}N$) [67, 68]. This points to the symmetrical compound, homospermidine (*6.93*), formed from two molecules of putrescine, being involved in the biosynthesis of retronecine (*6.94*). Appropriate experiments confirmed this to be so [69].

Further information was obtained with deuteriated cadaverine samples, notably (1*S*)-[1-$^2$H]- and (1*R*)-[1-$^2$H]-putrescine [as (*6.92*)]. Results with these precursors define the stereochemistry associated with the biochemical events involved in the elaboration of retronecine (*6.94*) (Scheme 6.22) [70–72].

A nice *in vitro* experiment was carried out in which homospermidine (*6.93*) was oxidized with a diamine oxidase enzyme to give trachelanthamidine (*6.91*). This is evidently what happens *in vivo*, subsequent hydroxylation at C-7 (with orthodox retention of configuration) and desaturation of (*6.91*) providing retronecine (*6.94*) [73].

Various carboxylic acids are found as the necic acid parts of the pyrrolizidine alkaloids, e.g. senecionine (*6.95*), monocrotaline (*6.96*) and heliosupine (*6.97*). Available evidence [74] points to the derivation of these acids from the branched-chain amino acids isoleucine and valine (see dotted lines). [In the case of monocrotaline C-4, C-5, and C-8 apparently derive

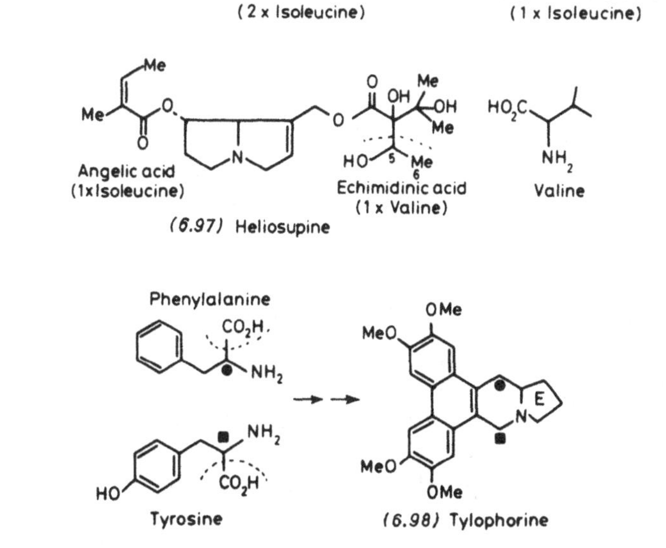

Isoleucine

(*6.95*) Senecionine
( 2 x Isoleucine)

(*6.96*) Monocrotaline
( 1 x Isoleucine)

Angelic acid
(1xIsoleucine)

Echimidinic acid
(1 x Valine)

Valine

(*6.97*) Heliosupine

Phenylalanine

Tyrosine

(*6.98*) Tylophorine

**Scheme 6.23**

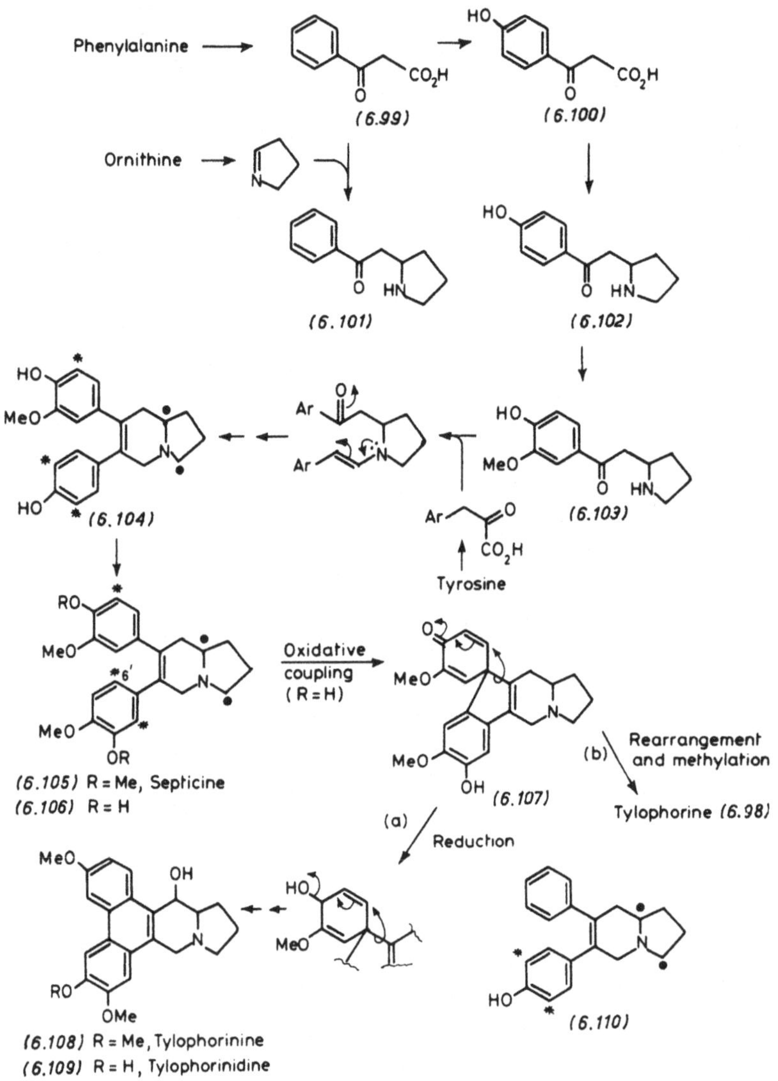

**Scheme 6.24**

from propionic acid. Carbon atoms 5 and 6 of the echimidinic acid fragment in (*6.97*) apparently derive from acetate.]

The phenanthroindolizidine alkaloids, e.g. tylophorine (*6.98*), are formed in *Tylophora asthmatica* from a molecule each of tyrosine (dihydroxyphenyl-alanine is also a precursor) and phenylalanine (Scheme 6.23), with ornithine apparently the source of ring E. The intact incorporation of the phenacyl-pyrrolidines (*6.101*), (*6.102*) and (*6.103*) provided important information on the steps of biosynthesis [75] (Scheme 6.24). [It is because of the involvement of these compounds in biosynthesis and the close structural similarity to, *inter alia*, hygrine (*6.2*), that discussion of the biosynthesis of these alkaloids appears here.]

In addition, the keto acids (*6.99*) and (*6.100*) were intact precursors [a common sequence *in vivo* is phenylalanine → cinnamic acid → (*6.99*)]. The late stages of biosynthesis were suggested by the structure (*6.105*) for the naturally occurring base, septicine, i.e. that linkage between the aromatic rings of the amino acid fragments to give the phenanthrene system of (*6.98*), would occur by oxidative phenol coupling (section 1.3.1), and at a late stage of biosynthesis. With this in mind examination of the structures (*6.98*), (*6.108*) and (*6.109*) of the *T. asthmatica* alkaloids indicated that (*6.106*) could be a key intermediate for the formation of all three alkaloids [76].

Samples of (*6.104*), (*6.106*) and (*6.110*), double labelled as shown (\* = $^3$H, ● = $^{14}$C), were found to be incorporated into (*6.98*), (*6.108*) and (*6.109*) at a similar level in each alkaloid, indicating a close biosynthetic relationship between the alkaloids. The changes in isotope ratio were consistent with necessary loss of tritium from sites in (*6.104*) and (*6.110*) which became hydroxylated in the course of biosynthesis and from C-6′ in (*6.106*) during phenol coupling. It follows from this that (*6.104*), (*6.106*) and (*6.110*) are intact precursors\* for (*6.98*), (*6.108*) and (*6.109*), but the much lower incorporation observed for (*6.110*) compared with the other two compounds indicated that it was utilized along a minor pathway. The major route to the alkaloids must be as shown (Scheme 6.24). The indolizidine (*6.106*) can only give (*6.98*), (*6.108*) and (*6.109*) via the dienone (*6.107*), alternative courses of rearrangement (path b, Scheme 6.24), and reduction and rearrangement (path a) affording the three alkaloids after further minor modification. In the rearrangement of (*6.107*) (and the dienol derived from it), the unique opportunity, in alkaloid biosynthesis, of styryl (as against aryl, section 6.3) migration must be taken in affording (*6.108*) and (*6.109*), and it may well be taken in the formation of (*6.98*).

## 6.3 ISOQUINOLINE AND RELATED ALKALOIDS

### 6.3.1 Phenethylamines and simple isoquinolines

The simplest of the phenethylamine alkaloids are *N*-methyltyramine (*6.113*) and hordenine (*6.114*), which are found in barley roots. Their genesis is an appropriately simple one from the aromatic amino acid tyrosine (*6.111*) via tyramine (*6.112*), with methionine serving as the source for the *N*-methyl groups [77, 78]. The biosynthesis of mescaline (*6.124*), the hallucinogen from the peyote cactus, has similar beginnings in tyrosine and tyramine. Extensive experimentation, including that with an *O*-methyltransferase

---

\*For discussion of the use of doubly-labelled precursors to test for intact incorporation see sections 2.2.1 and 6.4.2.

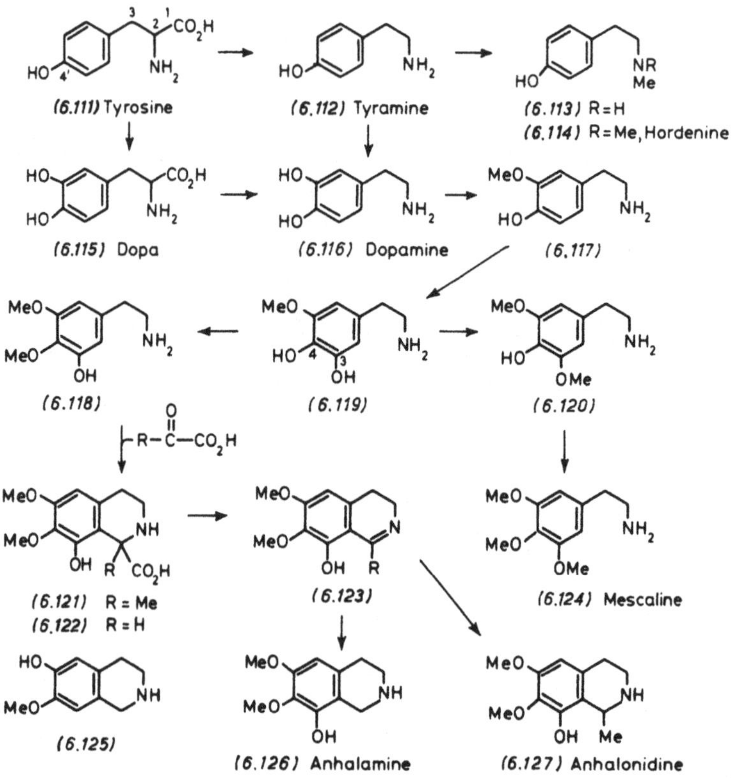

**Scheme 6.25**

from peyote, has allowed mapping of the biosynthetic pathways in detail (Scheme 6.25) [79, 80]. A key stage is reached with the formation of (*6.119*). Methylation of the hydroxy-group at C-3 leads to mescaline (*6.124*), whereas methylation of the alternative at C-4 leads to the isoquinoline alkaloids anhalonidine (*6.127*) and anhalamine (*6.126*).

All except one and two carbon atoms in the skeletons of, respectively, (*6.126*) and (*6.127*), are accounted for by phenethylamine (*6.118*). The origins of these almost trivial $C_1/C_2$ units proved quite elusive until a suggestion of some thirty years' standing, namely that (*6.121*) and (*6.122*) could be precursors for (*6.127*) and (*6.126*), respectively, was examined with positive results. The amine (*6.118*) and pyruvic acid, both precursors for anhalonidine (*6.127*), reacted together readily *in vitro* to give (*6.121*) (Scheme 6.26), which was a specific precursor *in vivo* for anhalonidine (*6.127*). It was shown to be a constituent of peyote and was decarboxylated by fresh peyote slices to the imine (*6.123*) which was then established as a further biosynthetic intermediate [81]. A similar route accounts for the

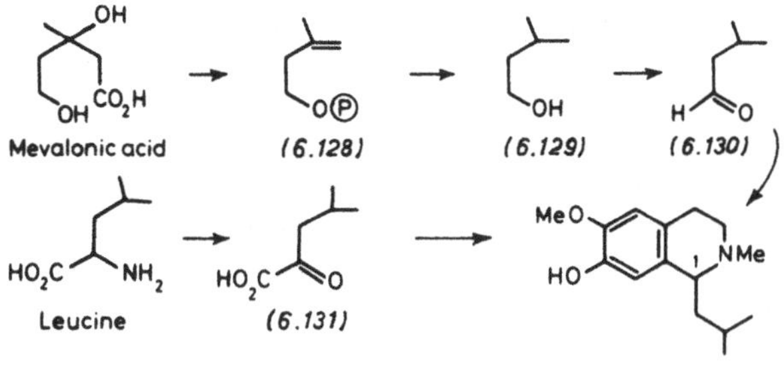

Mevalonic acid    (6.128)    (6.129)    (6.130)

Leucine    (6.131)

(6.132) Lophocerine

formation of anhalamine (6.126), via (6.122) which is derived from glyoxylic acid rather than pyruvic acid [81]. Salsoline (6.125) was subsequently shown to be derived along a related pathway. A similar pathway may be adduced for the unusual alkaloid lophocerine (6.132) from leucine, which is a proven precursor. A separate derivation for the $C_5$ unit (C-1 plus isobutyl group) in this alkaloid has been deduced [82, 83] to be mevalonate by way of (6.128), (6.129) and (6.130), all of which serve as lophocerine precursors. Isoquinoline formation along this route involves perforce an aldehyde in contrast to the α-carbonyl acids utilized for the biosynthesis of the other alkaloids discussed above.

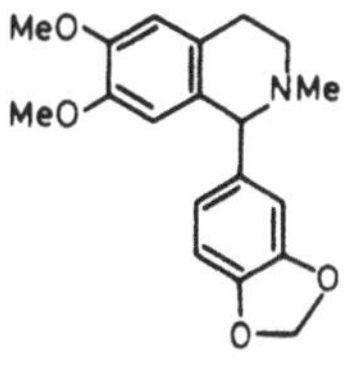

(6.133) Cryptostyline-I 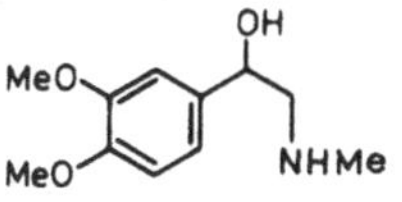    (6.134) Normacromerine 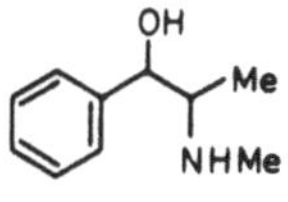    (6.135) Ephedrine

Cryptostyline-I (6.133) is formed from two molecules of tyrosine, and an aldehyde intermediate must again be involved [84]. From the preceding discussion and that which follows it is apparent that there are two alternatives for the biosynthesis of isoquinoline ring systems, [as (6.136)], that is to say via a keto-acid intermediate or, chiefly for the more complex phenyl-, benzyl-, and phenethyl-isoquinolines, via an aldehyde intermediate (Scheme 6.26).

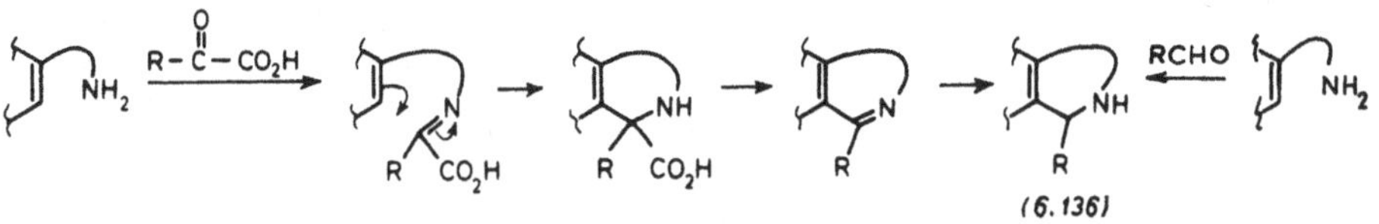

(6.136)

**Scheme 6.26**

Some naturally occurring phenethylamines, e.g. normacromerine (6.134), bear a hydroxy group in the β-position of the side-chain. Their origins are, however, unexceptionally in tyrosine. It appears that the distinguishing β-hydroxylation step may be an early one [85]. Ephedrine

(*6.135*) has a *C*-methyl group in its side-chain in addition to the β-hydroxy-group of bases like normacromerine. Curiously, in spite of the structural kinship of (*6.135*) to bases like (*6.134*), the genesis of ephedrine (*6.135*) is quite different. Phenylalanine was incorporated (via cinnamic acid) but only as a $C_6$—$C_1$ unit. Benzoic acid and benzaldehyde were incorporated at a higher level and it seems possible that normal derivation of the $C_6$—$C_1$ unit in ephedrine (*6.135*) is directly from shikimic acid rather than via its metabolite phenylalanine [86] (section 5.1). Methionine serves as the source of the *N*-methyl group of ephedrine, but the origin of the remaining $C_2N$ group is obscure, except that aspartate or close relative may be involved.

The 1-benzyltetrahydroisoquinoline skeleton is one of preeminence in the elaboration of plant alkaloids. Only the terpenoid indole skeleton (section 6.6.2) appears in more diversely modified form. Among the benzylisoquinolines, reticuline (*6.149*) is of outstanding importance as a substrate for modification. It is therefore important to know how this and other benzylisoquinolines are biosynthesized. It is well established through examining diverse examples that the benzylisoquinoline skeleton is elaborated from two molecules of tyrosine (*6.111*). Tyramine (*6.112*) and dopamine (*6.116*) serve as successive intermediates for the 'top' half of the skeleton [see (*6.142*)]; 3,4-dihydroxyphenylalanine (dopa) is also incorporated into this 'top' half (i.e. via dopamine) and essentially not at all into the 'bottom' half.

This latter observation has been puzzling because good evidence pointed to dihydroxyphenylacetaldehyde (*6.138*) as an intermediate for the 'bottom' half (expected to derive from tyrosine via dopa): an enzyme was obtained from plant tissue cultures which catalysed the stereospecific condensation of dopamine (*6.116*) with (*6.138*) to give (*S*)-norlaudanosoline (*6.140*) which can serve as a precursor for reticuline (*6.142*). It turns out, however, that, despite the evidence bolstered by prejudice, (*S*)-norlaudanosoline (*6.140*) is not an intermediate after all. The true intermediate is (*S*)-norcoclaurine (*6.139*) which is formed by condensation of dopamine with 4-hydroxyphenylacetaldehyde (*6.137*) (formed directly from tyrosine). This condensation is catalysed with similar facility by the synthase which (misleadingly) yields (*S*)-norlaudanosoline (*6.140*) [87, 88]. The (*6.139*) apparently undergoes *O*-methylation at C-6 to give (*S*)-coclaurine (*6.141*) which like reticuline (*6.142*) is a key benzylisoquinoline (see below). Hydroxylation at C-3′ in (*6.141*) followed by methylation yields (*S*)-reticuline (*6.142*) (Scheme 6.27). Work with tissue cultures has led impressively to isolation of the methyltransferases which catalyse, in the presence of *S*-adenosylmethionine, the sequence of methylations required [89].

One of the simplest benzylisoquinolines is papaverine (*6.143*) which is formed from *N*-norreticuline [as (*6.149*)] [90–92]. The most complex are the bisbenzylisoquinolines, e.g. tiliacorine (*6.144*). Most of those studied originate from coclaurine (*6.145*) and often also its *N*-methyl derivative

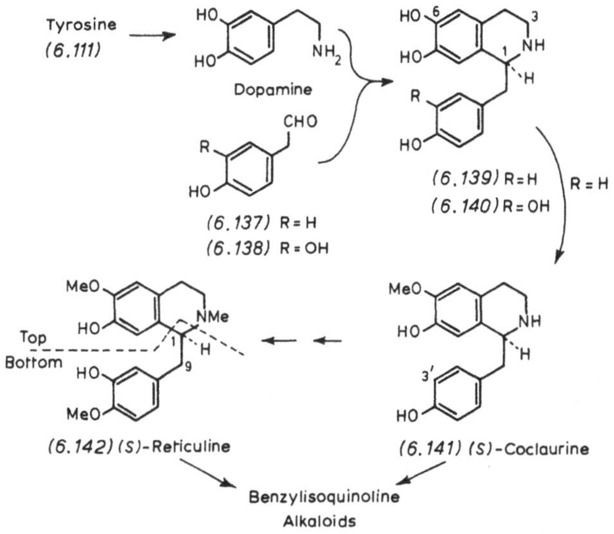


**Scheme 6.27**

(*6.146*). The results are exemplified by those found for the diastereoisomeric bases tiliacorine and tiliacorinine (*6.144*). (*R*)-*N*-Methylcoclaurine gives the 'right-hand' half of tiliacorine and the *S*-isomer [as (*6.141*)] gives the 'left-hand' half. Consequently, and since it is known that the two isomers of *N*-methylcoclaurine are not interconverted *in vivo*, the stereochemistry at C-1 and C-1′ in (*6.144*), which was previously unknown, is neatly determined *S* and *R* respectively. Tiliacorinine (*6.144*) is formed from two molecules of (*S*)-*N*-methylcoclaurine [as (*6.141*)] so the configurations at C-1 and C-1 are both *S* [93, 94].

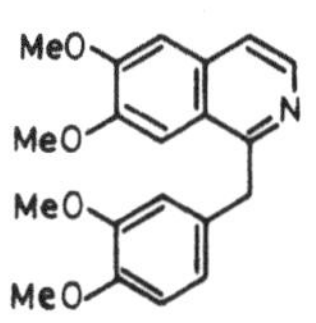

(*6.143*) Papaverine

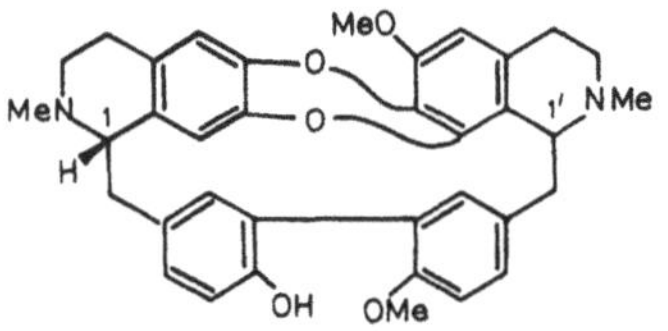

(*6.144*)

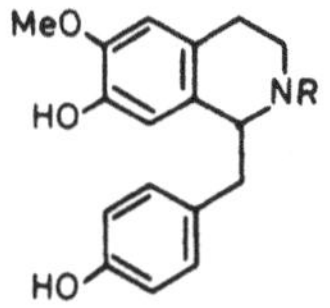

(*6.145*) R=H, Coclaurine
(*6.146*) R = Me

The bisbenzylisoquinolines are notable for the linking of two aromatic rings by way of one or more oxygen bridges [see (*6.144*)]. Such linkages are rationalized as occurring by phenol oxidative coupling (section 1.3.1). A single similar linkage is to be found in the much simpler alkaloid cularine (*6.147*). This alkaloid has been shown to be formed via (*6.148*) [95].

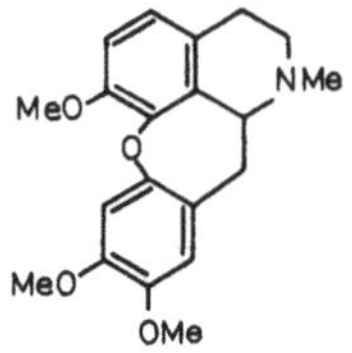

*(6.147)* Cularine 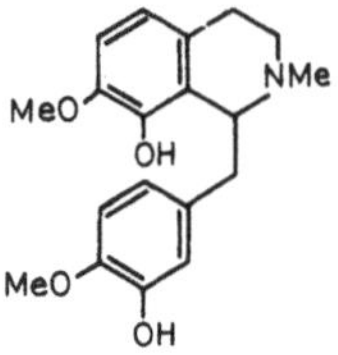        *(6.148)*

## 6.3.2 Aporphines

All that is required for the transformation of the benzylisoquinoline skeleton [as *(6.149)*] into that of the aporphines, e.g. bulbocapnine *(6.151)*, is a single new bond. Clearly phenol oxidative coupling (see section 1.3.1) is involved here, but there are several possible routes to a particular alkaloid. Interestingly examples of most of these possibilities have been found for the biosynthesis of one or more of the aporphine alkaloids, and in one case, that of boldine *(6.153)*, biosynthesis takes a different course in two different plants. Methylation pattern in the alkaloid produced does not provide a reliable guide to the course of biosynthesis, as witness that of *(6.153)* in Scheme 6.28 [96]. The methylation pattern suggests a different pathway (see Scheme 6.30), which is followed in another plant.

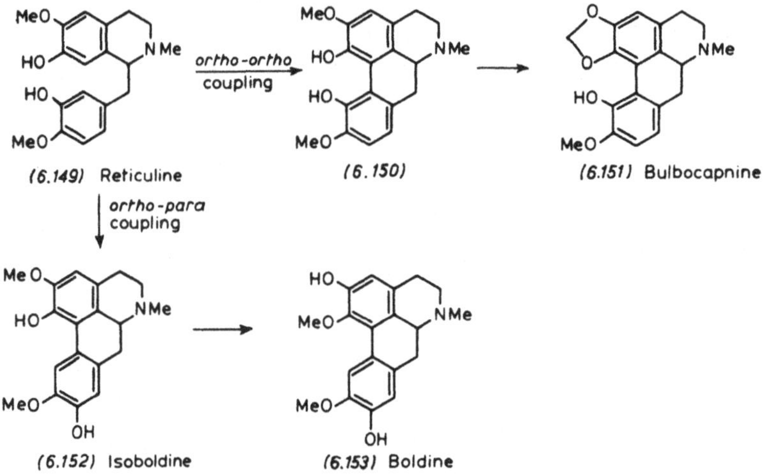

*(6.149)* Reticuline          *(6.150)*          *(6.151)* Bulbocapnine

*(6.152)* Isoboldine          *(6.153)* Boldine

**Scheme 6.28**

The simplest biosynthetic sequence is found for bulbocapnine *(6.151)*. Oxidative coupling occurs here between the sites *ortho* to the two hydroxy groups in reticuline *(6.149)* to give *(6.150)* which with minor modification affords *(6.151)* [97] (for the mechanism of methylene-dioxy-group formation see section 6.4.1, Scheme 6.39). Reticuline (*6.149*) is a proven precursor for two other aporphines [98, 99].

The importance of *O*-methylation pattern in deciding the course of coupling is well illustrated by results of a study on the biosynthesis of thebaine (*6.175*) and isothebaine (*6.157*) which are both natural constituents of *Papaver orientale*. Whereas reticuline (*6.149*) is modified to give thebaine (*6.175*) (section 6.3.4) orientaline (*6.154*), differing from reticuline in the location of one methyl group, has a quite different fate in isothebaine (*6.157*). In other cases a precursor, with the wrong methylation pattern, fails to yield alkaloid.

Because the benzylisoquinoline skeleton derives from tyrosine (*6.111*) which has a phenolic hydroxy-group at C-4', the aporphines should have hydroxy-groups at equivalent positions. Isothebaine (*6.157*) lacks such a group at one expected site, C-10. It follows that a hydroxy-group is lost from this site in the course of biosynthesis, plausibly via 'dienol-benzene' rearrangement in (*6.156*) (see section 1.3.1). This suggests the route shown in Scheme 6.29, which experiments have shown is correct [100]. Orientaline (*6.154*), and the key dienone (*6.155*) were shown to be precursors. The latter was also found as a natural constituent of *P. orientale*. One of the dienols obtained on chemical reduction of (*6.155*) was a much more efficient precursor for isothebaine (*6.157*) than the other indicating stereospecificity, and hence enzyme catalysis, in the dienol-benzene rearrangement.

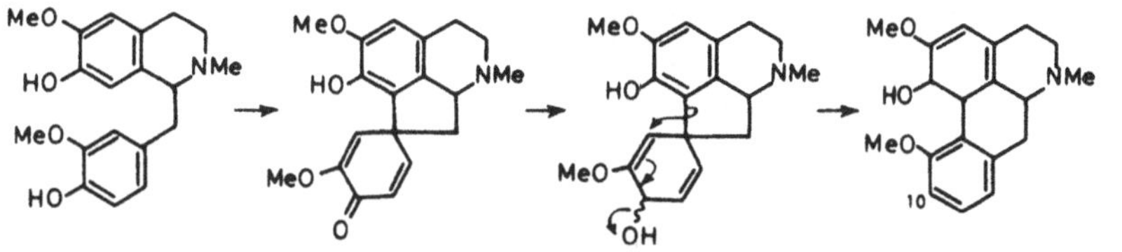

(*6.154*) Orientaline      (*6.155*)      (*6.156*)      (*6.157*) Isothebaine

**Scheme 6.29**

The structures of the *Dicentra eximia* bases, glaucine (*6.160*), dicentrine (*6.162*) and, particularly, corydine (*6.161*) indicate a similar pattern of biosynthesis to that already discussed. Quite unexpectedly the biosynthetic route turned out to be different. Painstaking work established [101] that biosynthesis is from norprotosinomenine (*6.158*); the likely following steps are indicated in Scheme 6.30 (the point of *N*-methylation is unknown). Boldine (*6.153*) is a logical, and proven, intermediate in the sequence to (*6.160*) and (*6.162*). Interestingly in the plant *Litsea glutinosa* boldine is formed from reticuline (*6.149*) via isoboldine (*6.152*) (Scheme 6.28) [96] [the conversion of (*6.152*) into (*6.153*), as observed in other examples, largely involves methyl loss and remethylation rather than methyl transfer].

A key role for dienones, e.g. (*6.155*) and (*6.159*), in aporphine alkaloid

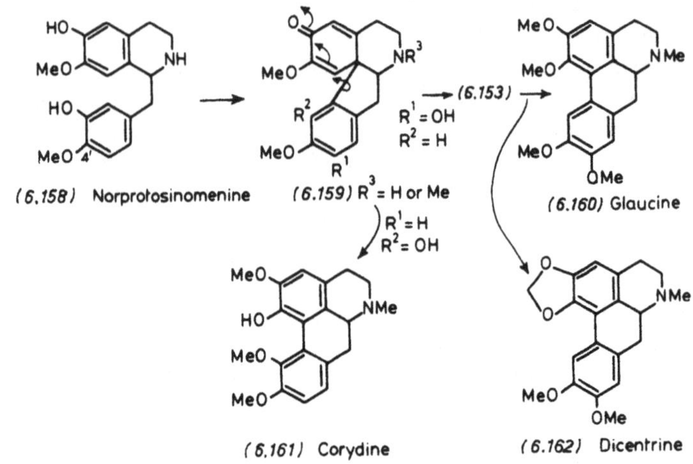

(6.158) Norprotosinomenine    (6.159) R³ = H or Me    (6.160) Glaucine

(6.161) Corydine    (6.162) Dicentrine

**Scheme 6.30**

biosynthesis is apparent. Their importance is emphasized by their natural occurrence. One such is crotonosine (6.163) formed from coclaurine (6.145) [102] which is also involved in the formation of related alkaloids [103, 104].

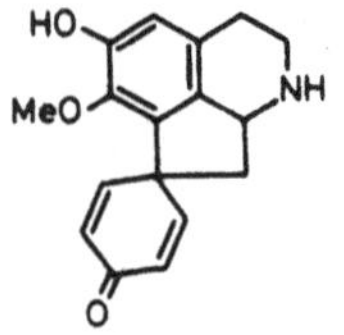

(6.163) Crotonosine

Aristolochic acid (6.165) is an interesting plant metabolite with a nitro substituent. Its structure suggests a biosynthesis via an aporphine, and this has been confirmed. Both (R)-orientaline [as (6.154)] and the aporphine, stephanine (6.164) are precursors [105].

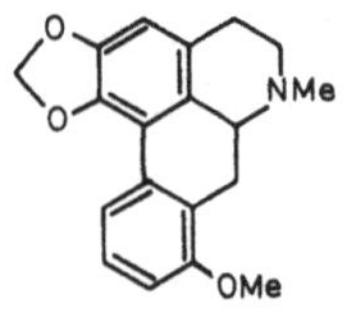

(6.164) 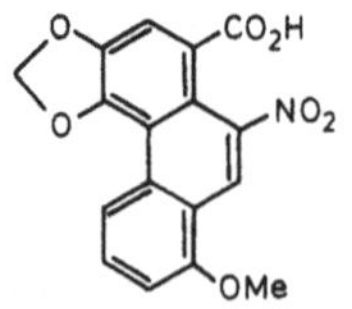    (6.165) Aristolochic acid

### 6.3.3 Erythrina alkaloids

The unusual structures of the *Erythrina* alkaloids, e.g. erythraline (*6.168*), suggest an unusual biogenesis. Although the later steps of biosynthesis are unusual, the first key intermediate is surprisingly a benzylisoquinoline: *N*-norprotosinomenine (*6.158*) (*S*-isomer), which is involved along with a dienone [as (*6.159*) = (*6.166*)] in the biosynthesis of both erythraline and some aporphine alkaloids (see above).

The pathway [106, 107] must involve a symmetrical intermediate because [2-$^{14}$C]tyrosine [as (*6.111*)] gave β-erythroidine (*6.171*) with equal labelling of C-8 and C-10, and more importantly because *N*-[4′-*methoxy*-$^{14}$C]norprotosinomenine [as (*6.158*)] gave erythraline (*6.168*) with equal labelling of methoxy- and methylenedioxy-groups.

Other intermediates in erythraline biosynthesis were proved to be (*6.167*), (*S*)-erysodienone (*6.170*), and 2-epierythratine [as (*6.169*)], and all the results are consistent with the suggested pathway (Scheme 6.31). The lactonic

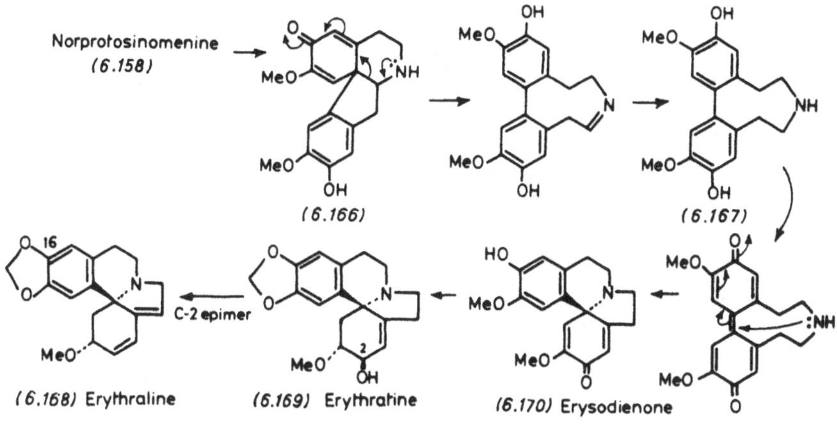

**Scheme 6.31**

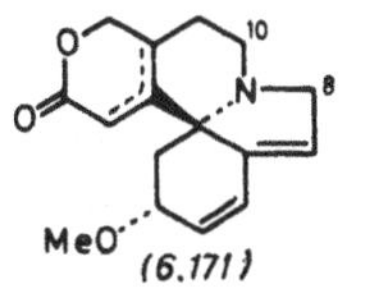

erythroidines, e.g. (*6.171*), arise from bases of the erythraline type, in a sequence which manifestly involves aromatic ring scission—a rare thing in the biosynthesis of plant bases (cf betalains, section 6.8).

Study of the biosynthesis of some alkaloids related to (*6.168*) but which lack an oxygen substituent at C-16 has shown that they also arise from (*6.158*). The pathway which must be followed is a modification of that shown in Scheme 6.31 [108].

### 6.3.4 Morphine and related alkaloids

The idea that the opium alkaloid morphine is a modified benzylisoquinoline provided the key to the structure (*6.176*) for the alkaloid; twisting of the benzylisoquinoline skeleton into that shown for reticuline (*6.149*) in (*6.172*) illustrates this relationship and suggests a possible biogenesis. This was later proved correct in the course of a study which is one of the classics of alkaloid biosynthesis. In the first experiments ever to be carried out with complex plant-alkaloid precursors, [1-$^{14}$C]- and [3-$^{14}$C]-norlaudanosoline [as (*6.140*)] were fed to opium poppies [109, 110]. They were found to label morphine (*6.176*), codeine (*6.177*) and thebaine (*6.175*) specifically, thus establishing that these alkaloids are indeed benzylisoquinoline derivatives (see discussion in section 6.3.1 on the biosynthesis of the benzylisoquinoline skeleton). The key intermediate [111] proved to be reticuline which as we have seen is biosynthesized as the (*S*)-isomer (*6.142*) (section 6.3.1). (*R*)-Reticuline (*6.172*) is, however, the isomer with the same relevant configuration as thebaine, codeine, and morphine (*6.176*). Both the (*R*)- and (*S*)-isomers of reticuline were found to serve as precursors and it is apparent that (*S*)-reticuline (*6.142*) is converted into (*R*)-reticuline (*6.172*) [by way of (*6.179*)] as an important step in the elaboration of these alkaloids [111, 112].

*Ortho-para* oxidative coupling of the diphenol, reticuline (*6.192*), can be conceived of as giving the dienone (*6.173*) which could afford thebaine (*6.175*) as shown [111]. This hypothesis is strongly supported by the observation that this dienone, (*6.173*), called salutaridine, is a constituent of opium poppies, is formed from radioactive tyrosine (*6.111*), and is a highly efficient and specific precursor for the opium alkaloids. The transformation of salutaridine (*6.173*) into thebaine (*6.175*) requires a further ring-closure, which occurs chemically when the two epimeric alcohols (*6.174*) are treated with acid. In contrast to the purely chemical reaction, only one of the alcohols was efficiently converted into thebaine *in vivo*, indicating that the reaction is enzyme mediated (and therefore part of normal biosynthesis).

Experiments, particularly with $^{14}CO_2$ [113, 114], have established that in the biosynthesis of the opium alkaloids thebaine (*6.175*) is the first alkaloid to be formed and the following sequence is essentially one of demethylation (Scheme 6.32). It is interesting to note that the demethylation at C-6 which converts thebaine (*6.175*) into (*6.178*), does not involve loss of the oxygen atom as expected of normal enol ether hydrolysis; a possible mechanism is shown in Scheme 6.33 [115].

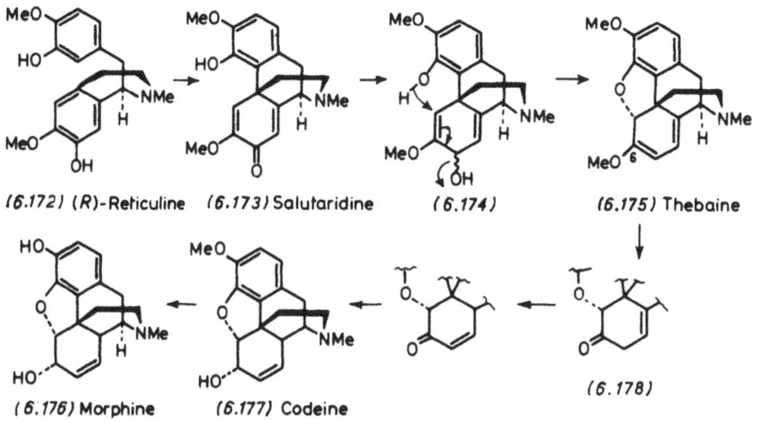
MeO
HO
NMe
H
MeO
OH
(6.172) (R)-Reticuline
MeO
HO
NMe
H
MeO
O
(6.173) Salutaridine
MeO
O
H
NMe
H
MeO
OH
(6.174)
MeO
O
NMe
H
MeO 6
(6.175) Thebaine
HO
O
NMe
H
HO
(6.176) Morphine
MeO
O
NMe
HO
(6.177) Codeine
O
O
(6.178)

**Scheme 6.32**

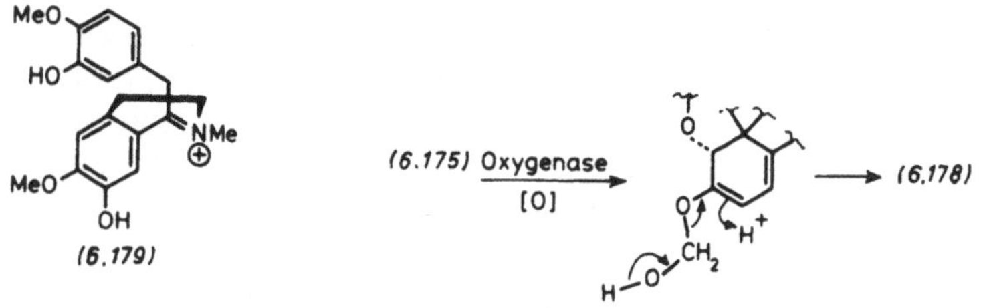
MeO
HO
NMe
MeO
OH
(6.179)
(6.175) Oxygenase
[O]
O
O
H+
CH2
H–O
(6.178)

**Scheme 6.33**

### 6.3.5 Hasubanonine and protostephanine

Hasubanonine (*6.182*) and protostephanine (*6.181*) have structures which in the light of the foregoing discussion appear to be benzylisoquinoline in origin. It took, however, really extensive, careful experiments to establish this [116], because the benzylisoquinoline substrate for oxidative coupling, i.e. (*6.180*), needed to have *two* hydroxy groups on ring B. This is quite unprecedented. In all other known examples (with one further, possible exception, see Lythraceae alkaloids, section 6.2.1) only one hydroxy group per ring has been found to be necessary. The deduced pathway is illustrated in Scheme 6.34.

### 6.3.6 Protoberberine and related alkaloids

Elaboration of the benzylisoquinoline skeleton in ways different from those already discussed involves the inclusion of an extra carbon atom (the 'berberine bridge') as seen in, e.g. berberine (*6.184*) (the extra atom is C-8).

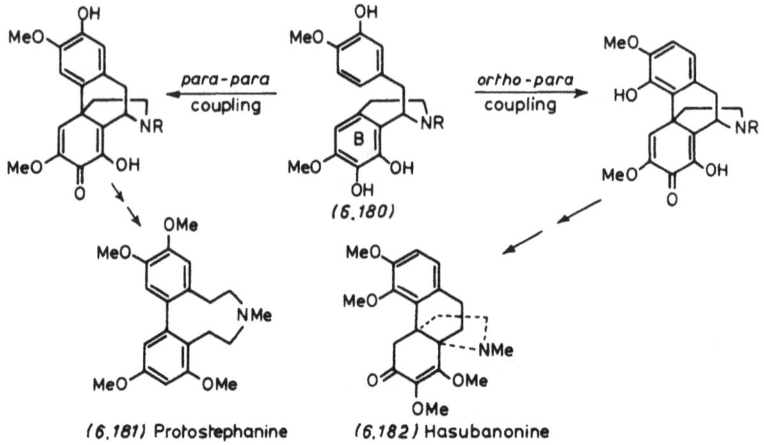

(6.181) Protostephanine     (6.182) Hasubanonine

**Scheme 6.34**

The question of how this occurs was answered simply in terms of an oxidative cyclization of the *N*-methyl group of a benzylisoquinoline precursor in a manner (Scheme 6.35) analogous to the formation of a methylenedioxy-group (section 6.4.1; Scheme 6.39) (both the 'berberine bridge' and

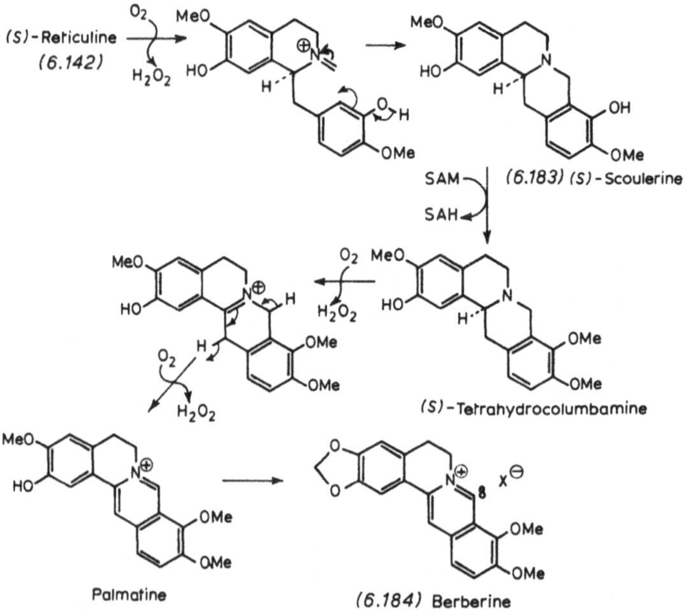

Palmatine     (6.184) Berberine

**Scheme 6.35**

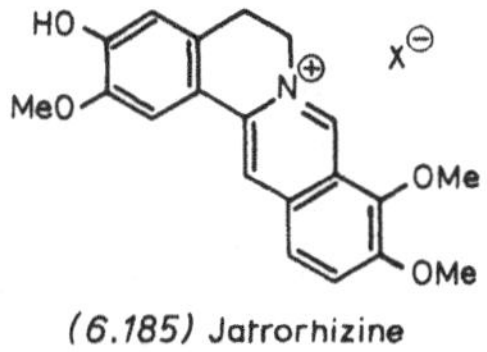

*(6.185)* Jatrorhizine

methylene-dioxy-carbons derive ultimately from methionine). The benzyl-isoquinoline precursor was established to be (*S*)-reticuline (*6.142*) which is utilized via (*S*)-scoulerine (*6.183*). Label in the *N*-methyl group of (*6.142*) was built intact into C-8 of berberine thus validating the hypothesis [117].

Further intricate detail on the course of berberine biosynthesis comes impressively through the isolation from plant tissue cultures, and characterization, of the enzymes which catalyse the steps involved (Scheme 6.35) [89, 118–120].

Jatrorrhizine (*6.185*) is biosynthesized via (*S*)-reticuline (*6.142*) and berberine (*6.184*). This conversion of (*6.184*) into (*6.185*) is a most unusual one since it involves the transfer of a methyl group from one oxygen to another (see palmatine; Scheme 6.35) by way of a methylenedioxy function [in (*6.184*)] [121].

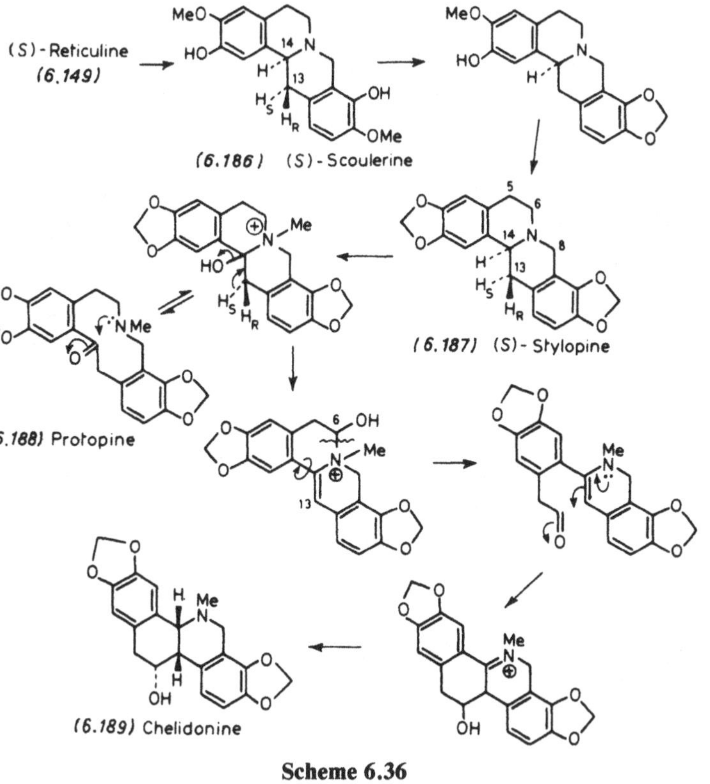

**Scheme 6.36**

The early suspicion that chelidonine (*6.189*) was a protoberberine [as (*6.186*)] variant has been confirmed [117, 122–125]. The deduced route, which includes another alkaloid protopine (*6.188*) and begins with the ubiquitous precursor reticuline, is illustrated in Scheme 6.36. Tritium-labelling results establish that C-1 and C-9 of reticuline (*6.149*) are unaffected by transformation through scoulerine (*6.186*) to stylopine (*6.187*). The conversion of the *N*-methyl group of reticuline into the 'berberine bridge' (C-8) of stylopine (*6.187*), which necessarily involves loss of one hydrogen/tritium atom, was found to occur with very high retention of tritium label on this carbon atom consistent with the expected operation of an isotope effect favouring loss of hydrogen rather than tritium. Subsequent transformation of stylopine (*6.187*) to chelidonine occurs with retention of tritium at C-5 and C-8, but loss of tritium from C-14 and the (*pro-S*)-hydrogen atoms from C-13 and C-6 (loss in each case of a proton from the α-face of the biosynthetic intermediates is notable).

The phthalideisoquinoline alkaloids, e.g. hydrastine (*6.190*), also derive from (*S*)-scoulerine (*6.186*). Entry of the C-13 oxygen atom occurs with removal of the *pro-S* proton (as in chelidonine biosynthesis), and the reaction thus proceeds with orthodox retention of configuration [122–124].

Corydaline (*6.191*), ochotensimine (*6.192*) [126] and alpinigenine (*6.194*) are further protoberberine variants; the labelling of (*6.191*) and (*6.192*) by [3-$^{14}$C]tyrosine [as (*6.111*); ●] and methionine (*; includes 'berberine-bridge' atom) is shown. Alpinigenine (*6.194*) is formed via the protopine relative (*6.193*) [127, 128].

(*6.190*) Hydrastine          (*6.191*) Corydaline          (*6.192*) Ochotensimine

(*6.193*)          (*6.194*) Alpinigenine

### 6.3.7 Phenethylisoquinoline alkaloids

Colchicine (*6.200*) is an alkaloid with a unique tropolone ring, and on the face of it appears to be quite unrelated to any other alkaloid.

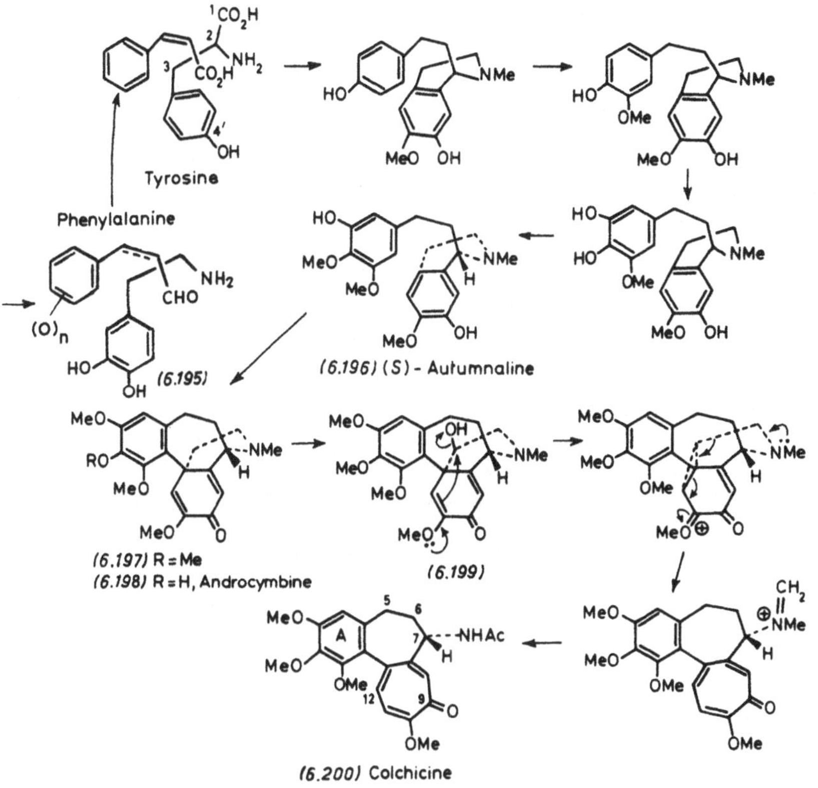

**Scheme 6.37**

Phenylalanine, by way of cinnamic acid, serves in *Colchicum* species as the source of ring A plus carbon atoms 5, 6, and 7. The seven carbons of the tropolone ring arise from tyrosine which loses C-1 and C-2 from the side-chain; labels from [4′-$^{14}$C]- and [3-$^{14}$C]-labelled amino acid appear at C-9 and C-12, respectively [129]. It follows that the tropolone ring arises by expansion of the tyrosine benzene ring with inclusion of the benzylic carbon atom, but through what series of intermediates?

The answer came from the elucidation of the structure of androcymbine, an alkaloid isolated from a relative of *Colchicum*. The dienone structure (*6.198*) was assigned to androcymbine. Androcymbine could then be thought of as arising from the phenethylisoquinoline skeleton [as (*6.196*)] which was quite unknown at the time [130]. A similar biogenesis for colchicine followed. Plausibly the transformation of the androcymbine skeleton (*6.198*) to that of colchicine could occur by hydroxylation to give (*6.199*). Homoallylic ring expansion would lead to colchicine (Scheme 6.37).

The crucial test for the hypothesis lay in the examination of compounds of type (*6.196*) and (*6.198*) as colchicine precursors. In the event [131–133], *O*-methylandrocymbine (*6.197*) was a spectacularly good precursor for colchicine. The phenethylisoquinoline (*6.196*), called autumnaline, was also clearly implicated in biosynthesis; only the (*S*)-isomer of autumnaline, with the same absolute configuration as colchicine is involved, and oxidative coupling of this base occurs in a *para-para* sense rather than the alternative, *ortho-para*. Results of other experiments, together with those discussed here, led to the pathway illustrated in Scheme 6.37.

The phenethylisoquinoline skeleton is formed in a similar way to that of the benzylisoquinolines (section 6.3.1), that is to say a phenethylamine dopamine (*6.116*) condenses with an aldehyde, in this case a cinnamaldehyde or dihydro-derivative [see (*6.195*)] [134].

Autumnaline (*6.196*) is a precursor for homoaporphine alkaloids, e.g. floramultine (*6.202*), found in *Kreysigia multiflora* [135]. The homo-Erythrina alkaloid, schelhammeridine (*6.203*) [136], and cephalotaxine 6.204) [137] are apparently also modified phenethylisoquinolines.

(*6.201*)          (*6.202*) Floramultine          (*6.203*)  Schelhammeridine

(*6.204*) Cephalotaxine

## 6.4 AMARYLLIDACEAE AND MESEMBRINE ALKALOIDS

The grouping of these two groups of alkaloids together can be justified on a certain structural resemblance and on a derivation from the same amino acids. But they are derived along apparently different pathways appropriate to the phylogenetically different plants which produce the two groups of alkaloids.

### 6.4.1 Amaryllidaceae alkaloids

Diverse though the structures of these alkaloids seem they may be classified into three main groups represented by (*6.210*), (*6.212*) and (*6.215*). The quite brilliant recognition that the three groups of alkaloids could arise simply by different modes of phenol oxidative coupling (section 1.3.1) within a mole-cule of type (*6.205*) has been of central importance to biosynthetic studies in this area, studies which were to prove the correctness of the original idea.

Examination of (*6.205*) indicates an assembly from a $C_6$—$C_1$ and a $C_6$—$C_2$ unit (thickened bonds) which can be traced through into the alkaloids (*6.210*), (*6.212*) and (*6.215*) (Scheme 6.38). It was established for representa-tive alkaloids, by the results of feeding radioactive compounds, that the $C_6$—$C_2$ unit arises from tyrosine (a common source for such units) via tyramine and that the $C_6$—$C_1$ unit arises from phenylalanine (a common source of $C_6$—$C_1$ and $C_6$—$C_3$ units) by way of cinnamic acid, its

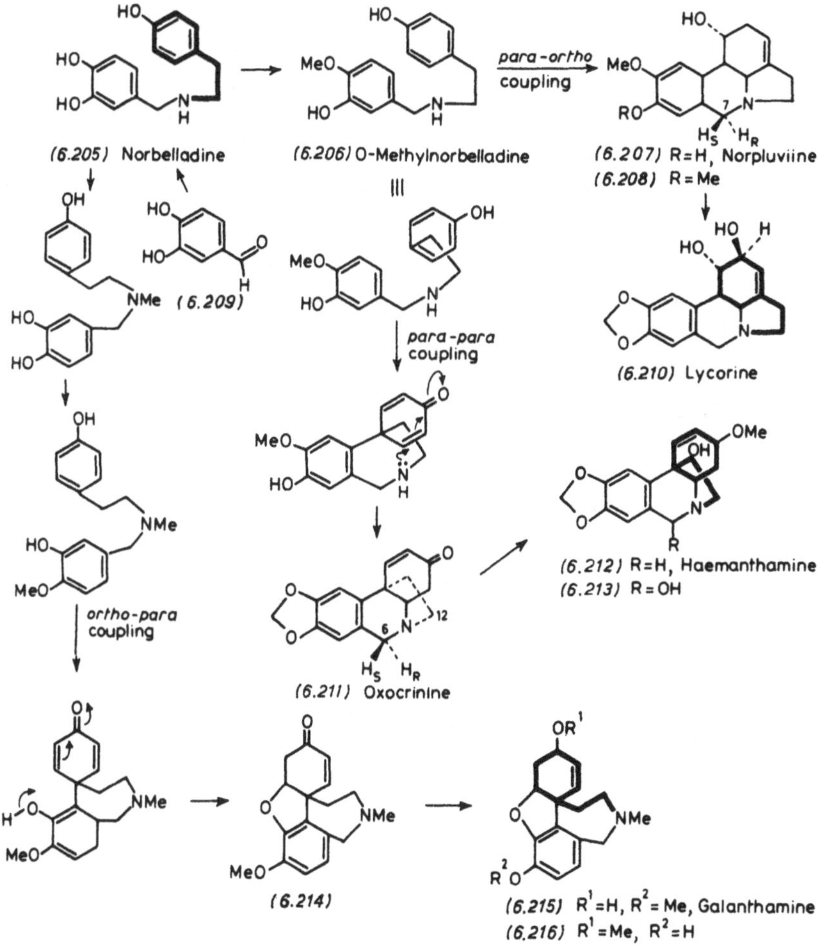

**Scheme 6.38**

3,4-dihydroxy-derivative, and protocatechualdehyde (*6.209*) [138–141]. It is interesting to note that, as is often the case in plant alkaloid biosynthesis, hydroxylation of phenylalanine to give tyrosine does not occur (as also colchicine, section 6.3.7, but not capsaicin [142, 143] or annuloline [144]).

Further experimental results established norbelladine (*6.205*) and some of its methylated derivatives (clearly not others) as key biosynthetic intermediates in the biosynthesis of, e.g., lycorine (*6.210*), haemanthamine (*6.212*) and galanthamine (*6.215*) [138–141, 145, 146]. As elsewhere (section 6.3) hydroxy-groups *ortho* and/or *para* to sites of new bond formation between aromatic rings are essential for biosynthesis to proceed, a telling set of examples in support of the phenol-oxidative coupling hypothesis. Of further interest is the reported isolation of an enzyme, from a plant of the Amaryllidaceae, which, when incubated with norbelladine and *S*-adenosylmethionine (source of methyl groups), yielded almost entirely the *O*-methylnorbelladine, (*6.206*), that is involved in alkaloid biosynthesis [147].

In the course of studies on the incorporation of *O*-[*methyl*-¹⁴C]methylnorbelladine [as (*6.206*)] into haemanthamine (*6.212*) it was demonstrated [141], for the first time, that a methylene-dioxy-group [as in (*6.212*)] arises by oxidative ring closure of an *o*-methoxyphenol [as in (*6.206*)]. Subsequent results on the biosynthesis of other alkaloids have confirmed the generality of this observation. The mechanism may be one involving radicals or cationic species (Scheme 6.39). A related oxidative cyclization is found in the conversion of an isoquinoline into a protoberberine (section 6.3.6) where an *N*-methyl group closes on to an aromatic ring.

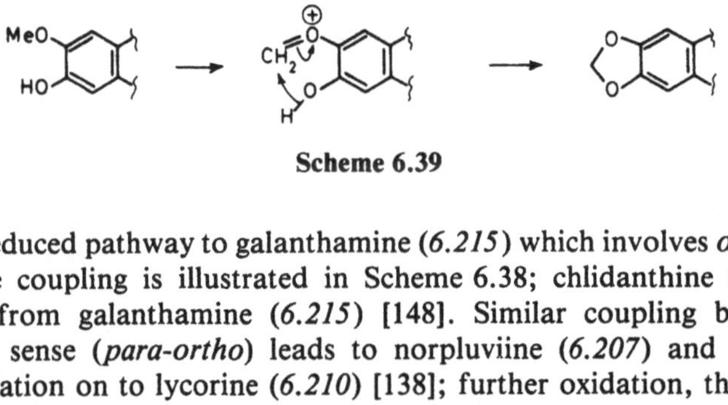

**Scheme 6.39**

The deduced pathway to galanthamine (*6.215*) which involves *ortho-para* oxidative coupling is illustrated in Scheme 6.38; chlidanthine (*6.216*) is formed from galanthamine (*6.215*) [148]. Similar coupling but in the opposite sense (*para-ortho*) leads to norpluviine (*6.207*) and by allylic hydroxylation on to lycorine (*6.210*) [138]; further oxidation, this time of galanthine (*6.217*), gives narcissidine (*6.218*) [149]. The retention of two out of four tritium atoms in the formation of norpluviine (*6.207*) from *O*-methyl-

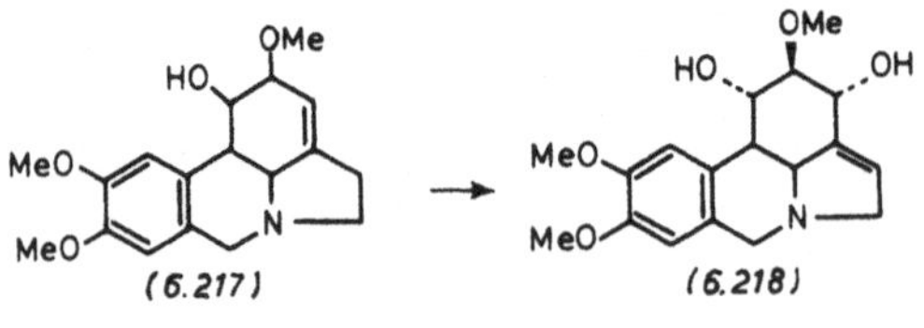

(*6.217*)          (*6.218*)

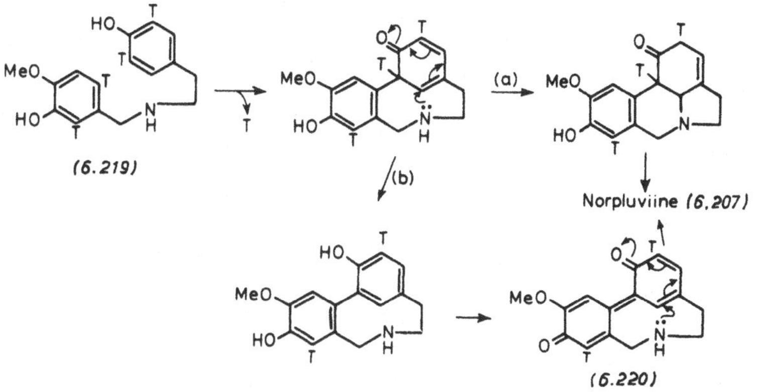

**Scheme 6.40**

norbelladine [labelled as shown in (*6.219*)] requires a complex pathway (b, Scheme 6.40) rather than a simpler one (path a) which would result in the retention of three tritium atoms [145].

The hydroxylation of (*6.207*) to give lycorine (*6.210*) would be expected to proceed with retention of configuration, the orthodox result (section 1.2.3). It has been found, however, that although this is true for biosynthesis in one plant, the opposite is true in another plant [150, 151].

The third group of alkaloids which arise from norbelladine (*6.205*), this time by *para-para* phenol oxidative coupling, is exemplified by haemanthamine (*6.212*), and biosynthesis is proved to be by way of compounds of type (*6.211*). Haemanthamine (*6.212*) shows an extra hydroxy-group at C-11, which has been shown to arise by hydroxylation with normal retention of configuration [152, 153].

Haemanthamine (*6.212*) is a precursor for haemanthidine (*6.213*). The conversion of protocatechualdehyde (*6.209*) through the sequence (*6.209*) → (*6.206*) → (*6.212*) → haemanthidine (*6.213*) has been shown by the results of tritium labelling to involve stereospecific hydrogen addition to, and removal from, what begins as the aldehyde carbonyl, the hydrogen added being later removed [154]. A similar sequence has been found for the conversion of protocatechualdehyde (*6.209*) through (*6.206*) and (*6.208*) into lycorenine (*6.221*); the proton added and removed is the (*7-pro-R*) one in (*6.208*) [155].

The structure (*6.225*) of narciclasine is such that it can *a priori* arise along a route involving intermediates of the norpluviine (*6.207*) or oxocrinine (*6.211*) types. Tracer experiments have shown that the latter route is utilized and involves *inter alia* oxocrinine (*6.211*) and vittatine (*6.222*). Loss of the two-carbon bridge in the conversion of (*6.222*) into (*6.225*) could plausibly occur as shown on 11-hydroxyvittatine (*6.223*). This is supported by the finding that (*6.223*) is an efficient precursor for narciclasine (*6.225*) [156, 157].

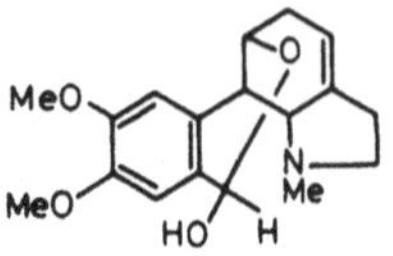

(6.221) Lycorenine

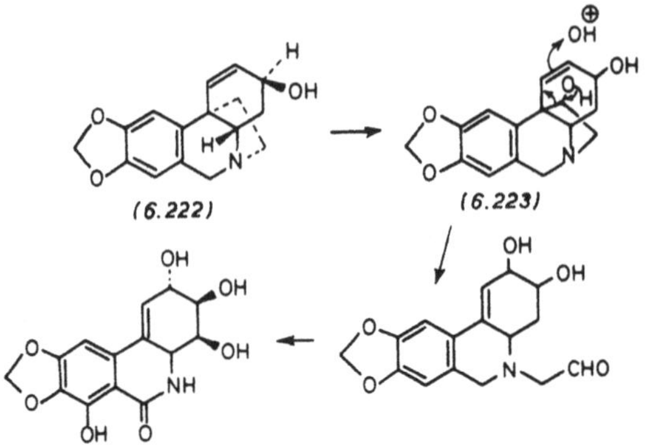

(6.222)

(6.223)

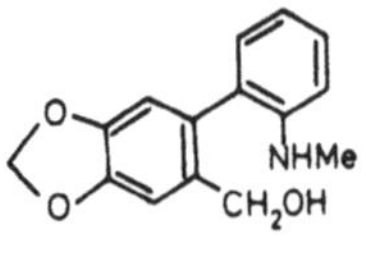

(6.224) Ismine

(6.225) Narciclasine

More extensive skeletal degradation than that seen in narciclasine biosynthesis is apparent in the genesis of ismine (*6.224*). Results of tracer experiments have shown [158] that it too arises from oxocrinine (*6.211*), with loss of C-12 (so it does not become the *N*-methyl group of ismine) and the (6-*pro-R*)-hydrogen atom. Interestingly this is a proton with the same stereochemistry as those lost in haemanthidine (*6.213*) and lycorenine (*6.221*) biosynthesis (see above).

Some evidence [159, 160] exists that alkaloids of the manthine (*6.226*) type are formed from those of the haemanthamine (*6.212*) kind.

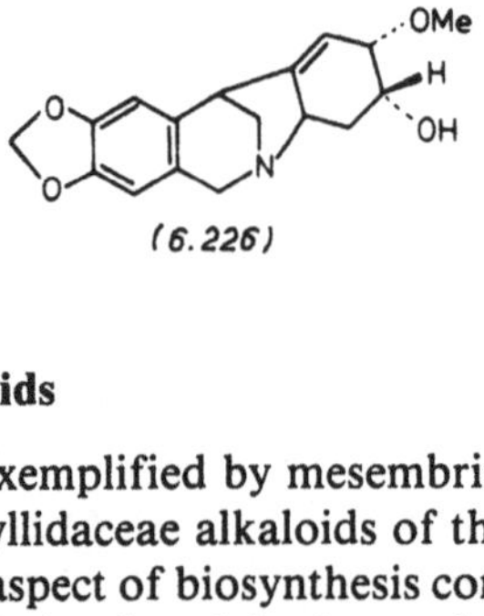

(6.226)

### 6.4.2 Mesembrine alkaloids

The group of alkaloids exemplified by mesembrine (*6.227*), shows a structural kinship with Amaryllidaceae alkaloids of the haemanthamine (*6.212*) type. However, the only aspect of biosynthesis common to these two groups of alkaloids is their origin in phenylalanine and tyrosine; results, crucially with doubly labelled precursors, showed that various norbelladine (*6.205*) derivatives were only incorporated after fragmentation [161].

The octahydroindole moiety in these alkaloids derives from tyrosine via tyramine, and *N*-methyltyramine [162]. Phenylalanine is the source of the unusual, if not unique, mesembrine $C_6$ unit (Scheme 6.41). As with other metabolites phenylalanine is utilized by way of cinnamic acid and, in this case, mono- and di-hydroxy-derivatives may be involved too. Late-stage aromatic hydroxylation is also possible, for skeletenone (*6.228*) is an

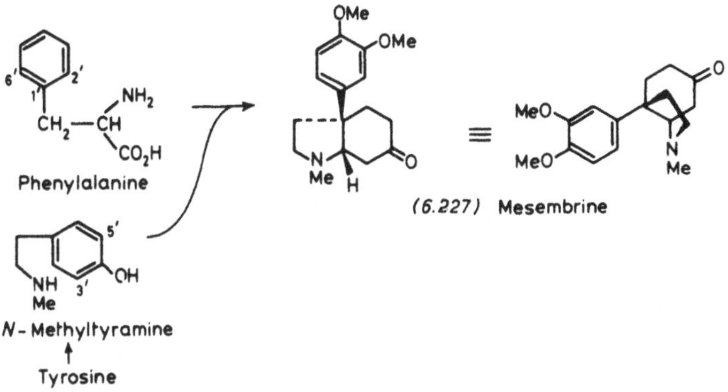

(6.227) Mesembrine

**Scheme 6.41**

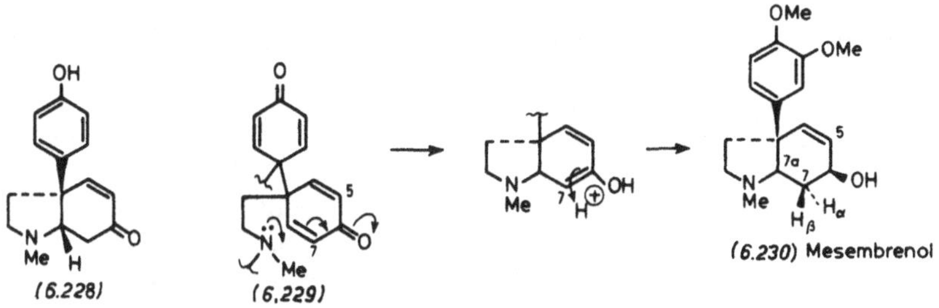

(6.228)

(6.229)

(6.230) Mesembrenol

efficient precursor for mesembrenol (*6.230*) [163] and appears to be on the major pathway.

The steps which lie beyond hydroxy-cinnamic acids and *N*-methyltyramine remain a mystery, but possible intermediates are in part circumscribed by the results of experiments with tritiated precursors. Thus tritium at C-2' and C-6' of phenylalanine is retained on formation of mesembrine (*6.227*) [161]. Formation of intermediates [as (*6.211*)] would require loss of one tritium atom, so further independent evidence for disparate pathways to these two groups of alkaloids is obtained. Moreover, the results indicate that bond formation between the phenylalanine- and tyrosine-derived units must occur at the carbon atom corresponding to C-1' of phenylalanine. This leads logically to a dienone of type (*6.229*) which can undergo fragmentation and aromatization as indicated without tritium loss. Cyclization of nitrogen on to C-7a [see (*6.230*)] must occur at some point. Half of the tritium from C-3' and C-5' of labelled tyrosine and *N*-methyl-tyramine was lost in mesembrenol (*6.230*) formation, residual tritium appearing at C-5 and C-7α (equal distribution of label between these sites indicates that tritium loss must occur whilst they are equivalent, i.e. before dienone formation). The results are consistent with the pathway shown, which involves internal conjugate addition of nitrogen to the enone followed by stereospecific β-protonation of the resulting enolate at C-7 [162].

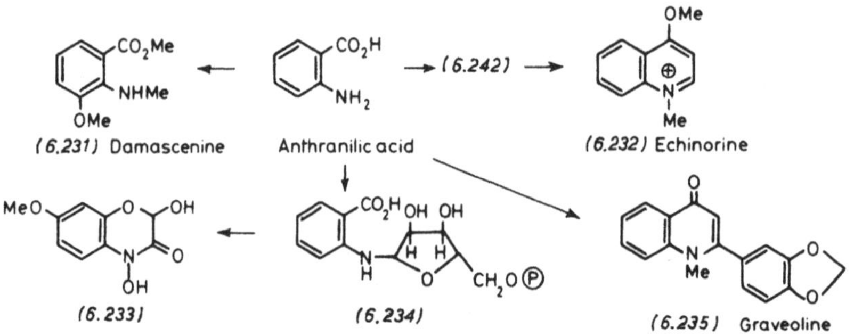


**Scheme 6.42**

## 6.5 QUINOLINE AND RELATED ALKALOIDS

A number of diverse alkaloids have a common anthranilic acid unit. This amino acid is an intermediate in tryptophan biosynthesis. The known pathways to some of the alkaloids, damascenine (*6.231*) [164], echinorine (*6.232*) [165], the benzoxazinone (*6.233*) [166], and graveoline (*6.235*) [167], are illustrated in Scheme 6.42. Further anthranilic acid derivatives are arborine (*6.236*) [168, 169], rutaecarpine (*6.237*) and evodiamine (*6.238*) (also derived from tryptophan) [170] and acridone alkaloids e.g. rutacridone (*6.239*). An intermediate in the biosynthesis of these last alkaloids is *N*-methylanthranilic acid and ring C is constructed from three molecules of acetate; a later intermediate in the biosynthesis of rutacridone is (*6.240*) [171].

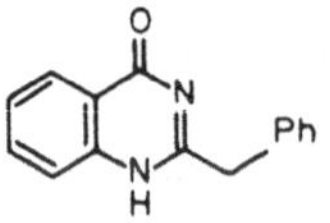

(*6.236*) Arborine

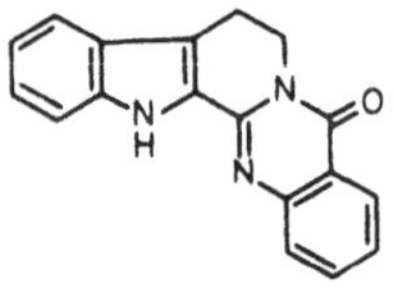

(*6.237*) Rutaecarpine

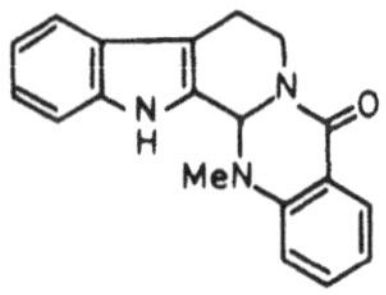

(*6.238*) Evodiamine

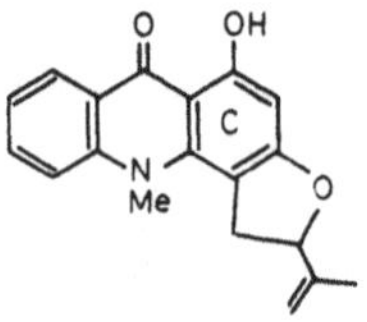

(*6.239*) Rutacridone

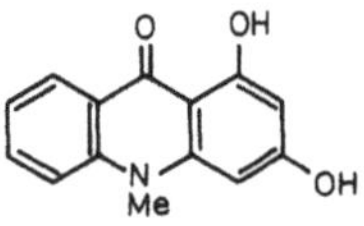

(*6.240*)

Peganine (≡ vasicine) (*6.241*) is interesting because its biosynthesis appears to be along quite different pathways in two plants from different families (Scheme 6.43) [172–174]. This is so far a unique observation in plant alkaloid biosynthesis.

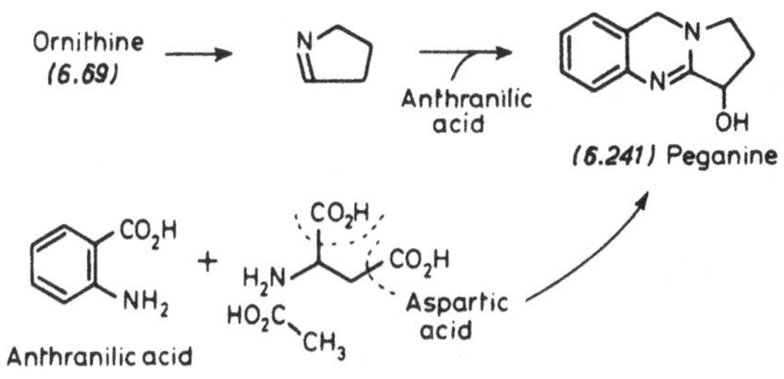

**Scheme 6.43**

The furoquinoline alkaloids, e.g. platydesmine (*6.246*) and dictamnine (*6.247*), are derived from anthranilic acid, acetate ($C_2$ unit), and mevalonate ($C_5$ unit) [see dotted lines in (*6.242*) and (*6.246*)] along the pathway [175] shown in Scheme 6.44; prenylation of the quinoline skeleton may occur as well with a methoxy-group at C-4 as with a hydroxy-group, because both (*6.242*) and (*6.243*), and (*6.244*) and (*6.245*) are efficient precursors [(*6.245*) being proven to be incorporated without loss of its methyl group], but methylation of the C-2 oxygen function prevents utilization in biosynthesis.

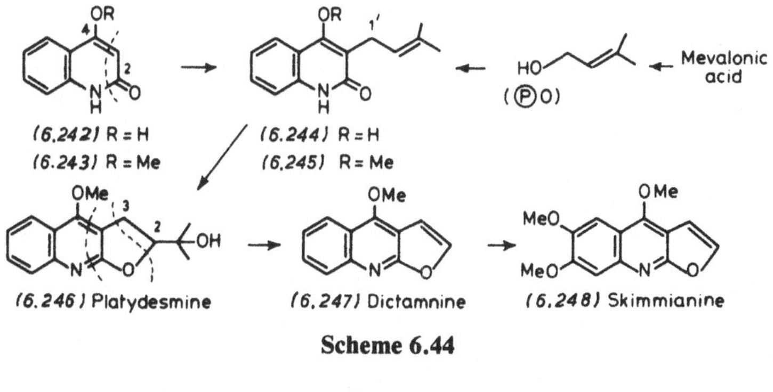

**Scheme 6.44**

The conversion of platydesmine (*6.246*) into dictamnine (*6.247*) involves loss of the isopropyl group from C-2. A mechanism involving stereospecific hydroxylation at C-3 [of (*6.246*)], followed by fragmentation (Scheme 6.45), was suggested by the observation that (*6.245*) was incorporated into skimmianine (*6.248*) with loss of half of a tritium label sited at C-1' [of (*6.245*)]. An alternative mechanism could be one initiated by hydride abstraction from (*6.246*) instead of hydroxyl loss from (*6.249*).

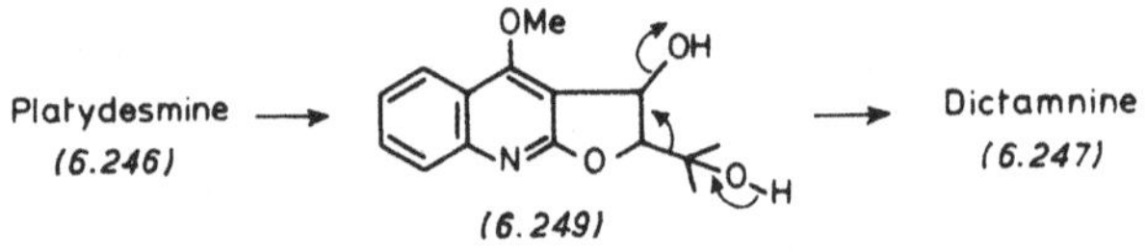

Platydesmine (6.246) → (6.249) → Dictamnine (6.247)

**Scheme 6.45**

Further results have been obtained with cell suspension cultures of *Ruta graveolens*, which produces furoquinoline alkaloids and edulinine (6.250). In particular, (6.242) was deduced [176] to be at the branch point for the biosynthesis of the two alkaloid types, with *N*-methylation of (6.242) causing diversion to edulinine biosynthesis. The pathway to (6.250) also involves the *N*-methyl derivative of (6.245).

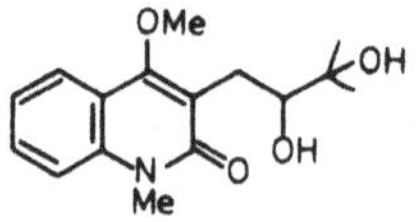

(6.250) Edulinine

## 6.6 INDOLE ALKALOIDS

### 6.6.1 Simple indole derivatives

The simple base gramine (6.253) is known to derive from tryptophan (6.251), in three plants from different families. C-2 is lost but apparently not the side-chain nitrogen [177].

Psilocybin (6.255) derives along the pathway: tryptophan (6.251) → tryptamine (6.252) → *N*-methyltryptamine → *N*,*N*-dimethyltryptamine → psilocin (6.254) = psilocybin (6.255) [178]. The biosynthesis of

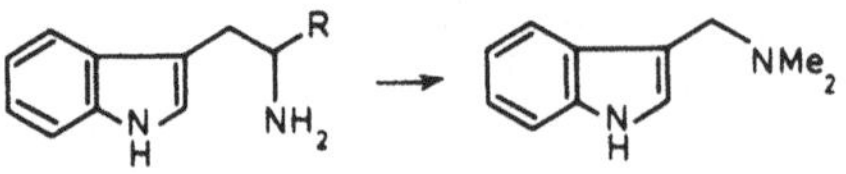

(6.251) R = $CO_2H$, Tryptophan
(6.252) R = H, Tryptamine
(6.253) Gramine

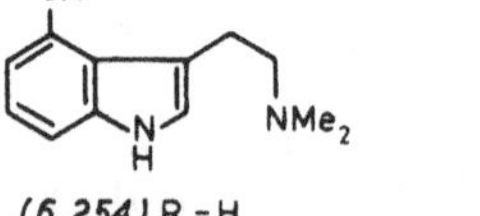

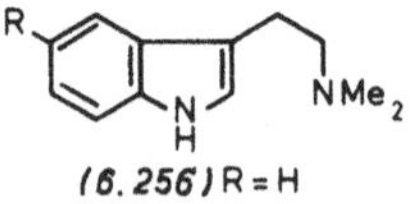

(6.254) R = H
(6.255) R = $PO_3^{2-}$, Psilocybin
(6.256) R = H
(6.257) R = OMe

(*6.256*) and (*6.257*) is more complex. Almost all the possible pathways from tryptophan (*6.251*) (made up of alternative sequences of decarboxylation, hydroxylation, and *O*- and *N*-methylation) appear to operate [179]. Such results are a caution against the simplistic belief in a single biosynthetic pathway to secondary metabolites.

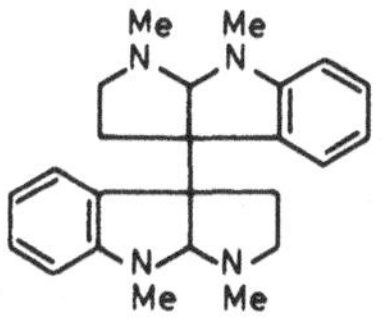

(*6.258*) Folicanthine 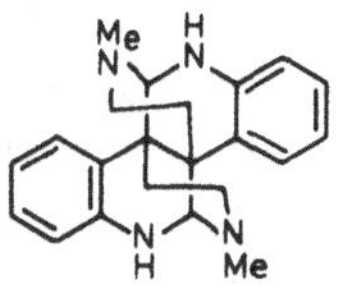          (*6.259*) Calycanthine

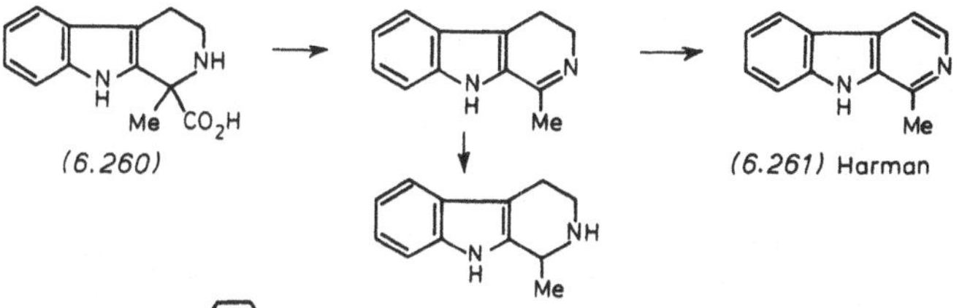

(*6.260*)          (*6.261*) Harman

(*6.262*) Eleagnine

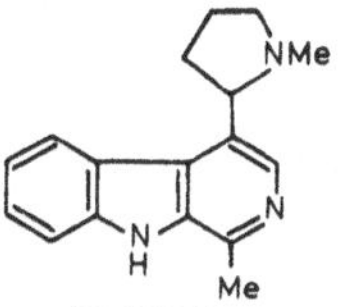

(*6.263*)

Folicanthine (*6.258*) and calycanthine (*6.259*) also derive from tryptophan (they are plausibly dimers of methyltryptamine) [180] as do the β-carboline alkaloids, e.g. harman (*6.261*) and eleagnine (*6.262*). The amino acid (*6.260*), plausibly formed by condensation of tryptamine (*6.252*) with pyruvic acid, is an intermediate in the biosynthesis of these β-carboline alkaloids (cf section 6.3.1) [181]. A similar origin is established for brevicolline (*6.263*): tryptophan, pyruvic acid and putrescine were incorporated [182].

### 6.6.2 Terpenoid indole and related alkaloids

The terpenoid indole skeleton is the most widely found amongst the plant alkaloids, the number of known examples exceeding one thousand. As seen in e.g. ajmalicine (*6.272*) these alkaloids are constituted from a tryptamine unit (proved to derive from tryptophan and tryptamine [183]) and a $C_9$ or $C_{10}$ unit which, after much experimentation, was proved to be terpenoid in origin. [Examination of (*6.272*) shows this is not immediately obvious.]

Significant variation in the alkaloidal tryptamine fragment is associated with loss of one or both side-chain carbon atoms [see (*6.301*)] or transformation of the aromatic nucleus into that of a quinoline [see (*6.308*)]. By far the greatest variation is associated with the terpenoid fragment.

The most important, indeed crucial, idea put forward on the biosynthesis of these alkaloids was that they are formed by fragmentation of a cyclopentane monoterpene [as (*6.266*)]. The pathway outlined in Scheme 6.46 indicates how the major skeletal types represented by ajmalicine (*6.272*) and akuammicine (*6.271*), catharanthine (*6.268*), and vindoline (*6.273*) may be formed; where one of these atoms is absent in derived alkaloids it appears always to be the one separated by dotted lines in (*6.267*), (*6.269*) and (*6.270*) [184, 185].

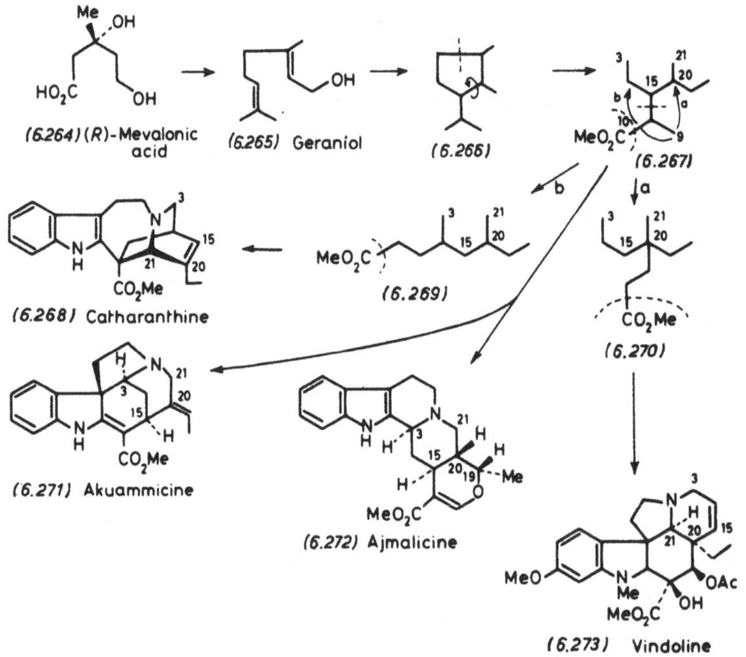

**Scheme 6.46**

The appropriate incorporations of mevalonic acid (*6.264*) [as elsewhere in biosynthesis it is the (3-*R*)-isomer which is utilized], and geraniol (*6.265*) and nerol (*6.274*) prove the essential correctness of the hypothesis; the hydroxy-derivatives (*6.275*) and (*6.276*) were also utilized [186]. Ensuing crucial experiments lay with identification of the cyclopentane monoterpene loganin (*6.278*) as a key intermediate in biosynthesis. Not only did (*6.278*) prove to be a specific alkaloid precursor, but its presence and biosynthesis from geraniol in *Catharanthus roseus*, a plant used for most of the experiments, could be demonstrated [187, 188].

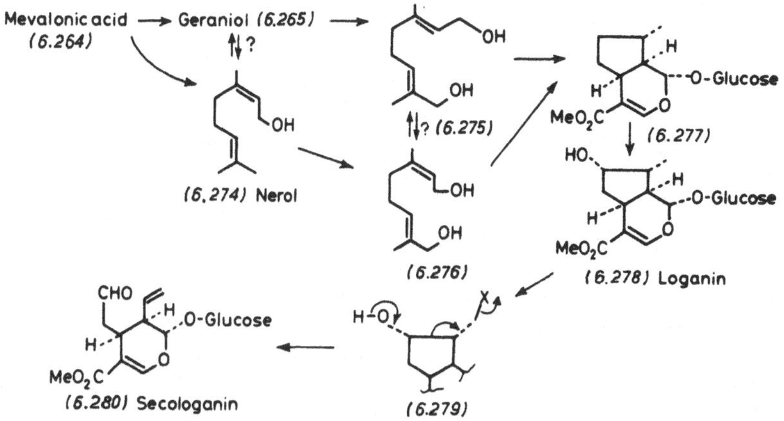

**Scheme 6.47**

Deoxyloganin (*6.277*) is sited as an intermediate before loganin (*6.278*) and cyclization of a geraniol precursor leading to it has been suggested to be as shown in Scheme 6.48; C-9 and C-10 become equivalent in this pathway as both are depicted as aldehydes, and this allows for the observation that label passing through (*6.265*) from mevalonate becomes equally divided between C-9 and C-10 of loganin (*6.278*) [189].

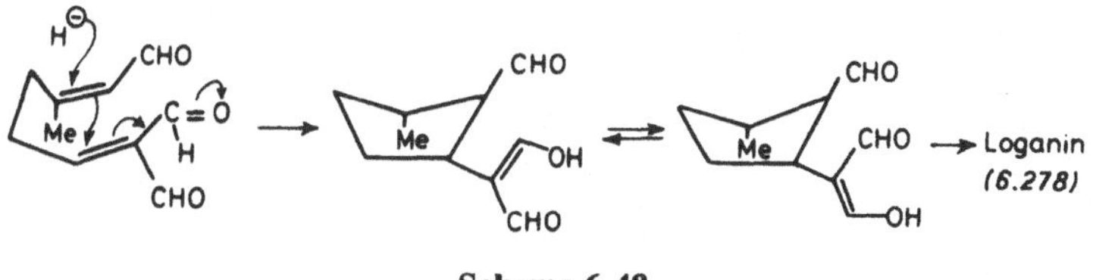

**Scheme 6.48**

Conversion of loganin (*6.278*) into alkaloids must involve cleavage of the cyclopentane ring [cf (*6.278*) → (*6.280*) in Scheme 6.47], which may be rationalized in terms of the mechanism shown in (*6.279*) (X is not a hydroxy group [190]) to give secologanin (*6.280*). This is corroborated by the firm identification of this terpene, (*6.280*), as a biosynthetic intermediate [191]. Consideration of the next step in biosynthesis leads to vincoside (*6.281*) and strictosidine (*6.282*) the epimeric products of chemical condensation between tryptamine (*6.252*) and secologanin (*6.280*). Strictosidine (*6.282*) with 3α-stereochemistry is the precursor for alkaloids like ajmalicine (*6.272*) with 3α-stereochemistry [and also vindoline (*6.273*) and catharanthine (*6.268*)] and alkaloids similar to (*6.272*) but with 3β-stereochemistry [192, 193]. For the former group tritium at C-3 in (*6.282*) is retained, for those with 3β-stereochemistry it is lost (cf ipecac alkaloids, section 6.7 for slightly different results).

The most recent results have been obtained chiefly using enzyme preparations from plant tissue cultures, and they are most impressive allowing clear definition of the early stages of biosynthesis [194, 195]. Thus only strictosidine (6.282) is the product of *enzymic* condensation between tryptamine and secologanin (6.280). It accumulates in the presence of a β-glucosidase inhibitor [which blocks removal of the glucose in (6.282) and thus prevents further biosynthesis]. In the absence of inhibitor biosynthesis can proceed normally and the alkaloids, ajmalicine (6.272), its C-19 epimer, (6.288), and tetrahydroalstonine (6.289) are formed. The enzyme preparation required reduced pyridine nucleotide (NADPH or NADH) and, in the absence of coenzyme, cathenamine (6.286) accumulated. It was moreover enzymically converted into (6.272), (6.288) and (6.289) in the presence of NADPH. It is thus clearly an intermediate in biosynthesis as is logically (6.283) and both compounds have been independently isolated from a plant [196]. The biosynthetic pathway outlined in Scheme 6.49 is nicely corroborated further by the

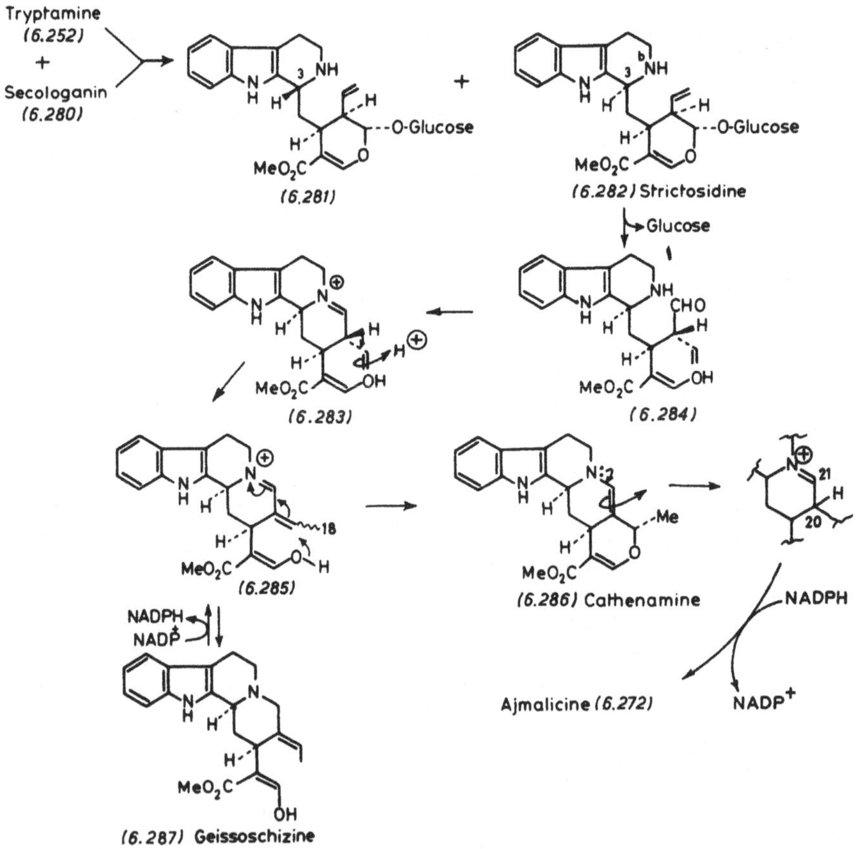

**Scheme 6.49**

results of a set of experiments in which (*6.282*) was incubated with a crude plant enzyme either in $D_2O$ with NADPH or in water with $NADP^2H$. Two protonations occur *en route* to (*6.272*), one each at C-18 and C-20 [see (*6.283*) and (*6.286*)], and these sites were found to be labelled by deuterium in the first experiment as required. Reduction of cathenamine (*6.286*) affords ajmalicine (*6.272*) with hydride addition at C-21: deuterium in the second experiment was incorporated at this site as required (it had the $\alpha$-configuration) [197].

The *Corynanthé* alkaloids represented by ajmalicine (*6.272*) and tetrahydroalstonine (*6.289*) are the simplest variations of the strictosidine skeleton. Formation of an extra carbocyclic ring [as shown in (*6.290*)] affords yohimbine (*6.291*).

Simple rearrangement of the *Corynanthé* skeleton [as (*6.272*)] affords the *Strychnos* type, exemplified by akuammicine (*6.271*), and strychnine (*6.292*). Like (*6.271*), strychnine has been found to have the expected origins in tryptophan and in mevalonate via geraniol; the extra $C_2$ unit (C-22 and C-23) arises from acetate. Late intermediates have been identified and the pathway deduced is shown in Scheme 6.50 [198].

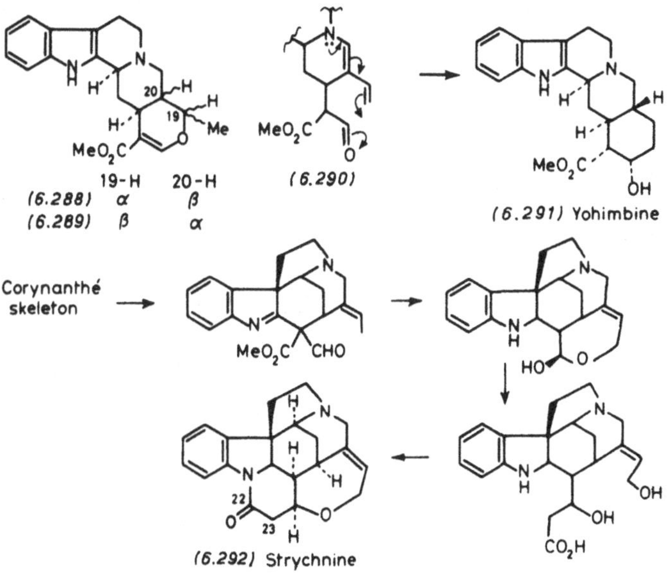

|  | 19-H | 20-H |
|---|---|---|
| (*6.288*) | $\alpha$ | $\beta$ |
| (*6.289*) | $\beta$ | $\alpha$ |

(*6.290*)

(*6.291*) Yohimbine

Corynanthé skeleton

(*6.292*) Strychnine

**Scheme 6.50**

It is, *a priori*, a reasonable hypothesis that alkaloids like catharanthine (*6.268*) (*Iboga* type) and vindoline (*6.273*) (*Aspidosperma* type) are formed by rearrangement of the *Corynanthé* skeleton [as (*6.272*)]. This is supported first by the observation that *Corynanthé* bases appear in *C. roseus* seedlings before *Iboga* and *Aspidosperma* alkaloids; second by the observation that

the *Corynanthé* base, geissoschizine (*6.287*) is an intact precursor for repre-
sentatives of the *Aspidosperma* and *Iboga* groups, and also those with the
*Strychnos* skeleton [199, 200]. [Experiments [201] with an enzyme preparation of *C. roseus* tissue cultures indicate that geissoschizine is a shunt from
the main biosynthetic pathway (in tissue cultures at least) (see Scheme 6.49).
The deduced relationship between the alkaloid groups still stands, however.]

Results of orthodox feeding experiments, and those where the sequence of
alkaloid formation is obtained by noting the appearance of precursor label in
individual alkaloids in relation to time, have led to the conclusion that
preakuammicine (*6.293*), stemmadenine (*6.294*) and tabersonine (*6.295*) are
further important intermediates in the biosynthesis of both *Aspidosperma*
[as (*6.273*)] and *Iboga* [as (*6.268*)] alkaloid types [202].

The formation of tabersonine (*6.295*) may be rationalized in terms of the
sequence shown in Scheme 6.51, central to which is the enamine (*6.296*)
formed by fragmentation of stemmadenine (*6.294*). *Iboga* alkaloids, e.g.
catharanthine (*6.268*), are reasonably derivable from the enamine (*6.296*)
also. If this is correct then the observed incorporation of labelled tabersonine
(*6.295*) into catharanthine (*6.268*) indicates that (*6.296*) → (*6.295*) is
reversible.

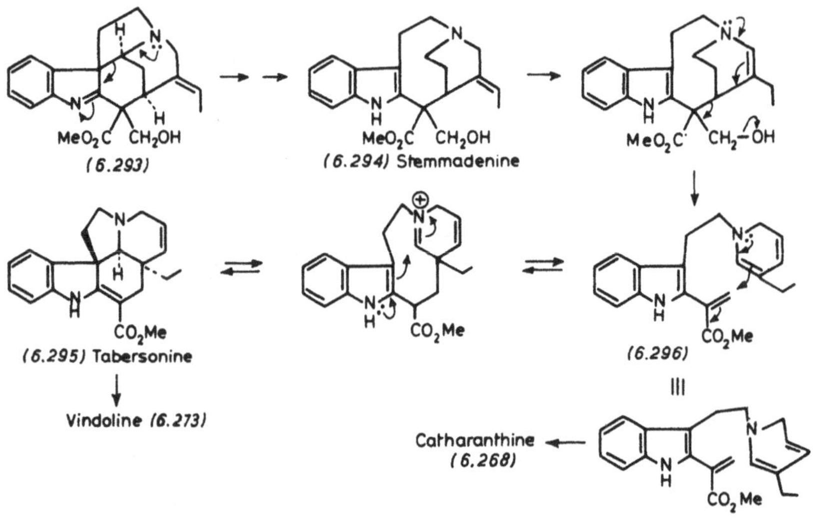

Scheme 6.51

The validity of (*6.296*) as a biosynthetic intermediate is supported by the
isolation of simple secodine (*6.297*) derivatives from plants and by the
specific incorporation of labelled secodine (*6.297*) into vindoline (*6.273*);
the tritium loss from (*6.297*) labelled in the dihydropyridine ring on trans-

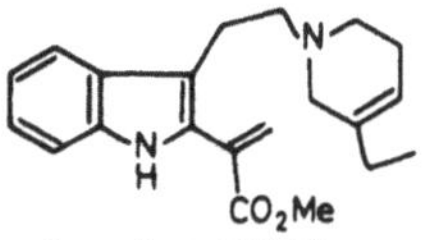

Secodine (6.297)

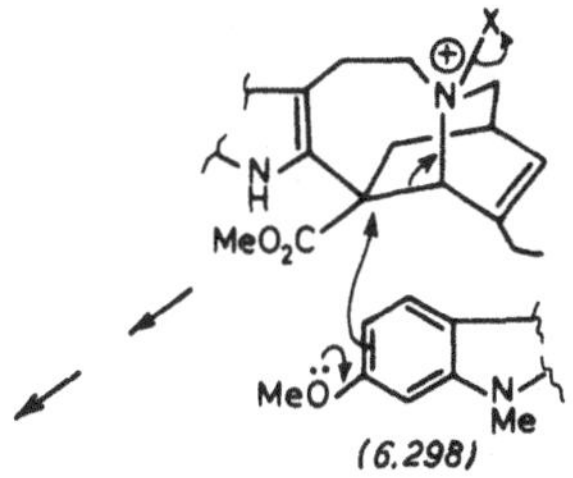

(6.298)

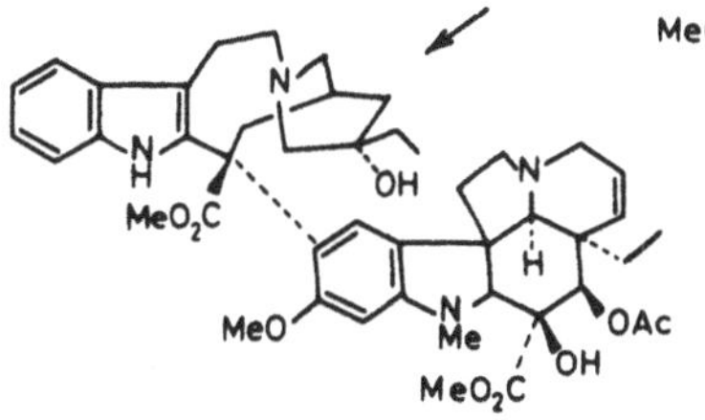

(6.299) Vinblastine

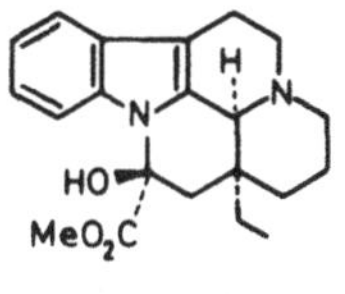

(6.300) Vincamine

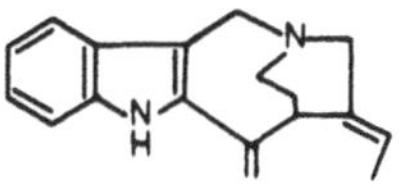

(6.301) Apparicine

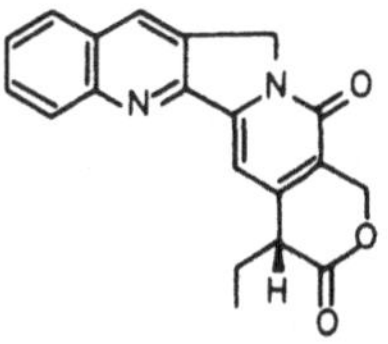

(6.302) Camptothecin

formation into vindoline (*6.273*) is consistent with involvement of secodine (*6.297*) via a more highly oxidized intermediate [as (*6.296*)] [203].

Investigation of the biosynthesis of the bisindole alkaloid, vinblastine (*6.299*), has shown that it does, as expected, arise from catharanthine (*6.268*) and vindoline (*6.273*) probably by the mechanism shown in (*6.298*) [204, 205].

Vincamine (*6.300*), with a skeleton different from that discussed so far, is derived via tabersonine (*6.295*) [206]. In another, different alkaloid, apparicine (*6.301*), loss of one of the tryptamine side-chain carbons is apparent, and this appears to happen at a stage beyond stemmadenine (*6.294*) [207].

The structure of polyneuridine aldehyde (*6.303*) is a minor modification of that seen in ajmalicine (*6.272*) in which C-16 has become joined to C-5. Work with plant enzyme preparations has provided a beautiful picture of the biosynthetic route which leads from (*6.303*) to ajmaline (*6.306*) (Scheme 6.52) and related alkaloids. The employment *in vivo* of the O-acetyl group is notable: it serves as a protecting group for (*6.304*) to hold the newly formed ring closed and the acetyl group remains until at a later stage the imine double bond has been reduced (the evidence indicates that N-methylation may occur before or after removal of the acetyl group) [208, 209].

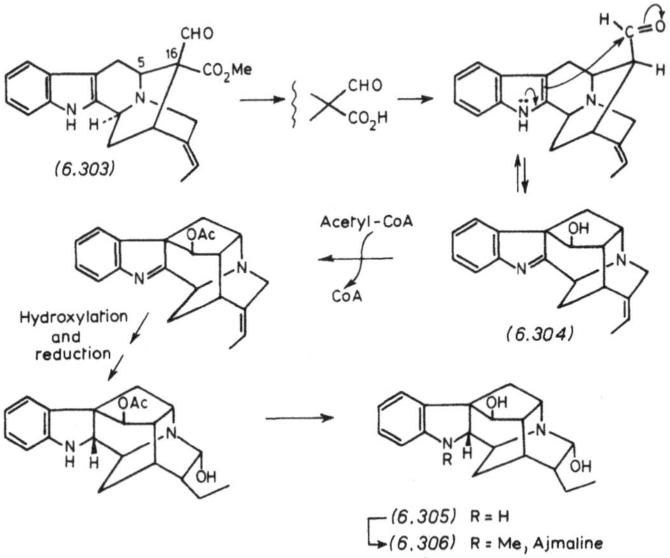

**Scheme 6.52**

The *Cinchona* alkaloid, quinine (*6.309*), does not immediately seem to be related to the alkaloids under discussion, but at quite an early date such a relationship was apparent since e.g. cinchonamine (*6.307*) was also identified as a *Cinchona* base. It seemed likely, moreover, that alkaloids of type (*6.307*)

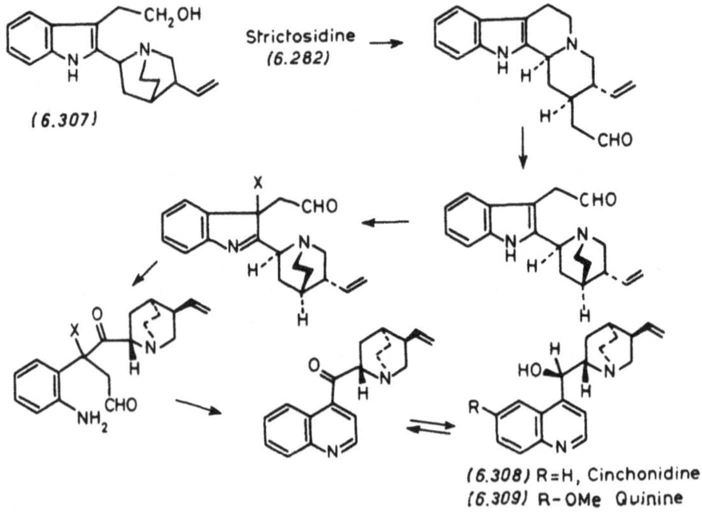

**Scheme 6.53**

would turn out to be intermediates in the formation of alkaloids like quinine (*6.309*). This has been proved to be correct. The pathway, which the experimental results indicate [210], is illustrated in Scheme 6.53. Another quinoline alkaloid is camptothecin (*6.302*). It is formed by diversion at the strictosidine (*6.282*) stage, by lactam formation [reaction of $N_b$ with the ester function of (*6.282*) rather than the aldehyde function] [211, 212].

## 6.7 IPECAC ALKALOIDS [193, 213, 214]

Part of the structure of the ipecac alkaloids, e.g. emetine (*6.314*) and cephaeline (*6.313*) [heavy bonding in (*6.313*)] is closely similar to the terpenoid fragment of the alkaloids just discussed (section 6.6.2); the remaining atoms are accounted for by two phenethylamine units. Investigations which have paralleled those of the terpenoid indole alkaloids have established the origins of these alkaloids (see Scheme 6.54). Desacetylisoipecoside (*6.310*) is the key intermediate for (*6.313*)

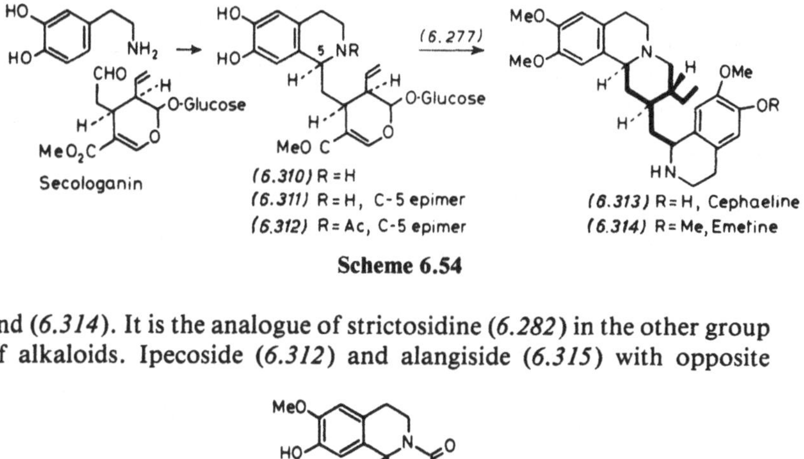

Secologanin

(*6.310*) R = H
(*6.311*) R = H, C-5 epimer
(*6.312*) R = Ac, C-5 epimer

(*6.313*) R = H, Cephaeline
(*6.314*) R = Me, Emetine

**Scheme 6.54**

and (*6.314*). It is the analogue of strictosidine (*6.282*) in the other group of alkaloids. Ipecoside (*6.312*) and alangiside (*6.315*) with opposite

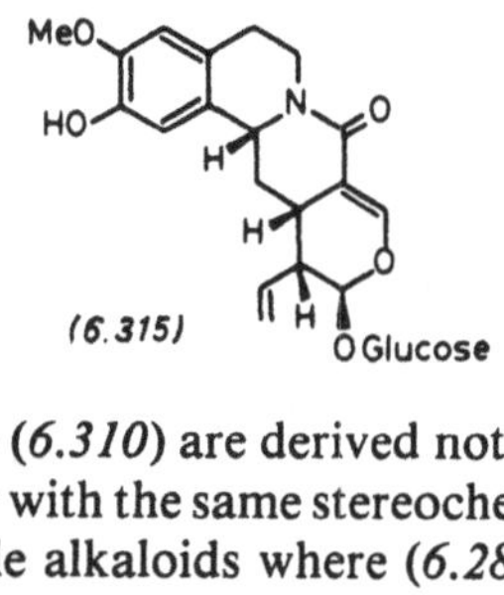

(*6.315*)

stereochemistry at C-5 to (*6.310*) are derived not from (*6.310*) but from (*6.311*), i.e. the precursor with the same stereochemistry unlike the situation with terpenoid indole alkaloids where (*6.282*) is the precursor for both of the corresponding epimeric series.

## 6.8 MISCELLANEOUS ALKALOIDS

The biosynthesis of steroidal and terpenoid alkaloids can be correlated with the appropriate steroid and terpenoid precursors (Chapter 4).

Alkaloids with unusual origins are dolichotheline (*6.316*), see Scheme 6.55 [215, 216], shihunine (*6.318*) which derives from (*6.317*) [217] an important precursor for 1,4-naphthoquinones (cf section 5.2), and pyrazol-1-ylalanine (*6.319*) which is formed from 1,3-diamino-propane via pyrazole which condenses with *O*-acetylserine [218].

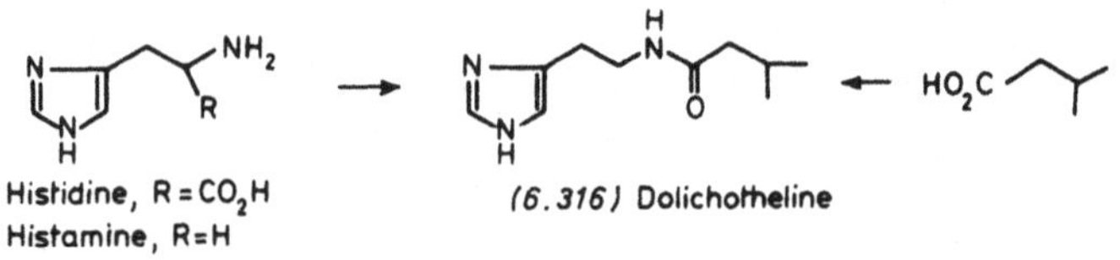

Histidine, R = CO₂H
Histamine, R = H

*(6.316)* Dolichotheline

**Scheme 6.55**

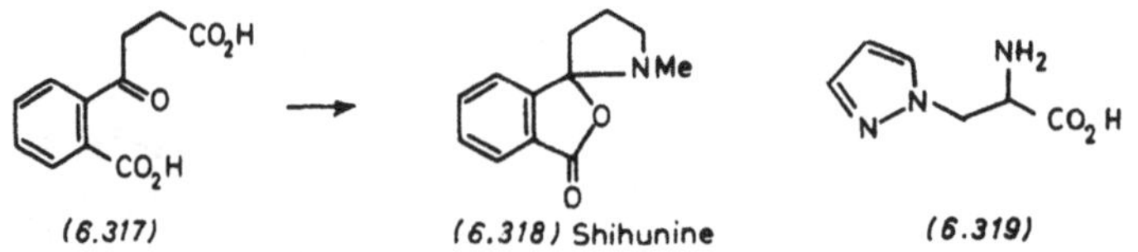

*(6.317)*          *(6.318)* Shihunine          *(6.319)*

The betalains, e.g. betanin (*6.321*) and indicaxanthin (*6.322*), are coloured bases found in plants of the Centrospermae order. They have a

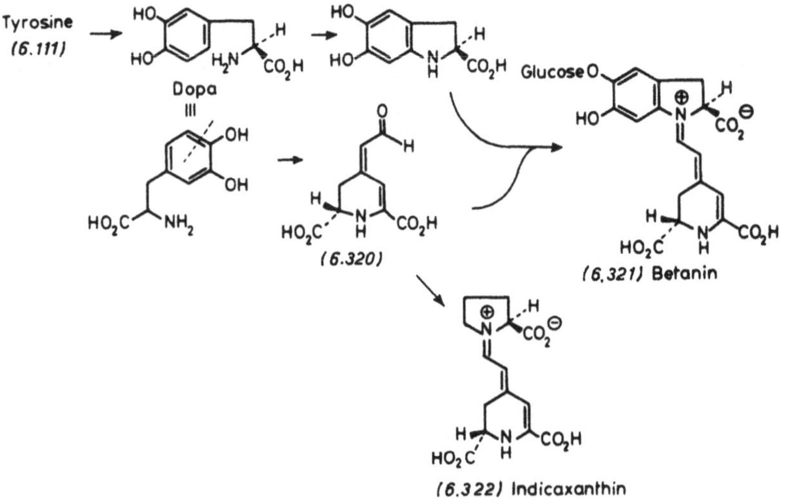

Tyrosine
(6.111)

Dopa
III

(6.320)

(6.321) Betanin

(6.322) Indicaxanthin

**Scheme 6.56**

well-defined function, e.g. as the pigments of some flowers. The pathway deduced for their biosynthesis is shown in Scheme 6.56. Notable is the exceptional extra-diol cleavage of the dopa unit which goes to make up the betalamic acid moiety (*6.320*) in (*6.321*) and (*6.322*) [219, 220].

# REFERENCES

Further reading: Herbert, R.B. (1979) In *Comprehensive Organic Chemistry* (eds D.H.R. Barton and W.D. Ollis), Pergamon, Oxford, vol. 5, pp. 1045–119. Herbert, R.B. (1980) In *Rodd's Chemistry of Carbon Compounds* (ed. S. Coffey), 2nd edn, Elsevier, Amsterdam, vol. IVL, pp. 291–455; (1988) Supplement (ed. M.F. Ansell), pp. 155–247. [2]. Spenser, I.D. (1968) In *Comprehensive Biochemistry* (eds M. Florkin and E.H. Stotz), Elsevier, Amsterdam, vol. 20, pp. 231–413. Dalton, D.R. (1979) *The Alkaloids, The Fundamental Chemistry, A Biogenetic Approach*, Dekker, New York. *Chemistry of The Alkaloids* (1970) (ed. S.W. Pelletier), van Nostrand Reinhold, New York.

1. *The Alkaloids* (ed. R.H.F. Manske), Academic Press, New York, now approx. twenty-nine volumes.
2. *Natural Product Reports*, The Royal Society of Chemistry, London.
3. Robinson, R. (1955) *The Structural Relations of Natural Products*, Oxford University Press, Oxford.
4. Leete, E. and Slattery, S.A. (1976) *J. Amer. Chem. Soc.*, **98**, 6326–30.
5. Leete, E. (1977), *Phytochemistry*, **16**, 1705–9.
6. Leete, E. and Olson, J.O. (1972) *J. Amer. Chem. Soc.*, **94**, 5472–7.
7. Roberts, M.F. (1978) *Phytochemistry*, **17**, 107–12.
8. Leete, E., Lechleiter, J.C. and Carver, R.A. (1975) *Tetrahedron Lett.*, 3779–82.
9. Keogh, M.F. and O'Donovan, D.G. (1970) *J. Chem. Soc.*, 1792–7.
10. Gupta, R.N. and Spenser, I.D. (1967) *Can. J. Chem.*, **45**, 1275–85.
11. Gupta, R.N. and Spenser, I.D. (1970) *Phytochemistry*, **9**, 2329–34.
12. Leete, E. (1956) *J. Amer. Chem. Soc.*, **78**, 3520–3.
13. Leete, E. (1969) *J. Amer. Chem. Soc.*, **91**, 1697–700.
14. Leete, E. and Friedman, A.R. (1964) *J. Amer. Chem. Soc.*, **86**, 1224–6.
15. Leete, E. and Chedekel, M.R. (1972) *Phytochemistry*, **11**, 2751–6.
16. Leistner, E. and Spenser, I.D. (1973) *J. Amer. Chem. Soc.*, **95**, 4715–25.
17. Gerdes, H.J. and Leistner, E. (1979) *Phytochemistry*, **18**, 771–5.
18. Gilbertson, T.J. (1972) *Phytochemistry*, **11**, 1737–9.
19. Leistner, E., Gupta, R.N. and Spenser, I.D. (1973) *J. Amer. Chem. Soc.*, **95**, 4040–7.
20. O'Donovan, D.G. and Creedon, P.B. (1971) *Tetrahedron Lett.*, 1341–4.
21. O'Donovan, D.G., Long, D.J., Forde, E. and Geary, P. (1975) *J. Chem. Soc. Perkin I*, 415–9.
22. Gupta, R.N. and Spenser, I.D. (1971) *Can. J. Chem.*, **49**, 384–97.
23. Sankawa, U., Ebizuka, Y. and Yamasaki, Y. (1977) *Phytochemistry*, **16**, 561–3.
24. Parry, R.J. (1978) *Bio-org. Chem.*, **7**, 277–88.
25. O'Donovan, D.G. and Creedon, P.B. (1974) *J. Chem. Soc. Perkin I*, 2524–8.
26. Braekman, J.-C., Gupta, R.N., MacLean, D.B. and Spenser, I.D. (1972) *Can. J. Chem.*, **50**, 2591–602.

27. Gupta, R.N., Horsewood, P., Koo, S.H. and Spenser, I.D. (1979) *Can. J. Chem.*, **57**, 1606-14.

28. Horsewood, P., Golebiewski, W.M., Wrobel, J.T. *et al.* (1979) *Can. J. Chem.*, **57**, 1615-30.

29. Hedges, S.H., Herbert, R.B. and Wormald, P.C. (1983) *J. Chem. Soc., Chem. Comm.*, 145-7.

30. Schütte, H.R., Hindorf, H., Mothes, K. and Hübner, G. (1964) *Annalen*, **680**, 93-104.

31. Schütte, H.R. and Hindorf, H. (1964) *Z. Naturforsch.*, **19b**, 855.

32. Rana, J. and Robins, D.J. (1986) *J. Chem. Soc. Perkin I*, 1133-7.

33. Fraser, A.M. and Robins, D.J. (1987) *J. Chem. Soc. Perkin I*, 105-9.

34. Golebiewski, W.M. and Spenser, I.D. (1985) *Can. J. Chem.*, **63**, 2707-18.

35. Fraser, A.M. and Robins, D.J. (1986) *J. Chem. Soc. Chem. Comm.*, 545-7.

36. Leeper, F.J., Grue-Sørensen G. and Spenser, I.D. (1981) *Can. J. Chem.*, **59**, 106-15.

37. Schütte, H.R., Nowacki, E., Kovacs, P. and Liebisch, H.W. (1963) *Arch. Pharm.*, **296**, 438-41.

38. Cho, Y.D., Martin, R.O. and Anderson, J.N. (1971) *J. Amer. Chem. Soc.*, **93**, 2087-9.

39. Cho, Y.D. and Martin, R.O. (1971) *Can. J. Biochem.*, **49**, 971-7.

40. Wink, M., Hartmann, T. and Witte, L. (1980) *Z. Naturforsch.*, **35c**, 93-7.

41. Anet, E.F.L., Hughes, G.K. and Ritchie, E. (1949) *Austral. J. Scientific Res., Ser. 2A*, 616-21.

42. Robinson, R. (1917) *J. Chem. Soc.*, 876-99.

43. Leete, E. (1962) *J. Amer. Chem. Soc.*, **84**, 55-7.

44. Leete, E. and Nelson, S.J. (1969) *Phytochemistry*, **8**, 413-8.

45. Liebisch, H.W., Schütte, H.R. and Mothes, K. (1963) *Annalen*, **668**, 139-44.

46. Ahmad, A. and Leete, E. (1970) *Phytochemistry*, **9**, 2345-7.

47. Hedges, S.H. and Herbert, R.B. (1981) *Phytochemistry*, **20**, 2064-5.

48. Leete, E. and McDonell, J.A. (1981) *J. Amer. Chem. Soc.*, **103**, 658-62.

49. Lamberts, B.L., Dewey, L.J. and Byerrum, R.U. (1959) *Biochim. Biophys. Acta*, **33**, 22-6.

50. Leete, E., Gros, G. and Gilbertson, T.J. (1964) *Tetrahedron Lett.*, 587-92.

51. Leete, E. and Yu, M.-L. (1980) *Phytochemistry*, **19**, 1093-7.

52. Wigle, I.D., Mestichelli, L.J.J. and Spenser, I.D. (1982) *J. Chem. Soc. Chem. Comm.*, 662-4.

53. Mizusaki, S., Tanabe, Y., Noguchi, M. and Tamaki, E. (1972) *Phytochemistry*, **11**, 2757-62.

54. Saunders, J.W. and Bush, L.P. (1979) *Plant Physiol.*, **64**, 236-40.

55. Leete, E. and Liu, Y.-Y. (1973) *Phytochemistry*, **12**, 593-6.

56. Zielke, H.R., Reinke, C.M. and Byerrum, R.U. (1969) *J. Biol. Chem.*, **244**, 95-8.

57. Leete, E. (1985) *Phytochemistry*, **24**, 953-5.

58. Leete, E. (1972) *Phytochemistry*, **11**, 1713-16.

59. Romeike, A. and Fodor, G. (1960) *Tetrahedron Lett.*, No. 22, 1-4.

60. Leete, E. and Lucast, D.H. (1976) *Tetrahedron Lett.*, 3401-4.

61. Leete, E. (1987) *Can. J. Chem.*, **65**, 226-8.

62. Beresford, P.J. and Woolley, J.G. (1974) *Phytochemistry*, **13**, 2143-4.

63. Beresford, P.J. and Woolley, J.G. (1974) *Phytochemistry*, **13**, 2511-3.

64. Leete, E. (1973) *Phytochemistry*, **12**, 2203–5.
65. Leete, E. (1982) *J. Amer. Chem. Soc.*, **104**, 1403–8.
66. Leete, E. (1983) *J. Amer. Chem. Soc.*, **105**, 6727–8.
67. Khan, H.A. and Robins, D.J. (1985) *J. Chem. Soc. Perkin I*, 101–5.
68. Grue-Sørensen, G. and Spenser, I.D. (1982) *Can. J. Chem.*, **60**, 643–62.
69. Khan, H.A. and Robins, D.J. (1985) *J. Chem. Soc. Perkin I*, 819–24.
70. Rana, J. and Robins, D.J. (1986) *J. Chem. Soc. Perkin I*, 983–8.
71. Kunec, E.K. and Robins, D.J. (1985) *J. Chem. Soc. Chem. Comm.*, 1450–4.
72. Grue-Sørensen, G. and Spenser, I.D. (1983) *J. Amer. Chem. Soc.*, **105**, 7401–4.
73. Robins, D.J. (1982) *J. Chem. Soc. Chem. Comm.*, 1289–90.
74. Cahill, R., Crout, D.H.G., Gregorio, M.V.M. *et al.* (1983) *J. Chem. Soc. Perkin I*, 173–80.
75. Herbert, R.B., Jackson, F.B. and Nicolson, I.T. (1984) *J. Chem. Soc. Perkin I*, 825–31.
76. Herbert, R.B. and Jackson, F.B. (1977) *J. Chem. Soc. Chem. Comm.*, 955–6.
77. Leete, E. and Marion, L. (1953) *Can. J. Chem.*, **31**, 126–8.
78. Leete, E. and Marion, L. (1954) *Can. J. Chem.*, **32**, 646–9.
79. Lundström, J. (1971) *Acta Pharm. Suecica*, **8**, 275–302.
80. Basmadjian, G.P. and Paul, A.G. (1971) *Lloydia*, **34**, 91–3.
81. Kapadia, G.J., Rao, G.S., Leete, E. *et al.* (1970) *J. Amer. Chem. Soc.*, **92**, 6943–51.
82. O'Donovan, D.G. and Barry, E. (1974) *J. Chem. Soc. Perkin I*, 2528–9.
83. Schütte, H.R. and Seelig, G. (1969) *Annalen*, **730**, 186–90.
84. Agurell, S., Granelli, I., Leander, K. and Rosenblom, J. (1974) *Acta Chem. Scand.*, **B28**, 1175–9.
85. Keller, W.J. (1981) *Phytochemistry*, **20**, 2165–7.
86. Yamasaki, K., Tamaki, T., Uzawa, S. *et al.* (1973) *Phytochemistry*, **12**, 2877–82.
87. Rueffer, M. and Zenk, M.H. (1987) *Z. Naturforsch.*, **42c**, 319–32.
88. Stadler, R., Kutchan, T.M., Loeffler, S. *et al.* (1987) *Tetrahedron Lett.*, **28**, 1251–4.
89. Zenk, M.H. Rueffer, M., Amann, M. *et al.* (1985) *J. Nat. Prod.*, **48**, 725–38.
90. Brochmann-Hanssen, E., Chen, C., Chen, C.R. *et al.* (1975) *J. Chem. Soc. Perkin I*, 1531–7.
91. Uprety, H., Bhakuni, D.S. and Kapil, R.S. (1975) *Phytochemistry*, **14**, 1535–7.
92. Battersby, A.R., Sheldrake, P.W., Staunton, J. and Summers, M.C. (1977) *Bioorg. Chem.*, **6**, 43–7.
93. Bhakuni, D.S. and Jain, S. (1981) *J. Chem. Soc. Perkin I*, 2598–603.
94. Bhakuni, D.S., Singh, A.N. and Jain, S. (1981) *Tetrahedron*, **37**, 2651–5.
95. Blaschke, G. and Scriba, G. (1985) *Phytochemistry*, **24**, 585–8.
96. Bhakuni, D.S., Tewari, S. and Kapil, R.S. (1977) *J. Chem. Soc. Perkin I*, 706–9.
97. Blaschke, G., Waldheim, G., von Schantz, M. and Peura, P. (1974) *Arch. Pharm.*, **307**, 122–30.
98. Brochmann-Hanssen, E., Chen, C.-H., Chiang, H.-C. and McMurtrey, K. (1972) *J. Chem. Soc. Comm.*, 1269.
99. Prakash, O., Bhakuni, D.S. and Kapil, R.S. (1978) *J. Chem. Soc. Perkin I*, 622–4.

100. Battersby, A.R., Brocksom, T.J. and Ramage, R. (1969) *J. Chem. Soc. Chem. Comm.*, 464–5.
101. Battersby, A.R., McHugh, J.L., Staunton, J. and Todd, M. (1971) *J. Chem. Soc. Chem. Comm.*, 985–6.
102. Haynes, L.J., Stuart, K.L., Barton, D.H.R. *et al.* (1965) *J. Chem. Soc. Chem. Comm.*, 141–2.
103. Barton, D.H.R., Bhakuni, D.S., Chapman, G.M. and Kirby, G.W. (1967) *J. Chem. Soc. (C)*, 2134–40.
104. Bhakuni, D.S. and Jain, S. (1981) *Tetrahedron*, **37**, 3175–81.
105. Sharma, V., Jain, S., Bhakuni, D.S. and Kapil, R. (1982) *J. Chem. Soc. Perkin I*, 1153–5.
106. Barton, D.H.R., Bracho, R.D., Potter, C.J. and Widdowson, D.A. (1974) *J. Chem. Soc. Perkin I*, 2278–83.
107. Bhakuni, D.S. and Singh, A.N. (1978) *J. Chem. Soc. Perkin I*, 618–2.
108. Bhakuni, D.S. and Jain, S. (1980) *Tetrahedron*, **36**, 2153–6.
109. Battersby, A.R. and Binks, R. (1960) *Proc. Chem. Soc.*, 360–1.
110. Battersby, A.R., Binks, R., Francis, R.J. *et al.* (1964) *J. Chem. Soc.*, 3600–10.
111. Battersby, A.R., Foulkes, D.M. and Binks, R. (1965) *J. Chem. Soc.*, 3323–32.
112. Borkowski, P.R., Horn, J.S. and Rapoport, H. (1978) *J. Amer. Chem. Soc.*, **100**, 276–81.
113. Parker, H.I., Blaschke, G. and Rapoport, H. (1972) *J. Amer. Chem. Soc.*, **94**, 1276–82.
114. Battersby, A.R. and Harper, B.J.T. (1960) *Tetrahedron Lett.*, No. 27, pp. 21–4.
115. Horn, J.S., Paul, A.G. and Rapoport, H. (1978) *J. Amer. Chem. Soc.*, **100**, 1895–8.
116. Battersby, A.R., Jones, R.C.F., Minta, A. *et al.* (1981) *J. Chem. Soc. Perkin I*, 2030–9.
117. Battersby, A.R., Francis, R.J., Hirst, M. *et al.* (1975) *J. Chem. Soc. Perkin I*, 1140–7.
118. Steffens, P., Nagakura, N. and Zenk, M.H. (1985) *Phytochemistry*, **24**, 2577–83.
119. Amann, M., Wanner, G. and Zenk, M.H. (1986) *Planta*, **167**, 310–20.
120. Rueffer, M. and Zenk, M.H. (1986) *Tetrahedron Lett.*, **27**, 923–4.
121. Rueffer, M., Ekundayo, O., Nagakura, N. and Zenk, M.H. (1983) *Tetrahedron Lett.*, **24**, 2643–4.
122. Battersby, A.R., Staunton, J., Wiltshire, H.R. *et al.* (1975) *J. Chem. Soc. Perkin I*, 1162–71.
123. Battersby, A.R., Staunton, J., Wiltshire, H.R. *et al.* (1975) *J. Chem. Soc. Perkin I*, 1147–56.
124. Battersby, A.R., Staunton, J., Summers, M.C. and Southgate, R. (1979) *J. Chem. Soc. Perkin, I*, 45–52.
125. Takao, N., Iwasa, K., Kamigauchi, M. and Sugiura, M. (1976) *Chem. and Pharm. Bull. (Japan)*, **24**, 2859–68.
126. Holland, H.L., Castillo, M., MacLean, D.B. and Spenser, I.D. (1974) *Can. J. Chem.*, **52**, 2818–31.
127. Rönsch, H. (1977), *Phytochemistry*, **16**, 691–8.
128. Tani, C. and Tagahara, K. (1977) *J. Pharm. Soc. Japan*, **97**, 93–102 (for the closely related alkaloid, rhoeadine).

129. Battersby, A.R., Dobson, T.A., Foulkes, D.M. and Herbert, R.B. (1972) *J. Chem. Soc. Perkin I*, 1730-6.

130. Battersby, A.R., Herbert, R.B., Pijewska, L. *et al.* (1972) *J. Chem. Soc. Perkin I*, 1736-40.

131. Battersby, A.R., Herbert, R.B., McDonald, E. *et al.* (1972) *J. Chem. Soc. Perkin I*, 1741-6.

132. Battersby, A.R., Sheldrake, P.W. and Milner, J.A. (1974) *Tetrahedron Lett.*, 3315-8.

133. Barker, A.C., Battersby, A.R., McDonald, E. *et al.* (1967) *J. Chem. Soc. Chem. Comm.*, 390-2.

134. Herbert, R.B. and Knagg, E. (1986) *Tetrahedron Lett.*, **27**, 1099-102.

135. Battersby, A.R., Böhler, P., Munro, M.H.G. and Ramage R. (1974) *J. Chem. Soc. Perkin I*, 1399-402.

136. Battersby, A.R., McDonald, E., Milner, J.A. *et al.* (1975), *Tetrahedron Lett.*, 3419-22.

137. Schwab, J.M., Chang, M.N.T. and Parry, R.J. (1977) *J. Amer. Chem. Soc.*, **99**, 2368-70.

138. Battersby, A.R., Binks, R., Breuer, S.W. *et al.* (1964) *J. Chem. Soc.*, 1595-609.

139. Zulalian, J. and Suhadolnik, R.J. (1964) *Proc. Chem. Soc.*, 422-3.

140. Wightman, R.H., Staunton, J., Battersby, A.R. and Hanson, K.R. (1972) *J. Chem. Soc. Perkin I*, 2355-64.

141. Barton, D.H.R., Kirby, G.W., Taylor, J.B. and Thomas, G.M. (1963) *J. Chem. Soc.*, 4545-58.

142. Bennett, D.J. and Kirby, G.W. (1968) *J. Chem. Soc. (C)*, 442-6.

143. Leete, E. and Louden, M.C.L. (1968) *J. Amer. Chem. Soc.*, **90**, 6837-41.

144. O'Donovan, D.G. and Horan, H. (1971) *J. Chem. Soc. (C)*, 331-4.

145. Kirby, G.W. and Tiwari, H.P. (1966) *J. Chem. Soc. (C)*, 676-82.

146. Fuganti, C. (1969) *Chim. Ind. (Milan)*, **51**, 1254-5.

147. Fales, H.M., Mann, J. and Mudd, S.H. (1963) *J. Amer. Chem. Soc.*, **85**, 2025-6.

148. Bhandarkar, J.G. and Kirby, G.W. (1970) *J. Chem. Soc. (C)*, 1224-7.

149. Fuganti, C., Ghiringhelli, D. and Grasselli, P. (1974) *J. Chem. Soc. Chem. Comm.*, 350-1.

150. Fuganti, C. and Mazza, M. (1972) *J. Chem. Soc. Chem. Comm.*, 936-7.

151. Bruce, I.T. and Kirby, G.W. (1968) *J. Chem. Soc. Chem. Comm.*, 207-8.

152. Battersby, A.R., Kelsey, J.W., Staunton, J. and Suckling, K.E. (1973) *J. Chem. Soc. Perkin I*, 1609-15.

153. Kirby, G.W. and Michael, J. (1973) *J. Chem. Soc. Perkin I*, 115-20.

154. Fuganti, C. and Mazza, M. (1971) *J. Chem. Soc. Chem. Comm.*, 1196-7.

155. Fuganti, C. and Mazza, M. (1973) *J. Chem. Soc. Perkin I*, 954-6.

156. Fuganti, C. (1973) *Gazzetta*, **103**, 1255-8.

157. Fuganti, C. and Mazza, M. (1972) *J. Chem. Soc. Chem. Comm.*, 239.

158. Fuganti, C. (1973) *Tetrahedron Lett.*, 1785-8.

159. Fuganti, C., Ghiringhelli, D. and Grasselli, P. (1973) *J. Chem. Soc. Chem. Comm.*, 430-1.

160. Feinstein, A.I. and Wildman, W.C. (1976) *J. Org. Chem.*, **41**, 2447-50.

161. Jeffs, P.W., Campbell, H.F., Farrier, D.S. *et al.* (1974) *Phytochemistry*, **13**, 933-45.

162. Jeffs, P.W., Johnson, D.B., Martin, N.H. and Rauckman, B.S. (1976) *J. Chem. Soc. Chem. Comm.*, 82–3.

163. Jeffs, P.W., Karle, J.M. and Martin, N.H. (1978) *Phytochemistry*, **17**, 719–28.

164. Munsche, D. and Mothes, K. (1965) *Phytochemistry*, **4**, 705–12.

165. Schröder, P. and Luckner, M. (1966) *Pharmazie*, **21**, 642.

166. Tipton, C.L., Wang, M.-C., Tsao, F.H.-C. *et al.* (1973), *Phytochemistry*, **12**, 347–52.

167. Blaschke-Cobet, M. and Luckner, M. (1973) *Phytochemistry*, **12**, 2393–8.

168. Johne, S., Waiblinger, K. and Gröger, D. (1970) *Eur. J. Biochem.*, **15**, 415–20.

169. O'Donovan, D.G. and Horan, H. (1970) *J. Chem. Soc. (C)*, 2466–70.

170. Yamazaki, M., Ikuta, A., Mori, T. and Kawana, T. (1967) *Tetrahedron Lett.*, 3317–20.

171. Baumert, A., Schneider, G. and Gröger, D. (1986) *Z. Naturforsch.*, **41c**, 187–92.

172. Liljegren, D.R. (1971) *Phytochemistry*, **10**, 2661–9.

173. Waiblinger, K., Johne, S. and Gröger, D. (1972) *Phytochemistry*, **11**, 2263–5.

174. Waiblinger, K., Johne, S. and Gröger, D. (1973) *Pharmazie*, **28**, 403–6.

175. Grundon, M.F., Harrison, D.M. and Spyropoulos, C.G. (1975) *J. Chem. Soc. Perkin I*, 302–4.

176. Boulanger, D., Bailey, B.K. and Steck, W. (1973) *Phytochemistry*, **12**, 2399–405.

177. Leete, E. and Minich, M.L. (1977) *Phytochemistry*, **16**, 149–50.

178. Agurell, S. and Nilsson, J.L.G. (1968) *Acta Chem. Scand.*, **22**, 1210–18.

179. Baxter, C. and Slaytor, M. (1972) *Phytochemistry*, **11**, 2767–73.

180. O'Donovan, D.G. and Keogh, M.F. (1966) *J. Chem. Soc. (C)*, 1570–2.

181. Herbert, R.B. and Mann, J. (1982) *J. Chem. Soc. Perkin I*, 1523–5.

182. Leete, E. (1979) *J. Chem. Soc. Chem. Comm.*, 821–2.

183. Battersby, A.R., Burnett, A.R. and Parsons, P.G. (1969) *J. Chem. Soc. (C)*, 1193–200.

184. Wenkert, E. (1962) *J. Amer. Chem. Soc.*, **84**, 98–102.

185. Thomas, R. (1961) *Tetrahedron Lett.*, 544–53.

186. Uesato, S., Kanomi, S., Iida, A., Inouye, H. and Zenk, M.H. (1986) *Phytochemistry*, **25**, 839–42.

187. Battersby, A.R., Hall, E.S. and Southgate, R. (1969) *J. Chem. Soc. (C)*, 721–8.

188. Battersby, A.R., Byrne, J.C., Kapil, R.S. *et al.* (1968) *J. Chem. Soc. Chem. Comm.*, 951–3.

189. Escher, S., Loew, P. and Arigoni, D. (1970) *J. Chem. Soc. Chem. Comm.*, 823–5 (and following two papers).

190. Battersby, A.R., Westcott, N.D., Glüsenkamp, K.-H. and Tietze, L.-F. (1981) *Chem. Ber.*, **114**, 3439–47.

191. Battersby, A.R., Burnett, A.R. and Parsons, R.G. (1969) *J. Chem. Soc. (C)*, 1187–92.

192. Rueffer, M., Nagakura, N. and Zenk, M.H. (1978) *Tetrahedron Lett.*, 1593–6.

193. Battersby, A.R., Lewis, N.G. and Tippett, J.M. (1978) *Tetrahedron Lett.*, 4849–52.

194. Scott, A.I., Lee, S.-L. and Wan, W. (1977) *Biochem. Biophys. Res. Comm.*, **75**, 1004–9.

195. Stöckigt, J., Husson, H.P., Kan-Fan, C. and Zenk, M.H. (1977) *J. Chem. Soc. Chem. Comm.*, 164–6.

196. Kan Fan, C. and Husson, H.-P. (1979) *J. Chem. Soc. Chem. Comm.*, 1015–6.
197. Stöckigt, J., Hemscheidt, T., Höfle, G. *et al.* (1983), *Biochemistry*, **22**, 3448–52.
198. Heimberger, S.I. and Scott, A.I. (1973) *J. Chem. Soc. Chem. Comm.*, 217–8.
199. Battersby, A.R. and Hall, E.S. (1969) *J. Chem. Soc. Chem. Comm.*, 793–4.
200. Scott, A.I., Cherry, P.C. and Qureshi, A.A. (1969) *J. Amer. Chem. Soc.*, **91**, 4932–3.
201. Stöckigt, J. (1978) *J. Chem. Soc. Chem. Comm.*, 1097–9.
202. Scott, A.I., Reichardt, P.B., Slaytor, M.B. and Sweeney, J.G. (1971) *Bioorg. Chem.*, **1**, 157–73.
203. Kutney, J.P., Beck, J.F., Eggers, N.J. *et al.* (1971) *J. Amer. Chem. Soc.*, **93**, 7322–4.
204. Scott, A.I., Gueritte, F. and Lee, S.-L. (1978) *J. Amer. Chem. Soc.*, **100**, 6253–5.
205. Stuart, K.L., Kutney, J.P., Honda, T. and Worth, B.R. (1978) *Heterocycles*, **9**, 1419–27.
206. Kutney, J.P., Beck, J.F., Nelson, V.R. and Sood, R.S. (1971) *J. Amer. Chem. Soc.*, **93**, 255–7.
207. Kutney, J.P., Beck, J.F., Ehret, C. *et al.* (1971) *Bioorg. Chem.*, **1**, 194–206.
208. Polz, L., Schübel, H. and Stöckigt, J. (1987) *Z. Naturforsch.*, **42c**, 333–42.
209. Pfitzner, A., Krausch, B. and Stöckigt, J. (1984) *Tetrahedron*, **40**, 1691–9.
210. Battersby, A.R. and Parry, R.J. (1971) *J. Chem. Soc. Chem. Comm.*, 31–2.
211. Hutchinson, C.R., Heckendorf, A.H., Daddona, P.E. *et al.* (1974) *J. Amer. Chem. Soc.*, **96**, 5609–11.
212. Sheriha, G.M. and Rapoport, H. (1976) *Phytochemistry*, **15**, 505–8.
213. Nagakura, N., Höfle, G. and Zenk, M.H. (1978) *J. Chem. Soc. Chem. Comm.*, 896–8.
214. Nagakura, N., Höfle, G., Coggiola, D. and Zenk, M.H. (1978) *Planta Med.*, **34**, 381–9.
215. Horan, H. and O'Donovan, D.G. (1971) *J. Chem. Soc. (C)*, 2083–5.
216. Rosenberg, H. and Paul, A.G. (1972) *Lloydia*, **34**, 372–6.
217. Leete, E. and Bodem, G.B. (1973) *J. Chem. Soc. Chem. Comm.*, 522–3.
218. Brown, E.G., Flayeh, K.A.M. and Gallon, J.R. (1982) *Phytochemistry*, **21**, 863–7.
219. Fischer, N. and Dreiding, A.S. (1972) *Helv. Chim. Acta*, **55**, 649–58.
220. Impellizzeri, G. and Piattelli, M. (1972) *Phytochemistry*, **11**, 2499–502.

# 7 Microbial metabolites containing nitrogen

## 7.1 INTRODUCTION

There is a wide variety of metabolites containing nitrogen, which is produced by micro-organisms. The nitrogen atom (or atoms) most frequently is derived ultimately from an α-amino acid which also provides at least part of the carbon skeleton. Modifications to cyclic dipeptides are commonly encountered as secondary metabolites (sections 7.3 and 7.4). In the case of the vastly important pharmaceuticals, the penicillins and cephalosporins (section 7.7.1), a linear tripeptide is involved in biosynthesis.

Tryptophan (*7.28*) serves as a precursor for a number of microbial metabolites. A notable example is the ergot alkaloids (section 7.5.1). Other common aromatic amino acids are involved in the biosynthesis of various metabolites and a number of quite unusual aromatic amino acids are implicated in the biosynthesis of some metabolites (sections 7.5.4, 7.6.2, 7.6.7). The origins of all of these amino acids are in the shikimic acid pathway (see section 5.1). Shikimic acid itself (*7.95*), or a close metabolic relative serves as a source for some metabolites, e.g. phenazines (section 7.6.1) and the ansamycins (section 7.6.2). For compounds like the latter, acetate and propionate are also important starting materials for biosynthesis. Another important part source for many metabolites is mevalonate, invariably as a $C_5$ (dimethylallyl) unit.

## 7.2 PIPERIDINE AND PYRIDINE METABOLITES

Pipecolic acid (*7.1*) occurs widely in plants, animals and micro-organisms. It has been found, in representative species of each, to derive from lysine with the D-isomer preferred (see section 6.2.1; Scheme 6.9). Slaframine (*7.5*), a toxin produced by the fungus *Rhizoctonia leguminicola*, has been firmly shown to be a pipecolic acid (*7.1*) metabolite. Results of experiments with likely precursors have allowed definition of the pathway shown in Scheme 7.1. This is supported by the discovery that a cell-free extract of *R. leguminicola* would catalyse the NADPH-

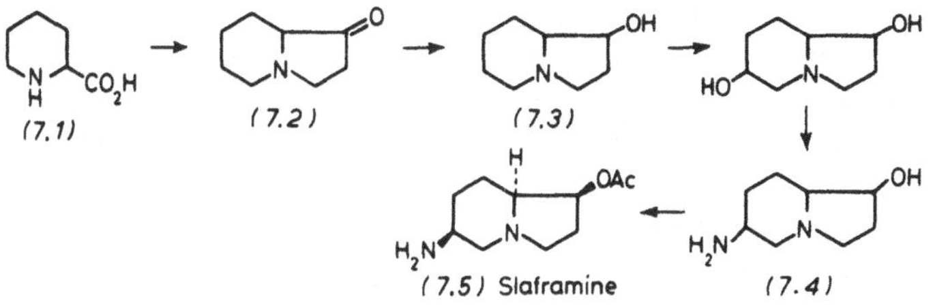


**Scheme 7.1**

dependent transformation of (*7.2*) into (*7.3*), and also the conversion of (*7.4*) into slaframine (*7.5*) in the presence of acetyl-CoA. Lysine is a precursor but curiously, in contrast to the evidence on pipecolic acid mentioned above (see also section 6.2.1), the L-isomer is preferred over its antipode [2].

Two possible primary metabolic pathways lead to lysine in different organisms [1]. An intermediate in one of these routes is dihydrodipicolinic acid (*7.6*). In *Bacillus megaterium* this is the point at which diversion to a secondary product, dipicolinic acid (*7.7*), may occur [3]. In the unrelated organism *Penicillium citreo-viride*, dipicolinic acid appears to be formed by diversion from the other route to lysine [4].

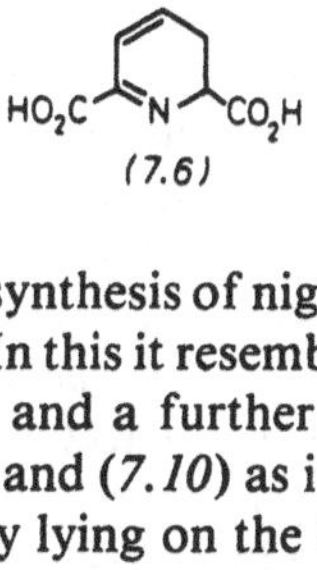

(7.6)

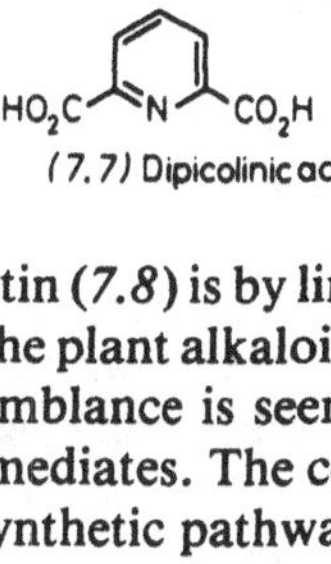

(7.7) Dipicolinic acid

By contrast, the biosynthesis of nigrifactin (*7.8*) is by linear combination of six acetate units. In this it resembles the plant alkaloid coniine (section 6.2.1; Scheme 6.5), and a further resemblance is seen in the apparent implication of (*7.9*) and (*7.10*) as intermediates. The compounds (*7.11*) and (*7.12*), plausibly lying on the biosynthetic pathway between (*7.10*) and nigrifactin (*7.8*), have also been found to be precursors for (*7.8*) [5].

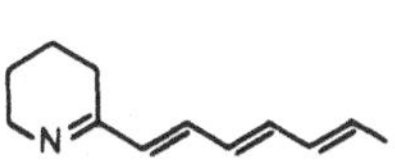

(7.8) Nigrifactin

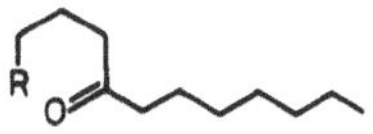

(7.9) R = $CO_2H$
(7.10) R = CHO

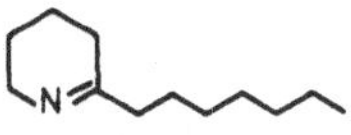

(7.11)

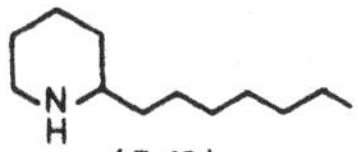

(7.12)

Three pyridine metabolites, fusaric acid (*7.13*), proferrorosamine A (*7.16*) and tenellin (*7.19*), have been found to have disparate origins.

(7.13) Fusaric acid        (7.14)        (7.15)        (7.16) Proferrorosamine A

(7.17)

Fusaric acid (*7.13*) derives from acetate (side-chain plus C-5 and C-6) and aspartic acid (*7.14*) (C-2, C-3, and C-4 plus the carboxy-group), and its origins thus resemble the biosynthesis of nicotinic acid from precursors other than tryptophan (section 6.2.2; Scheme 6.18) [6]. The picolonic part [as (*7.15*)] of proferrorosamine A (*7.16*) originates in lysine (and glutamic acid), being formed via picolinic acid (*7.15*) [7]. Ring A of caerulomycin A (*7.17*) is formed from lysine and (*7.16*) may be an intermediate in the biosynthesis of this *Streptomyces* metabolite; C-2 of (*7.17*) originates in C-1 of lysine and C-3 and C-4 have their origins in C-2 and C-1 of acetate, respectively, with the origin of the remainder being obscure (the *O*-methyl group is labelled by [3-$^{14}$C]serine [by way of methionine]) [8].

The biosynthesis of the fungal metabolite tenellin (*7.19*), has been studied in some detail [9, 10]. Notable use of $^{13}$C-labelled precursors gave results which establish the origins of this metabolite as those shown in (*7.19*); the intact acetate units indicated follow in the usual way (section 2.2.2) from experiments with [1,2-$^{13}$C$_2$]acetate. It is to be noted that the branched side chain originates from acetate plus methyl groups from methionine, rather than from the alternative, propionate, which is implicated in the biosynthesis of pyrindicin (*7.20*) [11], a *Streptomyces*

(7.18) Phenylalanine        (7.19) Tenellin        Methionine (■)

CH$_3$—CO$_2$H

CH$_3$—CH$_2$—CO$_2$H

(7.20) Pyrindicin

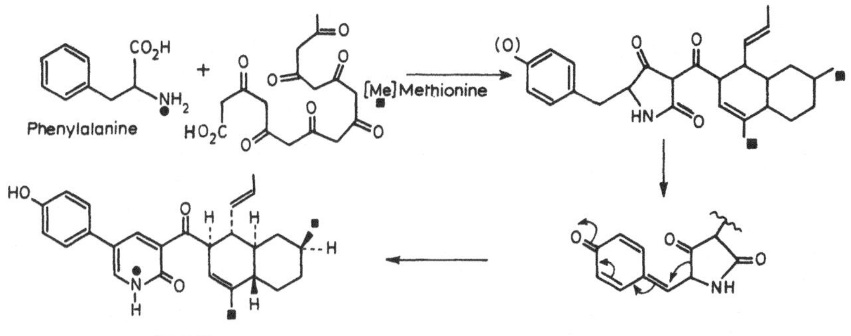

(7.21)

**Scheme 7.2**

metabolite. Incorporation of phenylalanine (7.18) into (7.19) is with rearrangement of the side-chain (C-1 of the amino acid appears at C-4 of tenellin, and C-2 at C-6). This is the same reordering of atoms as seen in tropic acid biosynthesis (section 6.2.2; Scheme 6.20) and it has been shown to involve a similar intramolecular 1,2-shift of the phenylalanine carboxy-group. Ilicicolin H (7.21) has the same origins as tenellin, i.e. phenylalanine, acetate, and methionine. The carboxy carbon of phenylalanine also undergoes a similar rearrangement; a similar mechanism is proposed (Scheme 7.2) [12].

Phomazarin (7.22) has been shown by use of [$^{13}$C]-labelled precursors to be formed from a single polyketide chain of nine intact acetate units [see (7.22)] with C-15 as the (only) starter acetate (see section 3.2.1 for further discussion of 'starter' acetate). Nitrogen intercalates in this chain at some point in biosynthesis, and the substrate for this is suggested to be (7.23) [13]. Fredericamycin A has a more extensive polyketide framework; it is formed like phomazarin with inclusion of a nitrogen atom [14].

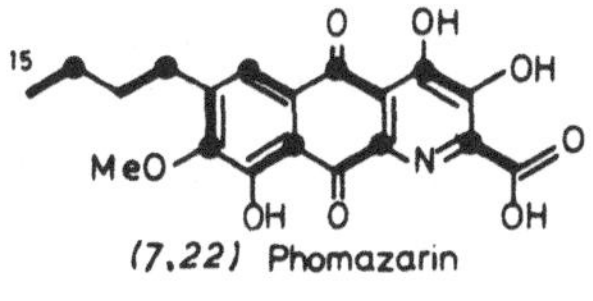

(7.22) Phomazarin

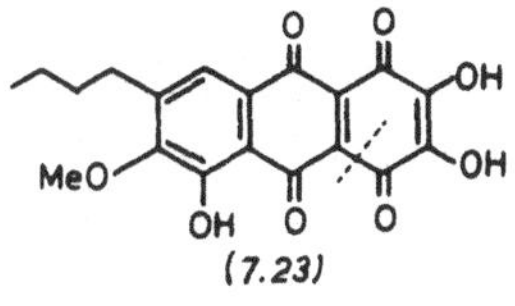

(7.23)

Cycloheximide (7.24) is biosynthesized from malonate (acetate) with three carbons in the glutarimide ring (either C-2, C-3, and C-4 or C-4, C-5, and C-6) being formed unusually and stereospecifically from an intact malonate unit. The two methyl groups (C-15 and -16) derive from methionine [15, 16].

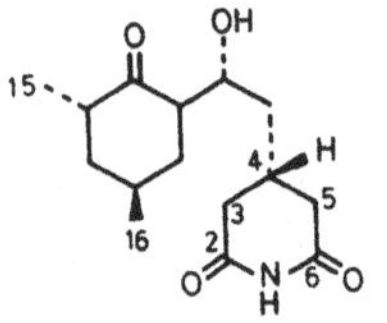

*(7.24)* Cycloheximide

## 7.3 DIKETOPIPERAZINES [17]

A variety of metabolites, isolated from fungi or actinomycetes, displays a diketopiperazine ring [as *(7.25)*] in modified form, e.g. echinulin *(7.27)*. (A related system is found in the benzodiazepines, see below.) The diketopiperazine metabolites can be seen as arising by combination of two α-amino acid residues through amide (peptide) linkages; in the case of echinulin these amino acids are tryptophan *(7.28)* and alanine *(7.29)*, and the diketopiperazine *(7.25)* is also a precursor for *(7.27)*.

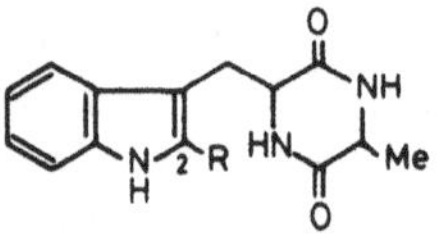

*(7.25)* R = H
*(7.26)* R = 

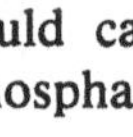

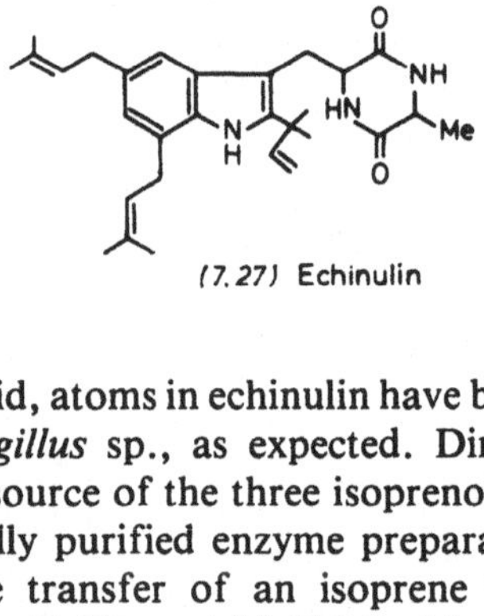

*(7.27)* Echinulin

The remaining, obviously isoprenoid, atoms in echinulin have been shown to arise from mevalonate in *Aspergillus* sp., as expected. Dimethylallyl pyrophosphate would follow as the source of the three isoprenoid residues and it has been found that a partially purified enzyme preparation from *A. amstelodami* would catalyse the transfer of an isoprene unit from dimethylallyl pyrophosphate *(7.30)* to *(7.25)* to yield *(7.26)*; this compound

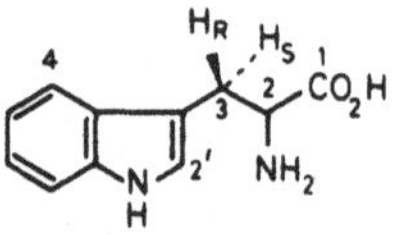

*(7.28)* Tryptophan

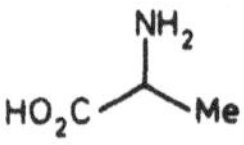

*(7.29)* Alanine

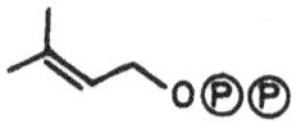

*(7.30)*

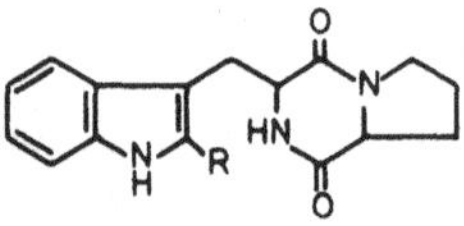

*(7.31)* R = H
*(7.32)* R = 

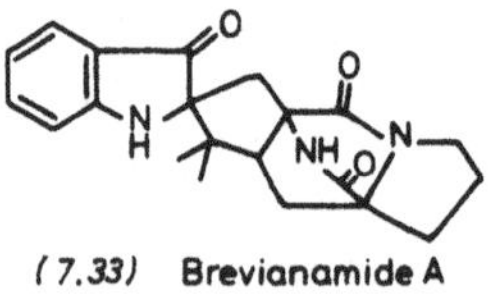

*(7.33)* Brevianamide A

was then found to be an intact precursor for echinulin (*7.27*). Prenylation of C-2 thus occurs before prenylation of the other two sites in echinulin biosynthesis [18, 19].

A similar pathway is deduced to lead to brevianamide A (*7.33*), in *Penicillium brevicompactum*, from tryptophan (*7.28*) and proline through the diketopiperazine (*7.31*), which has been isolated from *P. brevicompactum*. A further intermediate may be the *Aspergillus ustus* metabolite (*7.32*) (cf echinulin biosynthesis) [20].

Mycelianamide (*7.35*) is a diketopiperazine formed from tyrosine (*7.34*) rather than tryptophan. However, the appropriate diketopiperazines, *cyclo*-(L-alanyl-L-tyrosyl) and *cyclo*-(L-alanyl-D-tyrosyl), with radioactive labels, failed to act as precursors, suggesting possibly that the diketopiperazine when formed remains enzyme bound until after further reaction, and no equilibration occurs with unbound (radioactive) diketopiperazine [21].

(*7.34*) Tyrosine  (*7.35*) Mycelianamide

The desaturation of the tyrosine residue in mycelianamide (*7.35*) has been shown, with tyrosine samples chirally tritiated at C-3, to involve *cis* removal of hydrogen from L-tyrosine [22]. A (*Z*)-double bond similar to that in (*7.35*) is present in an echinulin (*7.27*) relative, and the desaturation proceeds again in the same stereochemical sense [18, 19].

Pulcheriminic acid (*7.36*) is, like mycelianamide, an *N,N*-dihydroxydiketopiperazine. It has been concluded that it derives by hydroxylation of the diketopiperazine (*7.37*) formed from two molecules of leucine [23].

(*7.36*) R = OH
(*7.37*) R = H

Other simple diketopiperazines examined are rhodotorulic and dimeric acids [24, 25], and roquefortine (*7.38*) [26]. Deuteriated tryptophan gave roquefortine (*7.38*) in which deuterium at C-2' [≡ C-5a in (*7.38*)] was lost. This indicates that the C$_5$ isoprene unit at C-10b of (*7.38*) may have arrived at this site via C-5a, with (*7.39*) as a possible intermediate [cf echinulin (*7.27*) above].

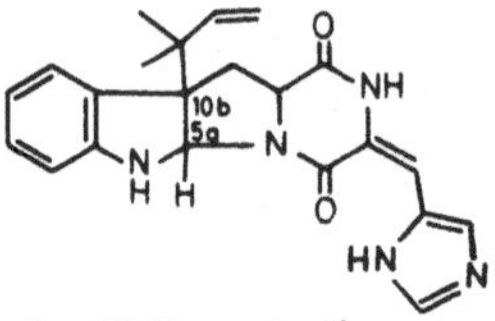

(7.38) Roquefortine

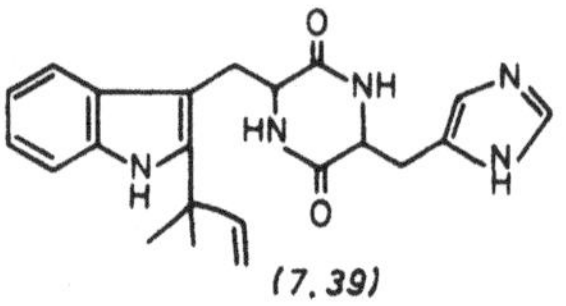

(7.39)

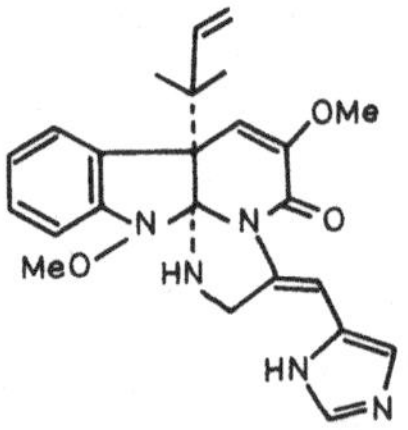

(7.40) Oxaline

Roquefortine (*7.38*) is a precursor for the more complex metabolite, oxaline (*7.40*). The double bond associated with the histidine unit in (*7.38*) and (*7.40*) [compare (*7.39*) and (*7.38*)] has been shown to be introduced with loss of two hydrogen atoms in a *syn* stereochemical sense as for mycelianamide (*7.35*) [27].

Gliotoxin (*7.44*) is unusual amongst diketopiperazine metabolites in having a disulphide bridge. Its biosynthesis though proceeds through a diketopiperazine [*cyclo*-(L-phenylalanyl-L-seryl) (*7.41*)]; the precursor amino acids are serine and phenylalanine (*7.18*). [Ar-$^2$H$_5$]phenylalanine was incorporated into gliotoxin with retention of all five deuterium atoms. This excludes aromatic hydroxy-intermediates, and biosynthesis has very reasonably been suggested to proceed via the arene oxide (*7.43*), as shown in Scheme 7.3 [28, 29]. A similar pathway leads to (*7.45*), ring expansion from an intermediate benzenoid ring being plausibly associated with another arene oxide (*7.46*). (For further discussion of arene oxides, see section 1.3.2)

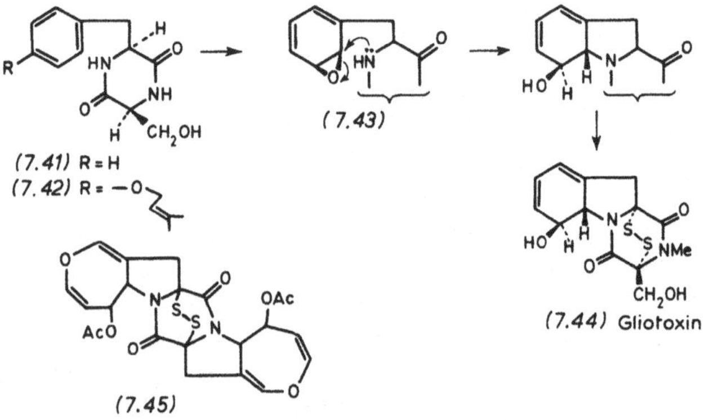

(7.41) R = H
(7.42) R = —O⟍

(7.43)

(7.44) Gliotoxin

(7.45)

**Scheme 7.3**

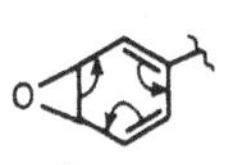

(7.46)

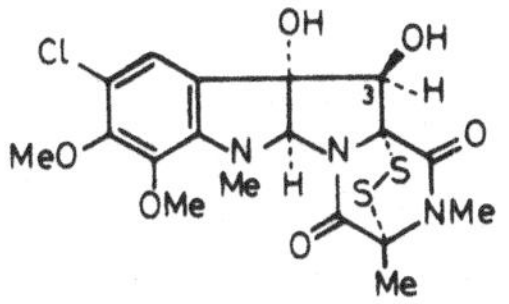

(7.47) Sporidesmin

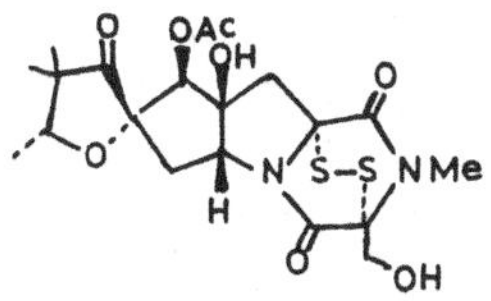

(7.48) Sirodesmin PL

Sporidesmin (*7.47*) is also a dithiodiketopiperazine. It derives from tryptophan and alanine; the hydroxy group at C-3 [≡ C-3 in tryptophan (*7.28*)] enters with normal retention of configuration [30]. An intriguing problem relates to the entry of sulphur into this and other metabolites above: the mechanism is unknown.

Sirodesmin PL (*7.48*) is formed in a related way to gliotoxin, but with extensive skeletal rearrangement, from tyrosine, serine and mevalonate via the diketopiperazine (*7.42*) [31].

## 7.4 BENZODIAZEPINES

The benzodiazepines, e.g. cyclopenin (*7.52*), are, formally at least related to the diketopiperazines in being formed from two amino acids. For cyclopenin (*7.52*) and cyclopenol (*7.53*), in *Penicillium cyclopium*, they are anthranilic acid (*7.49*) and phenylalanine (*7.18*). The first detectable intermediates are cyclopeptine (*7.50*) and dehydrocyclopeptine (*7.51*), which are interconvertible; desaturation proceeds in a *syn* sense (Scheme 7.4) (cf section 7.3). Earlier intermediates apparently only exist enzyme-bound, not being equilibrated with fed precursors.

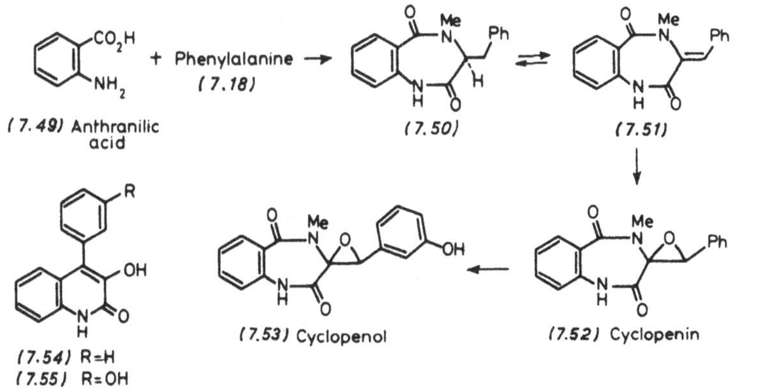

(7.49) Anthranilic acid

(7.18)

(7.50)

(7.51)

(7.54) R=H
(7.55) R=OH

(7.53) Cyclopenol

(7.52) Cyclopenin

**Scheme 7.4**

The epoxide and aromatic oxygens arise from molecular oxygen, and an enzyme which will effect the unusual aromatic *meta*-hydroxylation of

cyclopenin (*7.52*) to give cyclopenol (*7.53*) has been isolated from
*P. cyclopium*. It shows typical mixed-function oxidase properties, requiring
both oxygen and a hydrogen donating cofactor.

Also found in *P. cyclopium* cultures are viridicatin (*7.54*) and viridicatol
(*7.55*). They are enzymically derived from cyclopenin (*7.52*) and cyclopenol
(*7.53*), respectively, and the peculiar rearrangement is catalysed in a similar
way by base or acid [22, 32].

Although anthramycin (*7.56*) can be seen to contain a modified anthranilic
acid moiety (3-hydroxy-4-methyl-anthranilic acid; proved to have the
expected origins in tryptophan, cf actinomycin biosynthesis, below) the
origin of the remaining atoms ($C_8N$ unit) is not obvious. Results have been
obtained which show, however, that this $C_8$ unit originates in tyrosine [and
dopa (*7.59*)]. Incorporation of tyrosine (*7.34*) was with retention of seven of
the nine carbon atoms and the eighth carbon (C-14) derived from the methyl
group of methionine.

*(7.56)* Anthramycin

Sibiromycin (*7.58*) and 11-demethyltomaymycin (*7.57*) have a similar
biogenesis to anthramycin. The 1,11a double bond in (*7.58*) is introduced
with loss of the 3-*pro-S* proton from tyrosine (*7.34*).

*(7.57)* Demethyl tromamycin

*(7.58)* Sibiromycin

The results of experiments with $^2$H- and $^{13}$C-labelled precursors define the
ways in which the hygric acid moieties (*7.60*) and (*7.61*) seen in lincomycin A
(*7.62*) and B (*7.63*) are formed (Scheme 7.5). The tyrosine derived moieties
of (*7.56*) through (*7.58*) would plausibly derive by this common pathway
[33, 34]. The amino-octose portion of the lincomycins has a fairly complex
origin in sugar metabolism [35].

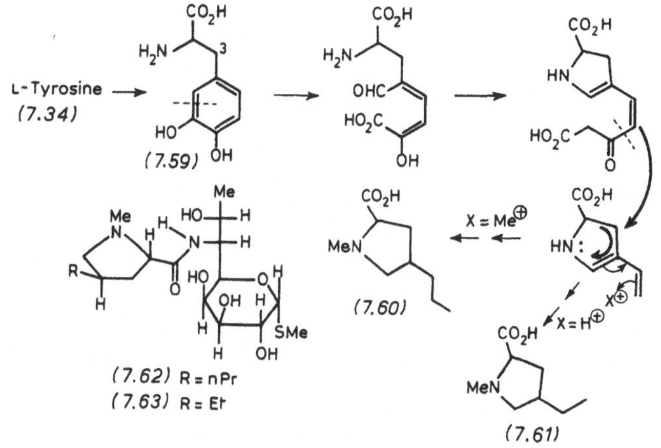

**Scheme 7.5**

## 7.5 METABOLITES DERIVED FROM TRYPTOPHAN PATHWAY

This section is concerned with diverse metabolites based on tryptophan (some also appear above). A large group is constituted by the ergot alkaloids.

### 7.5.1 Ergot alkaloids [36]

Ergot alkaloids, e.g. elymoclavine (*7.67*), are produced by a fungus, *Claviceps purpurea*, which infects rye and other grasses. They have been extensively studied and there are several fascinating aspects of their biosynthesis.

An impressive body of evidence has established that these 'alkaloids'* are formed from L-tryptophan (*7.28*) which condenses at C-4' with dimethylallyl pyrophosphate (*7.30*), derived in the usual way from (3-*R*)-mevalonic acid, to give dimethylallyltryptophan (*7.64*) [37]; this is then modified to give, e.g. elymoclavine (*7.67*). The main biosynthetic pathway is (*7.64*) → chanoclavine-I (*7.65*) → agroclavine (*7.66*) → elymoclavine (*7.67*) → lysergic acid (*7.68*) [36].

Most interestingly it has been found that the conversion of (*7.64*) through (*7.65*) into (*7.66*) and (*7.67*) involves two enigmatic changes in the ordering of the methyl groups about the double bond in (*7.64*) [38] (trace the labels ▲ and ● through the various compounds, Scheme 7.6); and it is logical to associate these isomerizations with the actual ring-closure reactions. The mechanism for closing ring C [formation of (*7.65*)] is unknown and the timing of the necessary allylic hydroxylation is uncertain. But the formation

---

*The term alkaloid only properly applies to bases isolated from plants.

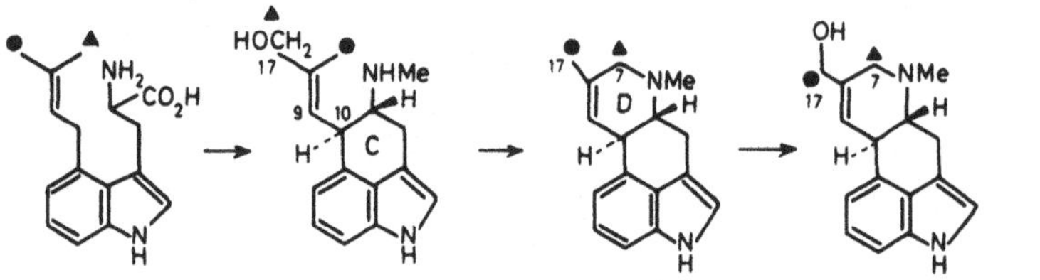

(7.64)        (7.65) Chanoclavine-I    (7.66) Agroclavine    (7.67) Elymoclavine

**Scheme 7.6**

of ring D and associated reactions can be accounted for satisfactorily [39]. In the conversion of chanoclavine-I (7.65) into elymoclavine (7.67) one of the C-17 hydrogens is lost; the other appears at C-7 as expected. This suggested that the aldehyde (7.69) might be an intermediate. This received strong support when radioactive (7.69) was shown to be a precursor for agroclavine (7.66) and elymoclavine (7.67); a tritium label from C-17 in the precursor appeared at C-7 in (7.67). Moreover, two deuterium atoms from a sample of mevalonic acid, labelled at C-3' were retained on formation of chanoclavine-I (C-3' of mevalonate becomes C-17 of chanoclavine-I); only one of these was retained into elymoclavine (7.67), consistent with the intermediacy of the aldehyde (7.69). Further, the retained deuterium atom (at C-7) was largely equatorial in (7.67) indicating proton addition to (7.69) would be stereospecific.

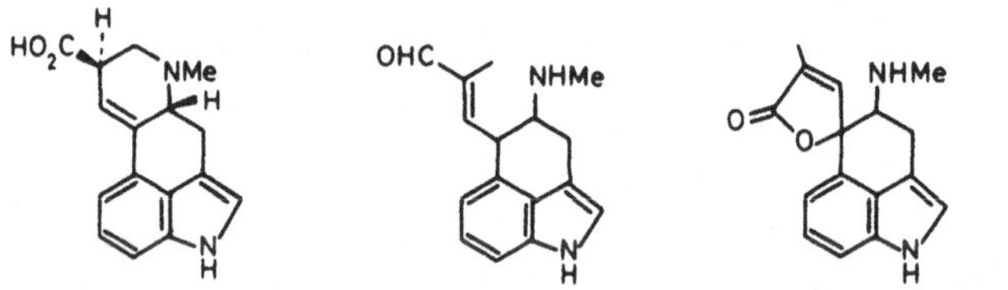

(7.68) Lysergic acid        (7.69)        (7.70) Rugulovasine

In considering the ring-closure of (7.65) to give (7.66) and the apparently connected double-bond isomerization, knowledge of the fate of the proton at C-9 in (7.65) is important. It was found to be incompletely retained ($\sim 70\%$) during biosynthesis (that at C-10 is completely retained). In an experiment with a 1:1 mixture of [2-$^{13}$C]- and [4-$^2$H$_2$]-mevalonate, the derived (7.65) showed the expected mixture of singly labelled species, but the elymoclavine (7.67) showed an appreciable percentage of double-labelled molecules, i.e. molecules which contained both $^{13}$C and deuterium. These results and the required double-bond isomerization are neatly accounted for in terms of the mechanism shown in Scheme 7.7 [39] in which there is an *inter*-molecular shift of the C-9 proton (hence the formation of doubly labelled elymoclavine). In the proposed sequence the aldehyde group is masked to allow

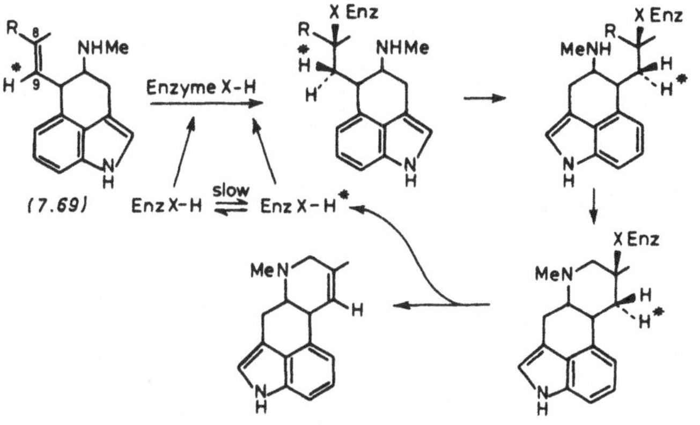

**Scheme 7.7**

anti-Michael addition of Enz-X-H to the double bond. Rotation around the 8,9-bond allows transfer of the original C-9 hydrogen to the enzyme in the subsequent elimination step. Assuming slow exchange of the hydrogen attached to the enzyme, most of it would be transferred to a molecule of (7.69) in the next cycle [because of dilution in the culture by unlabelled species, (7.69) might or might not be labelled].

Rugulovasine (7.70) and related *Penicillium* metabolites are biosynthesized via dimethylallytryptophan (7.64) but not chanoclavine-I (7.65) [40]. Of considerable interest is the finding that ergot alkaloids are elaborated by higher plants, also from tryptophan and mevalonic acid [41].

Lysergic acid (7.68), formed from elymoclavine (7.67), is a precursor for a range of peptidic ergot bases, e.g. ergotamine (7.71). Although the expected amino acids are incorporated into the complex peptidic alkaloids, attempts to obtain incorporation of larger fragments have met with little success and it is currently believed that assembly of the alkaloids from lysergic acid and the relevant amino acids occurs without desorption from an enzyme surface at any point or equilibration with exogenous (fed) material [42, 43].

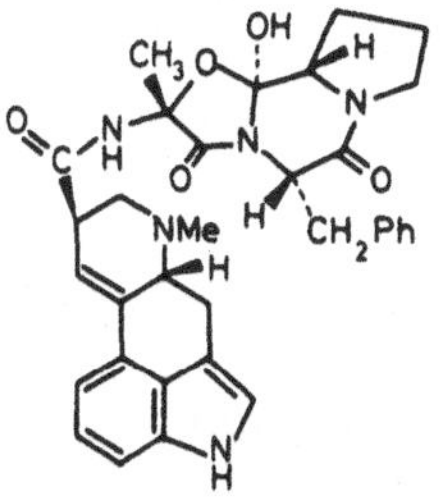

(7.71) Ergotamine

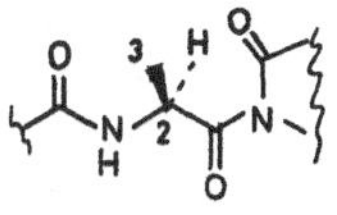

(7.72)

The biosynthesis of the cyclol unit in (*7.71*) involves the transformation of an alanine moiety [as (*7.72*)] to a α-hydroxy-α-amino acid fragment [as (*7.71*)]. This process does not affect the protons on C-3 of (*7.72*). The oxygen atom derives from molecular oxygen and hydroxylation which must occur directly on C-2 of (*7.72*) proceeds with normal retention of stereochemistry [44].

### 7.5.2 Cyclopiazonic acids and carbazomycin B

α-Cyclopiazonic acid (*7.75*) is formed in *Penicillium cyclopium* from tryptophan, mevalonate, and an acetate-derived C₄ unit [heavy bonding in (*7.75*)] by way of β-cyclopiazonic acid (*7.74*). An earlier intermediate is (*7.73*). [The ergot precursor, dimethylallyltryptophan (*7.64*) was not incorporated.]

(7.73)          (7.74)          (7.75) α-Cyclopiazonic acid

(7.76)

Tritium label at C-3 of tryptophan (*7.28*) is retained as far as (*7.74*), and subsequent conversion into (*7.75*) involves stereospecific removal of the tryptophan (3-*pro-S*)-proton. It follows from this, and the stereochemistry of (*7.75*), that C—C bond formation within (*7.74*) takes place from the opposite side of the molecule to proton removal. The retention of label from [2-$^3$H]tryptophan [label at C-5 of (*7.75*)] indicates that bond formation to C-4 cannot involve a 4,5-dehydrointermediate, but an intermediate with a double bond 1,4 is still possible [45].

The pattern of acetate labelling in the dimethylallyl unit in (*7.74*) and (modified) in (*7.75*) established that in the ring closure of (*7.74*) to give (*7.75*), addition to the double bond of the dimethylallyl unit in (*7.74*) occurs in a *syn* sense [46].

Carbazomycin B (*7.76*) derives from tryptophan (*7.28*) [C-2 appears as C-3 of (*7.76*) and the carboxy-group is lost], methionine and acetate; the origins of C-2 and C-11 are unknown [47].

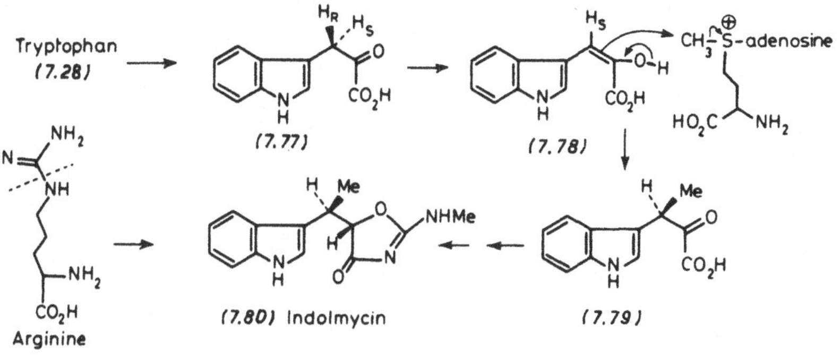

**Scheme 7.8**

### 7.5.3 Indolmycin

Indolmycin (*7.80*), a *Streptomyces griseus* metabolite, is derived as shown in Scheme 7.8. Of particular interest is the *C*-methylation at what is C-3 in tryptophan (*7.28*). Mechanistically indolepyruvic acid (*7.77*) is an attractive substrate for methylation via the enol (*7.78*) and enzymes have been isolated which will catalyse (*7.28*) → (*7.77*) and (*7.78*) → (*7.79*) (Scheme 7.8); removal of a tryptophan C-3 proton is stereospecific (*pro-R* proton) which indicates a stereospecific enolization reaction. Transfer of a methyl group from *S*-adenosyl-methionine to (*7.78*) occurs with inversion of configuration [48].

### 7.5.4 Streptonigrin and pyrrolnitrin

Streptonigrin (*7.81*) is a *Streptomyces flocculus* metabolite. It has been shown to have its genesis in part from tryptophan (via 3-methyltryptophan) (rings C and D). A notable point is the clever use of $^{13}C$—$^{15}N$ coupling: observation of this coupling in (*7.81*) derived from [2'-$^{13}C$, $^{15}N_{1'}$,]tryptophan indicates both atoms are retained and that the necessary cleavage of tryptophan occurs along the dotted line as shown. It is to be noted that this cleavage differs from the equally unique one in pyrrolnitrin (*7.87*) (below).

The use of [U-$^{13}C$]glucose and the labelling pattern (see section 2.2.2 for discussion of the method) which could be seen in the derived streptonigrin pointed to the solution of a difficult problem, namely where rings A and B originated (they were not labelled by phenylalanine, tyrosine, or shikimic acid) [49]. The intact glucose units are shown in (*7.81*). The pattern for rings C and D are those expected for metabolism through tryptophan and overall three erythrose (C₄) units could be discerned; labelled D-erythrose (*7.82*) was appropriately incorporated.

Further information came from the incorporation of $^{18}O_2$ into streptonigrin [labelling = * in (*7.81*); the labelling of the C-8 oxygen is uncertain]. In particular this indicated that in the formation of rings C and D of

(7.81) Streptonigrin

(7.87) Pyrrolnitrin          (7.86)

(7.81) tryptophan underwent oxidative ring opening; (7.83) is a plausible intermediate.

The evidence pointed to the previously unknown metabolites 4-amino-anthranilic acid (7.84) and 7-aminoquinoline-2-carboxylic acid (7.85) as possible precursors for streptonigrin. In the event they were found to be efficiently and specifically incorporated into (7.81) [50].

4-Aminoanthranilic acid is manifestly derived from the shikimate pathway (Chapter 5) but since shikimic acid is not a precursor the point of diversion from the pathway must be prior to shikimic acid (see section 7.6 for related results).

Pyrrolnitrin (7.87), a *Pseudomonas aureofaciens* metabolite, is formed from tryptophan with loss only of the carboxy-group in a biosynthetic sequence which obviously involves fracture of the indole nucleus. The proposed route is indicated by the results of feeding experiments and in part by the natural occurrence of (7.86) [51, 52].

(7.82)          (7.83)          (7.84)          (7.85)

## 7.6 METABOLITES DERIVED FROM THE SHIKIMATE PATHWAY

Tryptophan and other aromatic amino acids are products of the shikimate pathway (Chapter 5). Metabolites which are biosynthesized from tryptophan are discussed above, as are some which originate from phenylalanine and tyrosine. In the discussion which follows in this section other metabolites which are formed through these latter common amino acids are included.

In recent biosynthetic studies, and most interestingly, previously unknown aromatic amino acids, which are formed via the shikimate pathway, have been discovered to be key intermediates in the biosynthesis of some microbial metabolites. The biosynthesis of these metabolites is also discussed below. The intermediacy of an unusual amino acid, i.e. (*7.84*), in the biosynthesis of streptonigrin (*7.81*) has already been alluded to (section 7.5.4.)

### 7.6.1 Pseudans, phenoxazinones, phenazines, and chloramphenicol

A *Pseudomonas* sp. elaborates the pseudans, which are a group of simple quinoline derivatives with aliphatic side chains of varying length ($C_7$—$C_{12}$), e.g. pseudan-VII (*7.88*). The skeleton is accounted for by anthranilic acid (*7.49*), acetate and malonate, labelling by acetate and malonate being consistent with that expected for a fatty acid unit. Since 4-hydroxy-2-quinolone, a plant quinoline alkaloid precursor was not incorporated it follows that (*7.88*) is assembled from anthranilic acid and the complete fatty acid, which its incorporation establishes as 3-oxodecanoic acid for pseudan-VII [53].

(*7.88*) Pseudan - VII     (*7.89*) R = H     (*7.91*) Xanthommatin
                                          (*7.90*) R = OH

The insect ommochromes, e.g. xanthommatin (*7.91*), derive from the products of tryptophan catabolism, kynurenine (*7.89*) and hydroxy-kynurenine (*7.90*) [54] as do the *Streptomyces antibioticus* phenox-azinones, the actinomycins (*7.93*). A further intermediate in the biosynthesis of (*7.93*) is 3-hydroxy-4-methylanthranilic acid (*7.92*) (cf anthramycin biosynthesis, section 7.4), and aromatic methylation occurs on 3-hydroxykynurenine (*7.90*). Oxidative dimerization of (*7.92*) using a crude enzyme preparation from *S. antibioticus* gave actinocinin (*7.94*) [55].

It might be expected from the foregoing that bacterial phenazines, e.g.

(7.92)

(7.93) R = peptide, Actinomycin
(7.94) R = OH

phenazine-1,6-dicarboxylic acid (*7.98*) and phenazine-1-carboxylic acid (*7.99*), would derive from two molecules of anthranilic acid (*7.49*), but it has been proved that not only is this amino acid not involved, but also, using mutants, that biosynthesis along the pathway to aromatic amino acids gets no further than chorismic acid (*7.96*) (section 5.1). An earlier

(7.95) Shikimic acid    (7.96) Chorismic acid    (7.97)    (7.98)

intermediate on this pathway, shikimic acid (*7.95*), is, unlike chorismic acid, assimilated into phenazines when fed to bacterial cultures. It has been used to prove the orientation of precursor units in the phenazines. In particular, [2-$^2$H]shikimic acid [as (*7.95*)] gave iodinin (*7.101*), some molecules of which were dilabelled. This proved that two molecules of shikimic acid were involved in phenazine biosynthesis. These deuterium atoms were, moreover, located at C-2 and C-7, which established, in conjunction with $^{14}$C data, that the orientation of shikimic acid units is as shown in (*7.97*). This leads to phenazine-1,6-dicarboxylic acid (*7.98*) as the first phenazine to be formed and it has been shown to be a phenazine precursor [56, 57].

The nitrogen of the amide function of the amino acid L-glutamine is the primary source of the nitrogen atoms of iodinin (*7.101*) in *Brevibacterium iodinum* and of phenazine-1-carboxylic acid (*7.99*). *B. iodinum* also produces 2-aminophenoxazinone (*7.103*) and a close relationship between the biosynthesis of (*7.103*) and (*7.101*) is apparent.

Glutamine is the source of the amino group in anthranilic acid (an

(7.99) R = H
(7.100) R = OH

(7.101)

(7.102)

(7.103)          (7.104)

intermediate involved in the shikimate pathway prior to tryptophan) and
(7.104) is an intermediate in the biosynthesis of this amino acid. It seems
probable that (7.104) (but not anthranilic acid, see above) is an intermediate in the biosynthesis of phenazines and of 2-aminophenoxazinone
[58] (contrast the biosynthesis of actinomycin above).

Common reactions found in phenazine biosynthesis are hydroxylation, as in iodinin (7.101) biosynthesis, and this may be associated with
concomitant decarboxylation through (7.102) (solid arrows); iodinin
(7.101) is apparently formed in this way from (7.105) [59]. The latter
compound does not derive from phenazine-1-carboxylic acid (7.99) but
has an independent genesis presumably from phenazine-1,6-dicarboxylic
acid (7.98) again via a hydroxylative decarboxylation. It is proved using
deuterium-labelled precursors that pyocyanin (7.106) derives in a similar
way from (7.99) via (7.107); enzyme-catalysed nucleophilic substitution
of (7.107) by ammonia gives aeruginosin A (7.108). Opening of (7.102)
derived from (7.99) without decarboxylation (broken arrows) gives
2-hydroxyphenazine-1-carboxylic acid (7.100) [60].

Shikimic acid (7.95), rather than aromatic amino acids, is involved in
chloramphenicol (7.109) biosynthesis via chorismic acid (7.96) and
(7.110) [61].

(7.105)          (7.106) Pyocyanin          (7.107)

(7.110)          (7.109) Chloramphenicol          (7.108)

## 7.6.2 Ansamycins, mitomycins and antibiotic A23187

There is a common C₇N unit (heavy bonding) within the skeletons of the
ansamycins, e.g. rifamycin B (7.111), and the mitomycins, e.g. porfiromycin

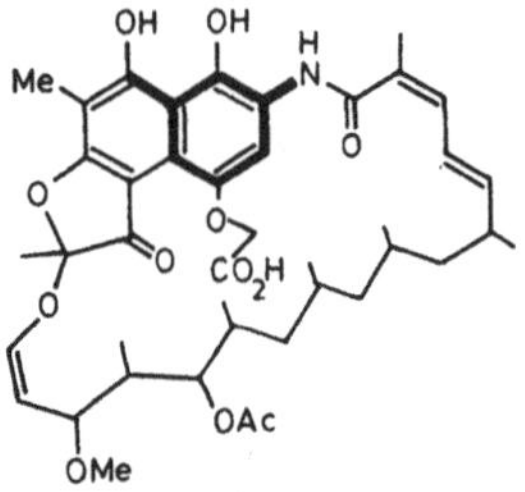

(7.111) Rifamycin B

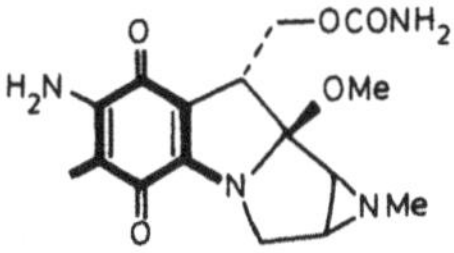

(7.112) Portiromycin

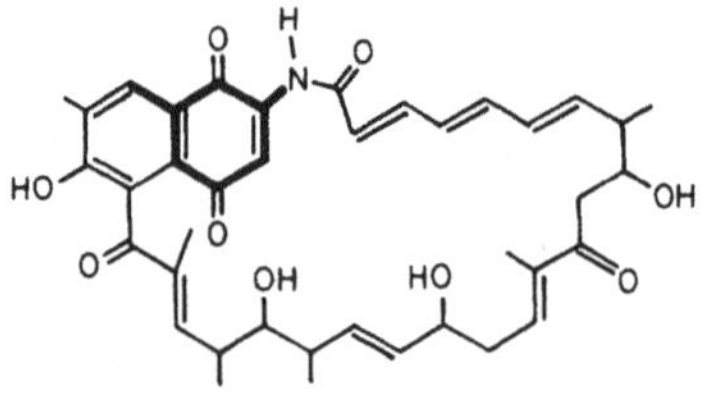

(7.113) Actamycin

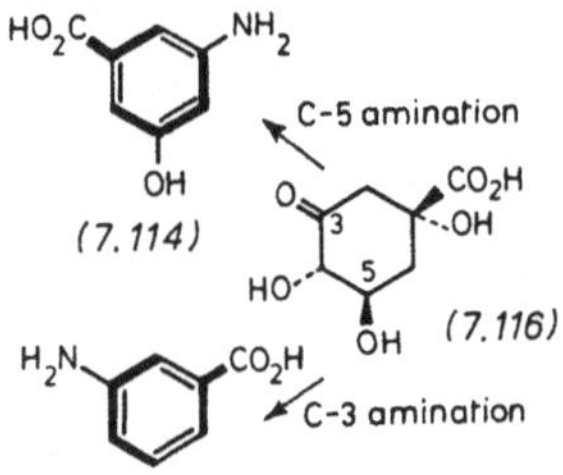

(7.114)

(7.115)

(7.116)

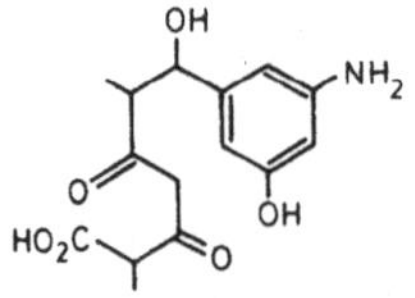

(7.117)

(7.112). The unit does not arise from shikimic acid (7.95) itself but is a product of the shikimate pathway, diverting from it between 3-deoxy-D-arabino-heptulosonic acid 7-phosphate and shikimic acid (Chapter 5). Initiated by inspiration (for the origin of the $C_7N$ unit was great puzzle) impressive and interlocking evidence has been obtained, chiefly with labelled precursors and mutants, which establishes that the key intermediate in the biosynthesis of rifamycin B (7.111), actamycin (7.113), ansamitocin P-3 (7.129), ansatrienin (7.127) and porfiromycin (7.112) is 3-amino-5-hydroxybenzoic acid (7.114). A further intermediate in rifamycin and actamycin biosynthesis is (7.117) [62–64].

The remaining atoms in the mitomycins are derived from citrulline (7.119)/arginine (7.120) and D-glucosamine (7.121): [1-$^{13}$C, $^{15}$N]glucosamine gave mitomycin B (7.118), the mass spectrum of which showed a fragment ion for C-1, C-2, C-3, and N-1a plus attached methyl group, with both stable isotopes present, indicating intact incorporation [65].

The pattern of precursor units in geldanomycin (7.123) follows clearly from experiments with [$^{13}$C]-labelled precursors. A sequence is indicated of acetate, propionate (not acetate + methyl from methionine) and glycerate units from C-14 through C-1 [66].

[U-$^{13}$C]Glucose was used in a notable study on the origins of the $C_7N$ unit in geldanomycin (7.123). The intact $C_3$ (pyruvate; C-15, -16, and -21) and $C_4$

(7.119) X = O
(7.120) X = NH

(7.121)

(7.118) Mitomycin B

(7.122) Pactamycin

[Me] Methionine

(erythrose phosphate; C-17, -18, -19, and -20) units arising from passage through the shikimate pathway could be discerned as those shown in (7.114) [67].

The origins of pactamycin are those illustrated in (7.122); the origin of both atoms of the C-ethyl group in methionine is to be noted. Interestingly the $C_7N$ unit in (7.122) is labelled differently to that of geldanomycin (7.123), i.e. (7.115) and (7.114), respectively. Thus notably in the course of biosynthesis the site of amination of a hypothetical dehydroquinate intermediate (7.116) from the shikimate pathway is different for the two compounds. The *m*-aminoacetophenone ($C_7N$) portion of (7.122) is formed via *m*-aminobenzoic acid (7.115), a compound which is, appropriately, not involved in the biosynthesis of geldanomycin [68].

Mixed acetate-propionate pathways (but not including glycerate in the chain) have been deduced for rifamycin S (7.124) and streptovaricin D (7.125). It seems likely that the streptovaricins and rifamycins derive from a common intermediate [69, 70]. The pattern deduced [see (7.126)] for rifamycin S (7.124) has two unusual features: loss of the propionate-derived methyl group from C-28 and insertion of an oxygen between C-12 and C-29 of the precursor. Isolation of rifamycin W with unmodified skeleton, its biosynthesis according to pattern (7.126) and its ready biotransformation into rifamycin B (7.111) confirms the rifamycin S pattern and its being a modification of the more orthodox arrangement in (7.126).

The origins of ansatrienin (7.127) are those indicated. Additional features are that the D-alanine residue is derived directly from D-alanine rather than the L-isomer and the components of the cyclohexylcarbonylalanine side chain are added one at a time. Importantly the cyclohexylcarbonyl unit was

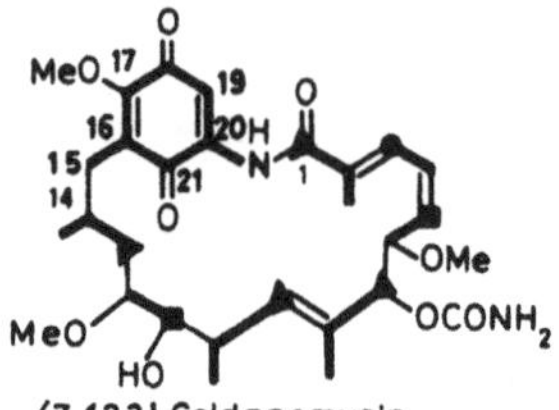

(7.123) Geldanomycin

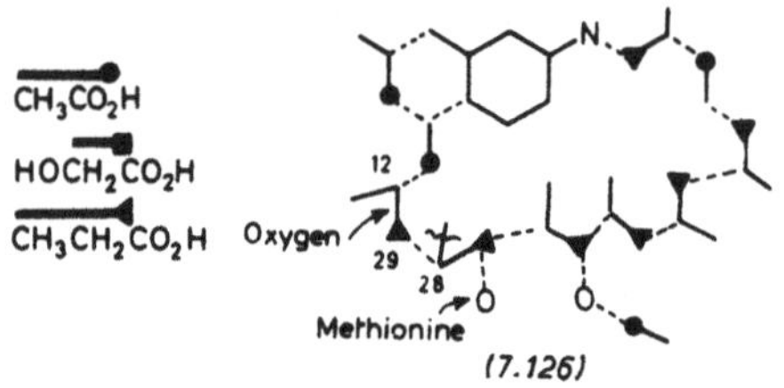

(7.126)

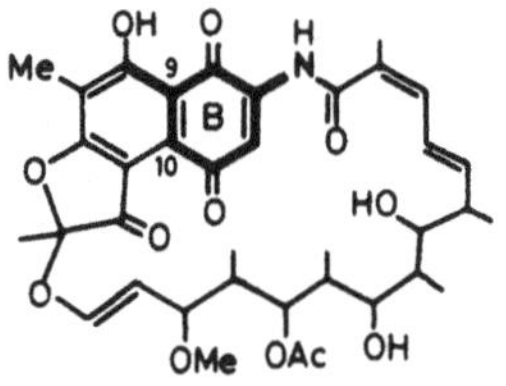

(7.124) Rifamycin S

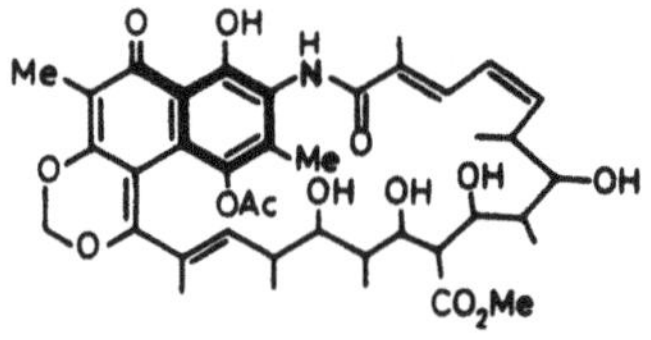

(7.125) Streptovaricin D

labelled by radioactive shikimate but the $C_7N$ unit [which was proved to arise from 3-amino-5-hydroxybenzoic acid (7.128)] was not. This confirms that (7.128) diverts from the shikimate pathway prior to shikimic acid (7.95) itself. The worry was that previous negative results with the metabolites discussed above could have been the result of permeability problems.

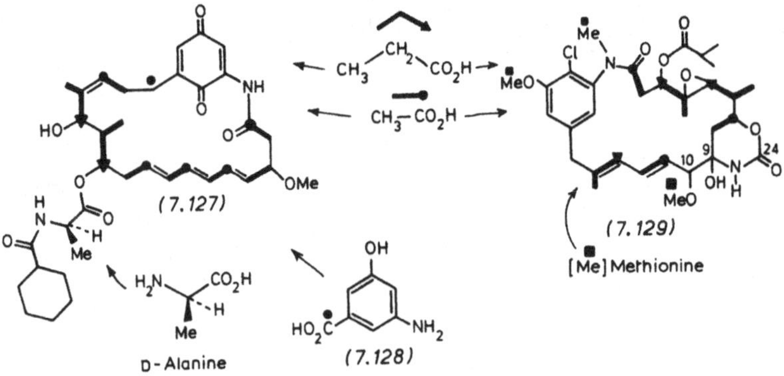

(7.127)

(7.129)

[Me] Methionine

D-Alanine     (7.128)

The seven atoms of the cyclohexylcarbonyl fragment in (7.127) are biosynthesized from the seven carbon atoms of shikimic acid (7.95) (but not benzoic acid or 2,5-dihydrophenylalanine). Cyclohexanecarboxylic acid and probably its 2,5-dihydro derivative are to be included as intermediates [71].

Ansamitocin P-3 (7.129) has similar origins to the other ansamycins, as illustrated. Interestingly C-9 and C-10 derive not from acetate but from glucose and C-24 derives from the amino acid citrulline (7.119) (cf geldanomycin above) [72].

There are similarities in the building blocks for antibiotic A23187 (7.130) and those discussed above, chiefly the acetate-propionate entities. Proline provides one end of the molecule (condensation occurs before dehydrogenation to yield the pyrrole ring) and another unusual aromatic amino acid provides the other. The [U-$^{13}$C]glucose labelling pattern is illustrated with heavy bonding (cf metabolites above). The lack of symmetry in this pattern excludes a symmetrical intermediate such as 2,6-diaminobenzoic acid [73].

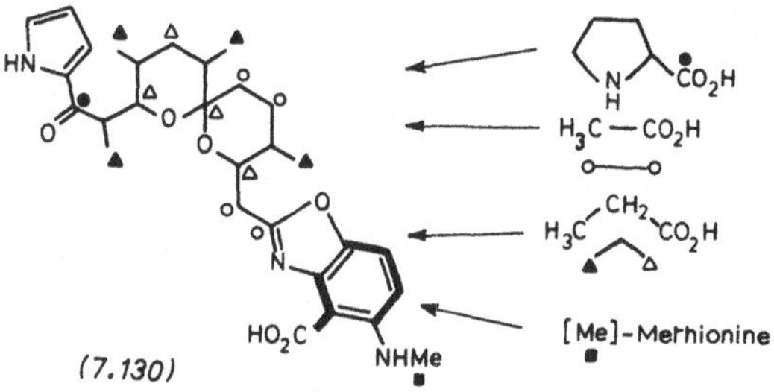

(7.130)

### 7.6.3 Cytochalasins and pseurotin A

The biologically interesting cytochalasins have an origin in phenyl-alanine (7.18) and a $C_{16}$ polyketide, as for example in cytochalasin D (7.132), or a $C_{18}$ polyketide, as for example in cytochalasin B (7.133). The chain in (7.133) is apparently fragmented by the insertion of an oxygen atom into the polyketide chain. This was proved to be so by showing that deoxaphomin (7.131) was a precursor for (7.133). It is to be

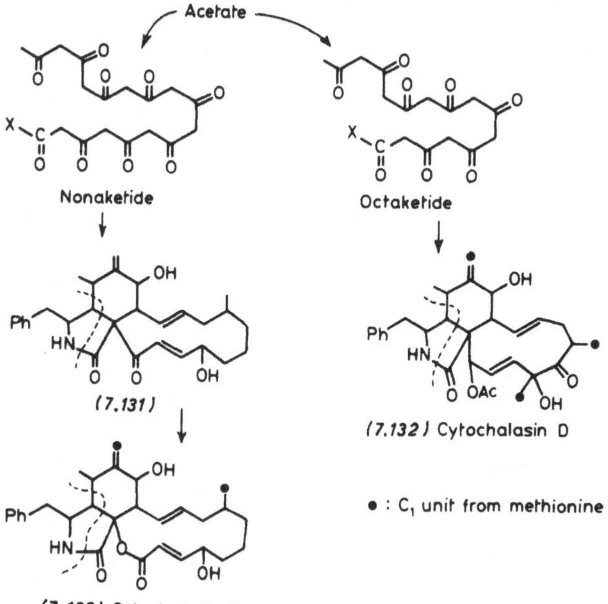

Nonaketide

Octaketide

(7.131)

(7.132) Cytochalasin D

(7.133) Cytochalasin B

● : $C_1$ unit from methionine

noted that, unlike some of the metabolites mentioned above, methionine plus acetate, rather than propionate, serve as precursors for the $C_3$ units [74]. A closely related metabolite is chaetoglobosin A (*Chaetomium*

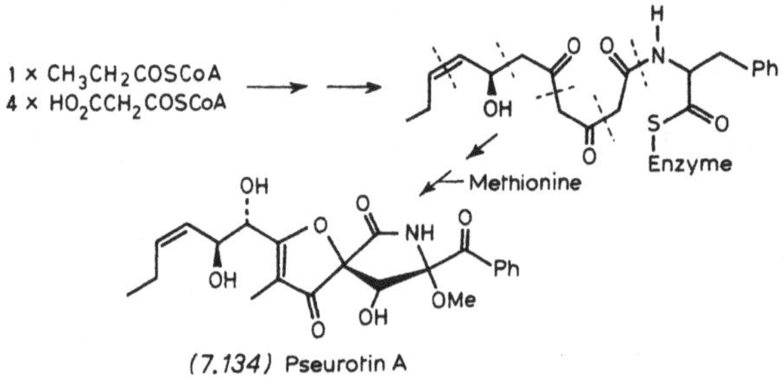

$1 \times CH_3CH_2COSCoA$
$4 \times HO_2CCH_2COSCoA$

Methionine

(7.134) Pseurotin A

**Scheme 7.9**

*globosum*); it has a related biosynthesis in which tryptophan replaces phenylalanine [75].

The biosynthesis of pseurotin A (*7.134*) (*Pseudorotium ovalis*) also follows a biosynthetic path involving an aromatic amino acid plus a methylated polyketide, but in addition one unit of propionate is also involved (as a starter unit) (Scheme 7.9) [76].

### 7.6.4 Nybomycin

The biosynthesis of nybomycin (*7.135*) has been revealed particularly by use of [1-$^{13}$C]acetate: C-4, C-6, C-8 and C-10 were found to be labelled and each to an equal extent. The exterior atoms of each pyridone ring therefore are formed from identical acetate-derived $C_4$ units ([2-$^{14}$C]acetate could be shown to label C-6′ and C-8′). Methionine is the source of C-2 and C-11′. The origin of the central ring is possibly shikimate or a close relative [77].

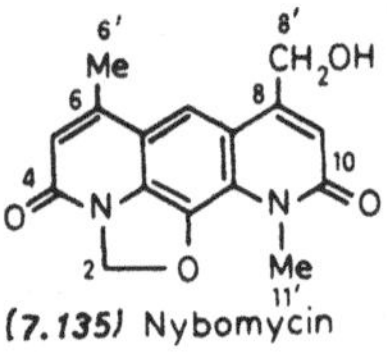

(7.135) Nybomycin

### 7.6.5 Naphthyridinomycin, saframycin A and CC-1065

The biosynthetic origins of naphthyridinomycin (*7.137*) have been mapped (Scheme 7.10) in experiments with $^{13}$C- and $^{14}$C-labelled precursors. C-1 and C-2 of the antibiotic derive, respectively, from C-2 and C-3 of serine. The previously unknown amino acid (*7.136*) is an intermediate formed first by C-methylation of tyrosine then hydroxylation [78].

Within (*7.137*) there is a $C_5$ unit which derives via the amino acid ornithine; the origins only of C-9 and C-9′ remain unknown.

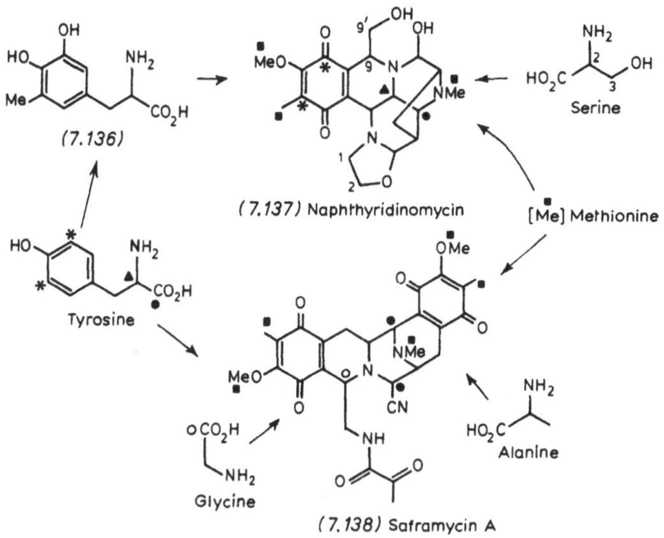

(7.137) Naphthyridinomycin

[Me] Methionine

Tyrosine

Glycine

(7.138) Saframycin A

Alanine

**Scheme 7.10**

As illustrated in Scheme 7.10 the biogenesis of saframycin A (*7.138*) resembles that of naphthyridinomycin in the implication of tyrosine and methionine [79]. These two amino acids together with serine account for the building blocks in the complex antibiotic CC-1065 (*7.139*) [80].

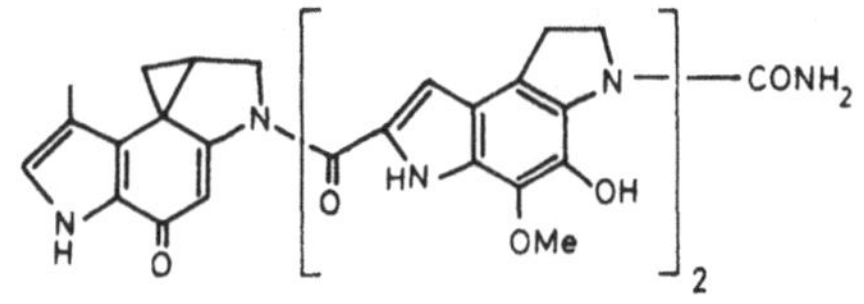

(*7.139*)

### 7.6.6 Isocyanides and tuberin

The isocyanide group is part of the structures of a small number of meta-bolites produced by micro-organisms on the one hand and marine sponges on the other. The microbial isonitrile (*7.141*) is biosynthesized from tyrosine (*7.140*) which undergoes aromatic ring scission leading to the cyclopentane ring of (*7.141*) [81, 82]. The hazimycins 5 (*R,R* + *S,S*) and 6 (*RS*) (*7.142*) are also formed as might be expected from tyrosine [83], two molecules of which are also the source of most of the skeleton of xanthocillin X (*7.143*) [84]. Results of experiments with $^{15}$N-labelled compounds show that the two nitrogen atoms of the isocyano groups in xanthocillin X monomethyl ether (*7.144*) derive from the amino group of tyrosine. The origin of the isonitrile carbon atoms remains unknown:

(7.140)    (7.141)

(7.142) Hazimycins    (7.143) R = H  Xanthocillin X
(7.144) R = Me

(7.145) Tuberin    (7.146) R = $H_R$
(7.147) R = OH

obvious sources of $C_1$ units in metabolism, notably $C_1$-tetrahydrofolate intermediates (cf section 1.3.3), have been excluded in experiments with labelled formate, serine, glycine, and methionine; in each case, only the O-methyl group of (7.144) was labelled [85]. However, some evidence was obtained that the isocyanide carbon atoms of the hazimycins (7.142) originate from the methyl group of methionine [83].

The structure of tuberin (7.145) resembles that of (7.144), and again the major part of the skeleton originates in tyrosine (7.146). In this case, however, not only the O-methyl group but also the N-formyl group (which might on chemical analogy be seen as a precursor, by dehydration, for an isocyano group) is biosynthesized through $C_1$-tetrahydrofolate intermediates (in section 1.3.3 further results involving tuberin are discussed) [86].

The way in which the double bonds in (7.144) and tuberin (7.145) are formed has been probed. Double bond formation in neither case involves C-3 hydroxylation of tyrosine or derivative; *threo*-3-hydroxytyrosine (7.147), the product of tyrosine hydroxylation, undergoes degradation by an aldol-type reaction to give p-hydroxybenzaldehyde and glycine [87].

In the case of tuberin the formation of the *trans*-double bond involves loss of the 3-*pro-R* proton in L-tyrosine (7.146) and concomitantly the carboxyl group (the C-2 proton is retained). The elimination thus occurs in a formal antiperiplanar sense. In the case of the xanthocillin (7.144) the formation of the *cis*-double bonds involves the loss of the 3-*pro-S* proton of tyrosine (7.146) [87].

Whilst the isocyanide functions in the microbial xanthocillin (7.144) do not originate in cyanide ion, those in the sponge isocyanide

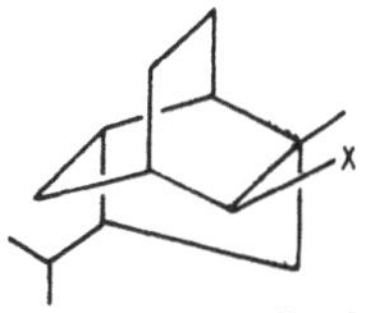

(7.149) X = N≡C
(7.150) X = N=C=S
(7.151) X = NHCHO

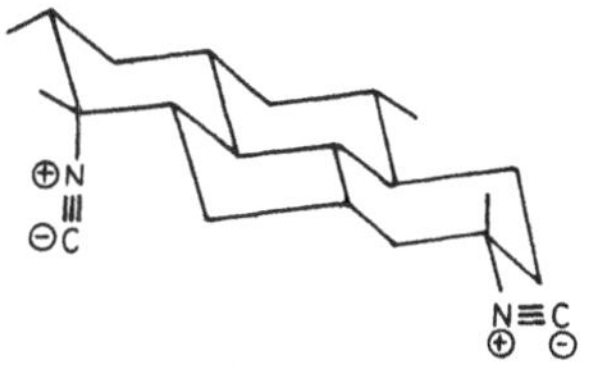

(7.148) Di-isocyanoadociane

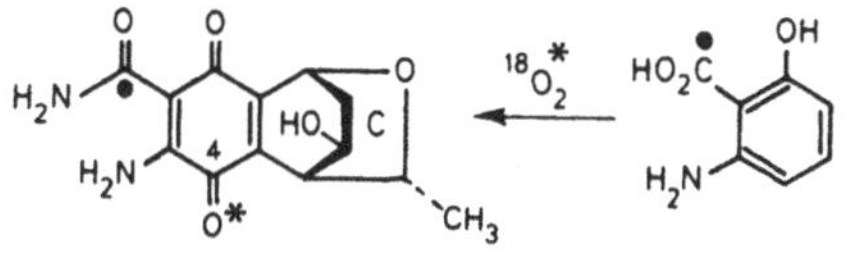

(7.152) Sarubicin A          (7.153)

di-isocyanoadociane (*7.148*) do indeed derive from cyanide ion [88]. In addition to isocyanides [as (*7.149*)] sponges also elaborate the corresponding isothiocyanates [as (*7.150*)] and *N*-formyl compounds [as (*7.151*)]. These latter metabolites are formed from the corresponding isocyanide and not the other way round [89].

### 7.6.7 Sarubicin A

Glucose is a direct precursor of the tetrahydropyran portion (C) of sarubicin A (*7.152*). The remainder of the skeleton is formed via the shikimate pathway (Chapter 5) and the unusual amino acid (*7.153*) which was previously unknown as a product of this pathway is implicated as an intermediate. (For other unusual aromatic amino acids see sections 7.5.4 and 7.6.2.) Since only the quinone oxygen at C-4 in (*7.152*) was labelled by $^{18}O_2$ the other quinone oxygen must be carried over from the shikimate pathway [90].

### 7.6.8 Arphamenines

The building blocks for arphamenine A (*7.154*; R = H) and B (*7.154*; R = OH) are arginine, acetate and phenylalanine (for A) or tyrosine (for B). It follows from the phenylalanine-tyrosine results that (*7.154*; R = H) is not a precursor for (*7.154*; R = OH). A detailed picture (Scheme 7.11) of the biosynthetic route to arphamenine A (*7.154*; R = H) has emerged beautifully from experiments with a cell-free preparation of the organism which produces the arphamenines [91].

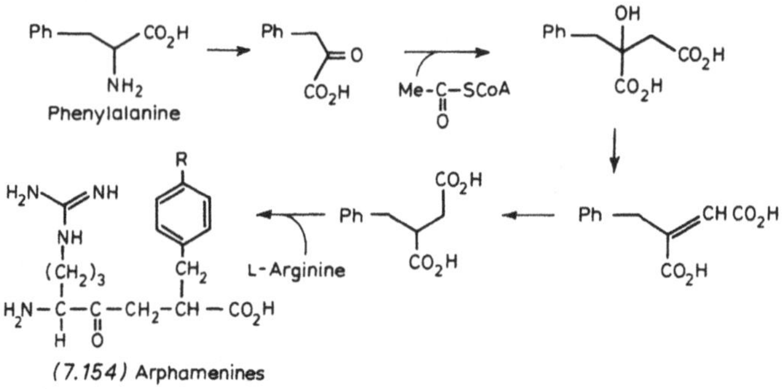

(7.154) Arphamenines

**Scheme 7.11**

## 7.7 β-LACTAMS

### 7.7.1 Penicillins and cephalosporins [92, 93]

The pre-eminence of penicillins and cephalosporins as clinically important antibiotics has spawned a prodigious amount of research into the biosynthesis of these antibiotics. The result of these labours is a clear and detailed picture of the course of biosynthesis and some definition of substrate acceptability to the key enzymes which are involved. The penicillins and cephalosporins are closely related and as we shall see the cephalosporins are biosynthesized through a common tripeptide (7.158), isopenicillin N (7.160) and penicillin N (7.162).

The penicillins [e.g. isopenicillin N (7.160)] and cephalosporins [e.g. cephalosporin C (7.165)] are elaborated from three amino acids: L-valine (7.157), L-serine (7.156) and L-α-aminoadipic acid (7.155) (Scheme 7.12) (which is constructed from 2-oxoglutarate plus acetyl coenzyme A [94]) in cultures of e.g. *Cephalosporium acremonium*, *Penicillium chrysogenum* and *Streptomyces clavuligerus*. These three amino acids are condensed together to give the pivotal tripeptide intermediate (LLD-ACV; 7.158). An enzyme, δ-(L-α-aminoadipyl)-L-cysteinyl-D-valine synthetase, has been identified in cell-free preparations of *C. acremonium* which catalyses this condensation in the presence of $Mg^{2+}$ and ATP [95]; no dipeptide is formed as an intermediate. During the course of the biosynthesis of LLD-ACV, the L-valine undergoes a curious inversion of configuration; the nitrogen atom is retained while the C-2 proton is lost and D-valine is not a substrate for the enzyme.

The LLD-ACV (7.158) undergoes a double cyclization to yield isopenicillin N (7.160) as the first penicillin to be formed. Using cell-free preparations of *C. acremonium* (eucaryote) and *S. clavuligerus* (procaryote) it has been clearly shown that the LLD-tripeptide (7.158), and this isomer only, is an intact penicillin precursor. Notably the tripeptide (7.159), labelled with $^{13}$C as shown, gave isopenicillin N (7.161), labelled as shown, on incubation with

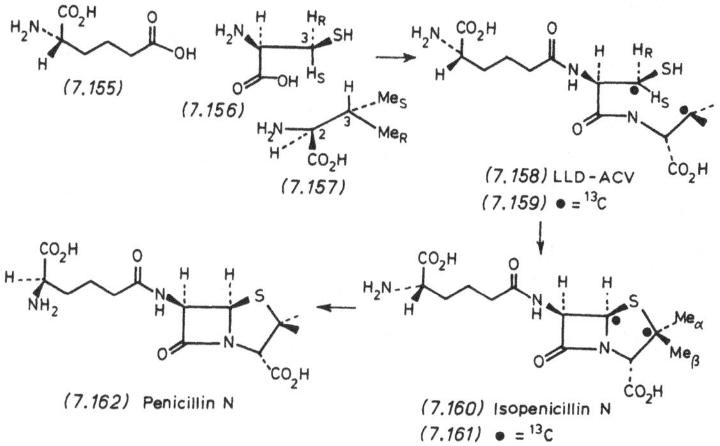

**Scheme 7.12**

a cell-free preparation of *C. acremonium*. The experiment was conducted in the probe of an n.m.r. spectrometer and no intermediate formed by the first of the two ring-closures could be detected [96–98]. Other evidence below indicates which ring closes first.

The conversion of the tripeptide (*7.158*) into isopenicillin N (*7.160*) is catalysed by isopenicillin N synthetase (IPNS) in the presence of ascorbic acid, ferrous ion and oxygen (the stoichiometry of the reaction is appropriate to the loss of four hydrogen atoms). The IPNS from several sources has been purified; IPNS genes from *C. acremonium* and *P. chrysogenum* have been cloned in the bacterium *Escherichia coli*. The protein-coding region of the genes from both sources are more than 74% homologous. Both genes code for two cysteine residues in exactly analogous regions; these sulphur-bearing residues may be involved in catalysis and/or iron binding in association with a histidine residue [99].

The cloned IPNS from *C. acremonium* has been purified. It displayed catalytic properties identical with those of fungally produced IPNS even though the cloned enzyme bears an additional glycine residue at the *N*-terminous of the enzyme [100].

In the course of the conversion of cysteine (*7.156*) into isopenicillin N (*7.160*) the 3-*pro-R* hydrogen is retained while the 3-*pro-S* hydrogen is lost [101] (for a discussion of prochirality see section 1.2.1). The two methyl groups in valine are diastereotopic. In the same way as the hydrogens on C-3 of cysteine are distinguished in the course of biosynthesis, so are the methyl groups in valine. Thus, the (*pro-R*)-methyl group becomes the *β*-methyl group of (*7.160*) and C-2 of cephalosporin C (*7.165*). These conversions of valine and cysteine are therefore stereospecific and the two ring closure reactions which lead to the penicillin skeleton both proceed with retention of stereochemistry. These results come from labelling experiments with the individual amino acids and the tripeptide (*7.158*). Finally, results of deuterium labelling experiments showed that penicillin formation occurred without proton loss from the methyl groups of valine and two out of the three protons on the (*pro-R*)-methyl group are retained at C-2 in cephalosporin biosynthesis [102–104].

The sequence of cephalosporin biosynthesis begins with the isomerization of isopenicillin N (*7.160*) to penicillin N (*7.162*), then ring expansion to give deacetoxycephalosporin C (DAOC; *7.163*). Hydroxylation (with normal retention of configuration [105, 106]) yields deacetylcephalosporin C (*7.164*). The oxygen present here derives from molecular oxygen as does the one present in cephamycin C. DAOC synthetase and DAOC hydroxylase, two of the enzymes involved here, require iron, molecular oxygen, ascorbate and $\alpha$-ketoglutarate for activity [107].

(*7.163*) R = H
(*7.164*) R = OH
(*7.165*) R = OAc, Cephalosporin C

A wide range of compounds based on the key tripeptide (*7.158*) have been examined as substrates for IPNS. To be effective these analogues, if modifications are within the $\alpha$-aminoadipyl moiety, need a six-carbon or equivalent structure terminating in a carboxyl group. (A similar requirement has been found for cephalosporin formation from (*7.162*) [108]). Wide variations are acceptable in the valine part of the molecule. Results here, where hydroxylative and desaturative ring-closures involving concomitant C—S bond formation are observed, point to the mechanism of ring-closure; it is hypothesized that in both the cyclization of ACV (*7.158*) and the ring-expansion of penicillin N (*7.162*) free radicals or equivalent, i.e. a very weak iron—carbon bond derived from an iron—oxo species, may be implicated. The chemical feasibility of a free radical process has been successfully modelled [109].

In the formation of isopenicillin N (*7.160*) from (*7.158*) which of the two rings shuts first? The most persuasive evidence comes from the following. When a 1:1 mixture of LLD-ACV (*7.158*) and LLD-A[3,3-$^2$H]CV (*7.166*) was

(*7.166*) X = $^2$H, Y = $^1$H
(*7.167*) X = $^1$H, Y = $^2$H

incubated with crude IPNS the protic substrate was preferentially converted into β-lactam product. In a similar experiment with a 1:1 mixture of (*7.158*) and LLD-AC[3-²H]V (*7.167*), no isotopic discrimination was observed. Both of the deuteriated tripeptides showed significant $V_{max}$ (maximum velocity) deuterium-isotope effects as measured in separate non-competitive experiments. In competitive mixed-label experiments isotopic discrimination is a $V_{max}/K_m$ effect ($K_m$ = Michaelis constant) and reflects events only up to the first irreversible step. From the above it can be seen that only one of the C—H bond cleavages shows a $V_{max}/K_m$ effect, i.e. cysteinyl C(3)-H cleavage. It follows therefore that this is the first chemical step in the reaction sequence, i.e. the β-lactam ring shuts first [110].

### 7.7.2 Clavulanic acid

Clavulanic acid (*7.169*) is constructed from a $C_3$ unit and a $C_5$ unit. The principal source for the latter unit is ornithine (*7.168*). C-5 of the amino acid becomes the hydroxymethyl group of (*7.169*). In this process the *5-pro-R* proton is retained whilst the *5-pro-S* proton is lost. Overall the process is one of inversion (Scheme 7.13) [111].

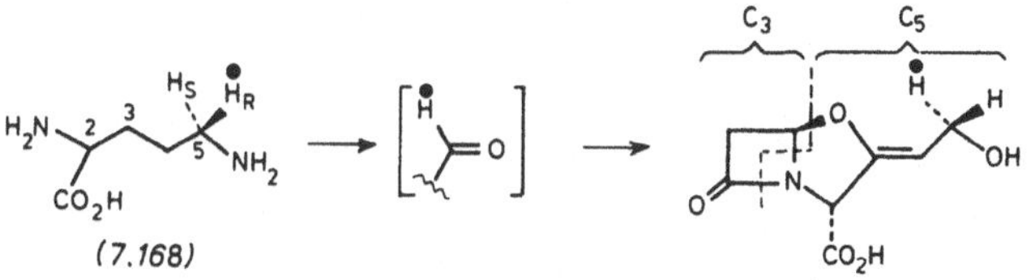

**Scheme 7.13**

D-glycerate (*7.170*) has been identified as a late intermediate in the biosynthesis of the $C_3$ unit in (*7.169*) [112]. 3-Hydroxypropionic acid is an alternative precursor [113], and glycerol may serve as a progenitor for both. Results of experiments with chirally tritiated samples of glycerol show that, as for β-lactam formation in penicillin biosynthesis, similar ring-closure occurs here with retention of configuration [114].

### 7.7.3 Nocardicins

Nocardicin A (*7.172*) is constructed from two molecules of tyrosine by way of L-(*p*-hydroxyphenyl)glycine (*7.173*), one molecule of L-serine (*7.171*) and one of methionine (homoserine was a less efficient precursor than methionine which indicates that it is sulphur rather than oxygen which is displaced in forming the ether link seen in nocardicin A). Both the nitrogen atoms in (*7.172*) appear to have their genesis from the amino group in (*7.173*).

C-3 of serine (*7.171*) becomes C-4 of the β-lactam of (*7.172*) and overall the hydroxy group at C-3 of the amino acid is displaced by nitrogen to form

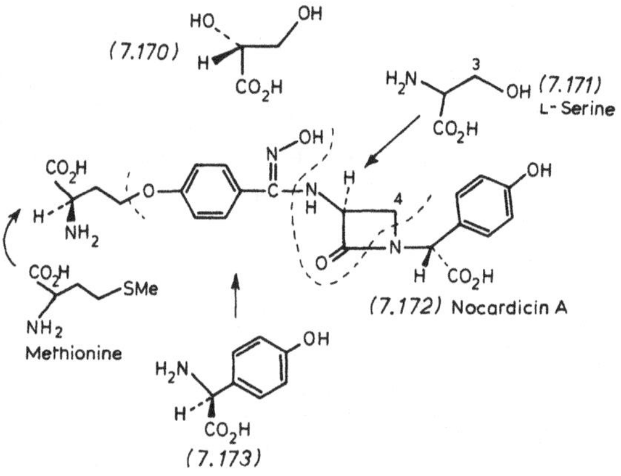

(7.170)

(7.171) L-Serine

(7.172) Nocardicin A

Methionine

(7.173)

the $\beta$-lactam ring. In this process serine is utilized without change of oxidation level at C-3 and displacement of the hydroxy-group occurs with inversion of configuration, i.e. a simple nucleophilic displacement by nitrogen of the (presumably) activated hydroxy group [115] (cf different results above for the biosynthesis of the $\beta$-lactam ring in the penicillins and clavulanic acid).

### 7.7.4 Thienamycin and tabtoxin

The origins of thienamycin (*7.174*) are as follows; C-6 and C-7 derive from C-2 and C-1 of acetate, with C-8 and C-9 originating in the methyl group of methionine. All of the protons present at C-8 and C-9 are carried over from the methyl group of methionine. The pyrrolidine ring has its origins in glutamate and cysteine is the source of the sulphur-bearing side chain.

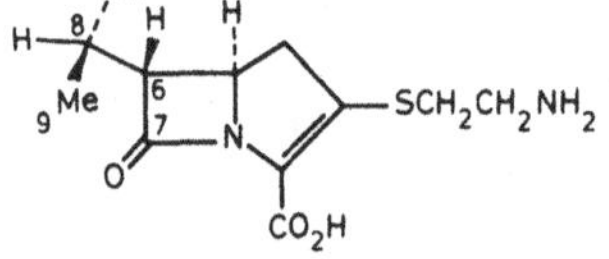

(7.174) Thienamycin

The stereochemical fate of the methyl group of methionine which becomes C-9 of (*7.174*) on attachment to C-8 was explored with methionine chirally labelled on the methyl group ($^1$H, $^2$H, $^3$H). It turns out that methylation of C-8 occurs with overall retention of configuration. Two steps each occurring with inversion seems likely and a methylated corrin intermediate is a possibility [116].

The building blocks for the $\beta$-lactam tabtoxin (*7.175*) were deduced using $^{13}$C-labelled precursors. The results are illustrated. The extensive oxidation of the methyl group of methionine is notable [117].

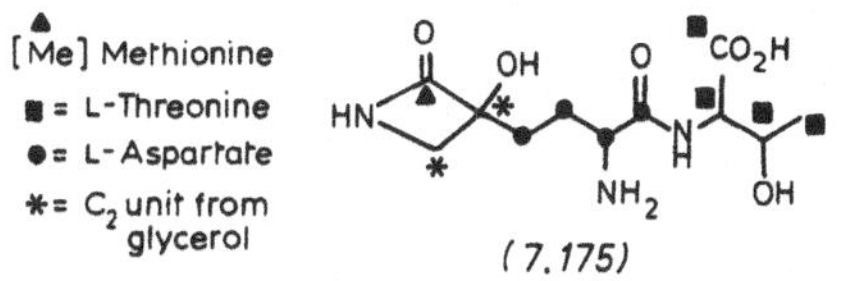

▲
[Me] Methionine
■ = L-Threonine
● = L-Aspartate
✱ = C$_2$ unit from glycerol

(7.175)

## 7.8 MISCELLANEOUS METABOLITES

### 7.8.1 Prodiginines

Particularly clear information on the biogenetic origins of prodigiosin (7.178) was obtained by use of [$^{13}$C]-labelled precursors. This tripyrrole, it turned out, is constituted in the unprecedented way shown (Scheme 7.14). Similar origins have been found for rings A and B of two other metabolites, (7.179) and (7.180). In ring C there are marked differences between (7.178) and the other two metabolites. Appropriately the origins are different (Scheme 7.14).

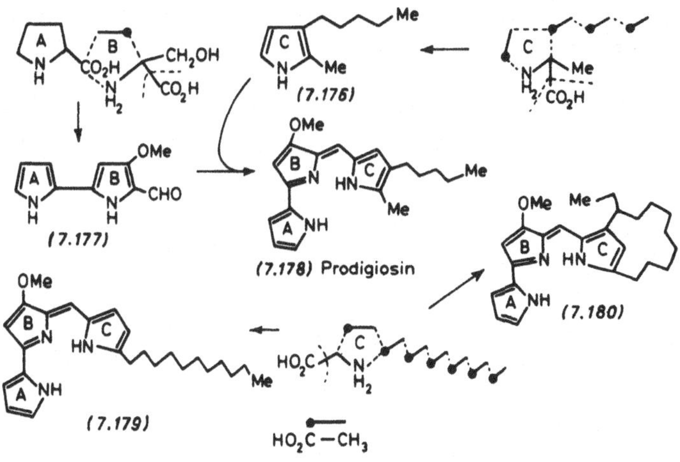

**Scheme 7.14**

The aldehyde (7.177) condenses with (7.176) in the ultimate step of prodigiosin (7.178) biosynthesis. It seems probable that (7.177) is a precursor for (7.179) and (7.180) also [118, 119].

### 7.8.2 Elaiomycin and valinimycin

Elaiomycin (7.181) and valinimycin (7.182) are unusual naturally occurring azoxy compounds. Carbons 5 to 12 and the β-nitrogen of (7.181)

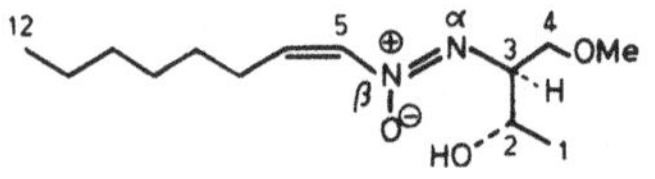

(7.181) Elaiomycin 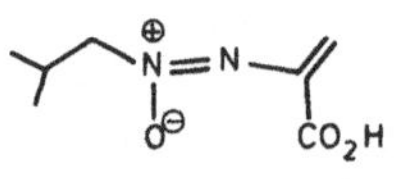     (7.182) Valinimycin

have their genesis through *n*-octylamine; introduction of the double bond involves the *syn* removal of two protons. Carbons 2 through 4 plus the α-nitrogen have their origin in serine (7.171); most remarkably the C-1 methyl group is derived from C-2 of acetate (see sections 7.8.4 and 7.8.5 for similar findings). Methionine labels only the *O*-methyl group [120].

The amino acids valine and alanine are the sources for the skeleton of valinimycin (7.182) [121].

### 7.8.3 Streptothricin, acivicin, reductiomycin and asukamycin

The pattern of labelling of streptothricin F (7.184) by [$^{13}$C$_2$]acetate provided the necessary clues to the origins of the metabolite's skeleton. These results pointed to metabolism by way of the tricarboxylic acid cycle through arginine (7.183) with the β-lysine moiety [as (7.186)] originating in lysine. Further detailed examination with arginine labelled with $^{13}$C, $^{15}$N and deuterium, nicely gave results which establish that this amino acid is an intact precursor for streptothricin and which indicate the pathway illustrated in Scheme 7.15. Loss of deuterium label from both C-2 and C-3 of arginine (7.183) on incorporation into (7.184) was observed [122].

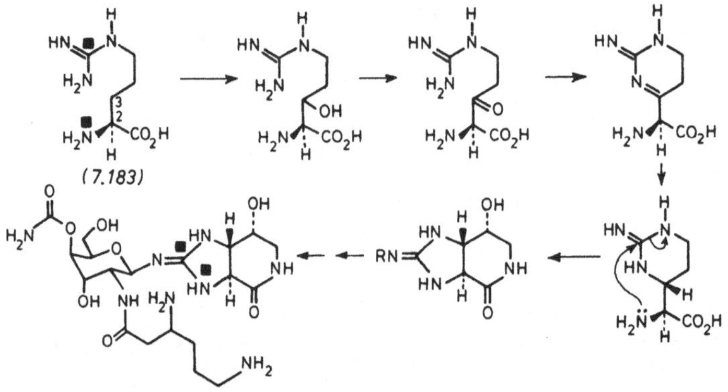

(7.184) Streptothricin F

**Scheme 7.15**

The β-lysine moiety (7.186) is formed from L-lysine (7.185) in a reaction catalysed by lysine-2,3-aminomutase. The 3-*pro-R* hydrogen of (7.185) is transferred stereospecifically to C-2. This occurs in an *inter-*

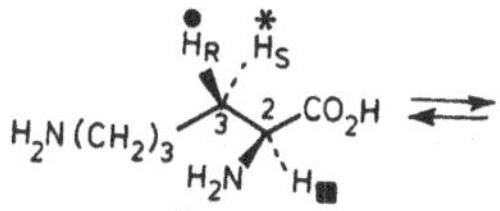

(7.187) Acivicin    (7.185) 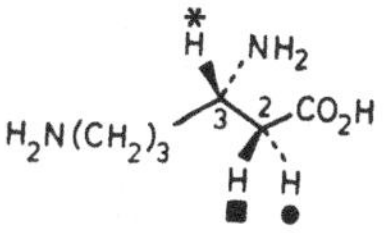    (7.186)

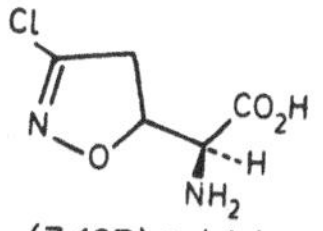

molecular sense while the nitrogen atom is transferred from C-2 to C-3 in an *intra*-molecular sense. Overall in this rearrangement both C-2 and C-3 suffer inversion of configuration [123].

Acivicin (*7.187*) with its unusual heterocyclic ring originates in ornithine (*7.168*). As for streptothricin (above) deuterium at C-2 of the amino acid was lost. Appropriately, half of a deuterium label at C-3 was also lost [124].

Both asukamycin and reductiomycin (*7.188*) bear a 2-amino-3-hydroxycyclopent-2-enone moiety which is formed through δ-amino-laevulinic acid (*7.189*). The remainder of the skeleton in (*7.188*) is biosynthesized from *p*-hydroxybenzoic acid which is elaborated directly from an intermediate in the shikimate pathway (i.e. not via an aromatic amino acid like phenylalanine) [125].

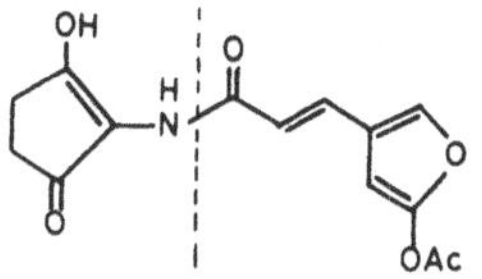

(7.188) Reductiomycin

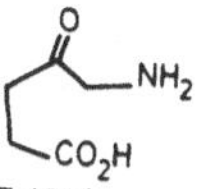

(7.189)

### 7.8.4 Myxopyronin A, myxothiazol, angiolam A, rhizoxin and malonimycin

The building blocks involved in the biosynthesis of myxopyronin A (*7.190*) are those illustrated. The origin of one of the methyl groups in C-2 of acetate is notable (cf section 7.8.2 and 7.8.5). The origin of C-13 is not clear [126].

A mixed amino acid-polyketide origin is apparent for myxothiazol (acetate, propionate, leucine, cysteine and the methyl groups of methionine) [127], for angiolam A (acetate, propionate and alanine) [128] and rhizoxin (acetate, methyl group of methionine and serine) [129].

The tetramic acid malonimycin (*7.191*) is constructed from one unit each of 2,3-diaminopropionic acid (C-3, -4, and -5), acetic acid (C-1 and -2), succinic acid (C-6, -7, -8, and -9/10), carbon dioxide (C-10/9) and L-serine (C-11, -12, and -13). The tetramic acid (*7.192*) is an intact precursor for (*7.191*) [130].

## 7.8.5 Virginiamycin antibiotics

The biosynthetic origins of virginiamycin M₁ (*7.193*) are those illustrated. Antibiotic A2315A is of similar structure and has appropriately

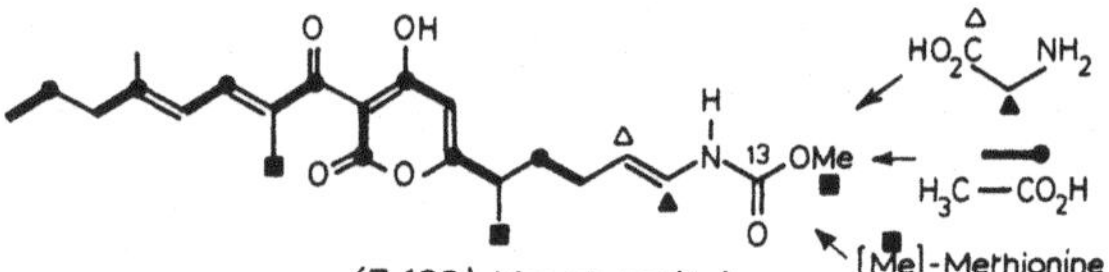

(7.190) Myxopyronin A

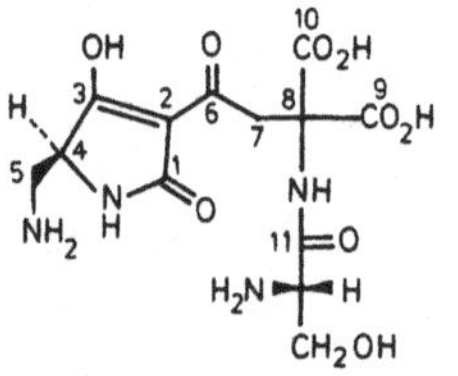

(7.191) Malonimycin

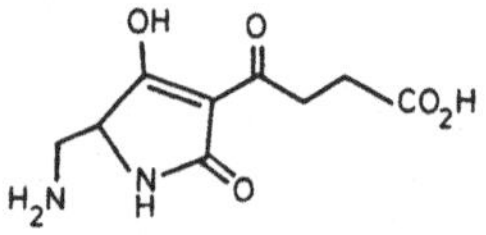

(7.192)

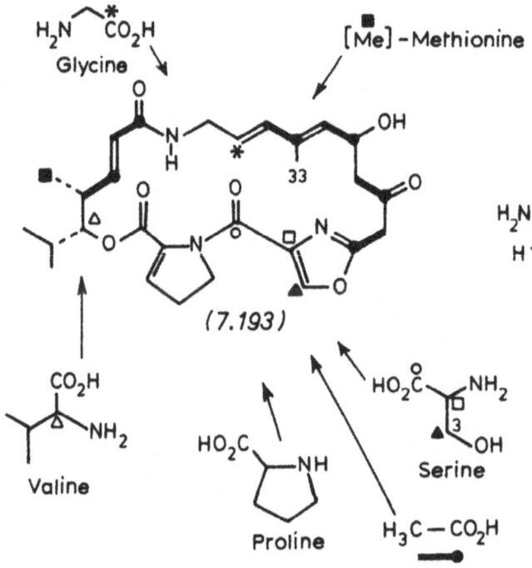

(7.193)

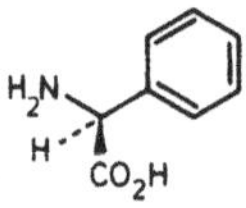

(7.194)

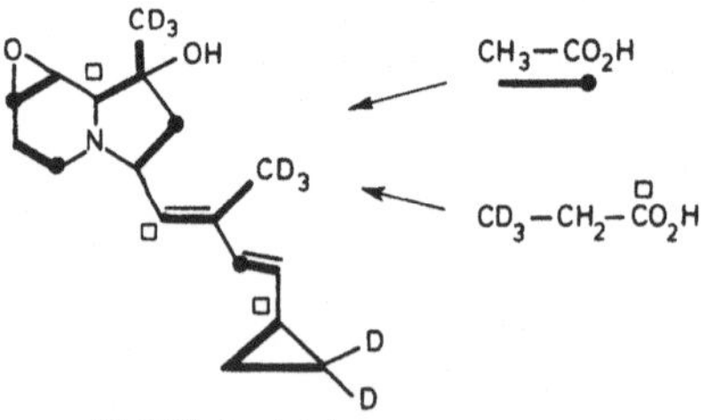

(7.195) Cyclizidine

similar origins. In both valine, presumably via isobutyryl CoA, serves as a starter unit for the 'western' side of the antibiotic, L-serine provides the oxazole units (with stereospecific loss of the 3-*pro-S* proton; experiments with virginiamycin $M_1$) and the C-33 methyl groups derive from C-2 of acetate (cf sections 7.8.2 and 7.8.4) [131, 132].

Virginiamycin $S_1$ contains an L-phenylglycine (*7.194*) moiety. The results of an experiment with DL-[3-$^{13}$C, $^{15}$N]phenylalanine [as (*7.18*)] show that this moiety is formed from phenylalanine with loss of the nitrogen atom [133] (cf nocardicins, section 7.7.3).

### 7.8.6 Cyclizidine

Notable in the unusual structure of cyclizidine [as (*7.195*)] is the cyclopropyl group at the chain terminus. The biosynthetic origins of this *Streptomyces* metabolite have been mapped with $^{13}$C-labelled acetate and [1-$^{13}$C]- and [3-$^2$H$_3$]-propionate [134].

## REFERENCES

Further reading: *Biosynthesis (Specialist Periodical Reports)* [ed. T.A. Geissman (vols. 1-3), ed. J.D. Bu'Lock (vols. 4 and 5)], The Chemical Society, London; Herbert, R.B. In *The Alkaloids (Specialist Periodical Reports)* [ed. J.E. Saxton (vols. 1-5), ed. M.F. Grundon (vols. 6-10) , The Chemical Society, London; [1]; *Natural Product Reports*.

1. Herbert, R.B. (1980) In *Rodd's Chemistry of Carbon Compounds*, 2nd edn, (ed. S. Coffey), Elsevier, Amsterdam, vol. IV L, pp. 291-455; (1988) Supplements ed. M.F. Ansell, pp. 155-247.
2. Schneider, M.J., Ungemach, F.S., Broquist, H.P. and Harris, T.M. (1982) *J. Amer. Chem. Soc.*, **104**, 6863-4.
3. Bach, M.L. and Gilvarg, C. (1966) *J. Biol. Chem.*, **241**, 4563-4.
4. Kanie, M., Fujimoto, S. and Foster, J.W. (1966) *J. Bact.*, **91**, 570-7.
5. Terashima, T., Idaka, E., Kishi, Y. and Goto, T. (1973) *J. Chem. Soc. Chem. Comm.*, 75-6.
6. Dobson, T.A., Desaty, D., Brewer, D. and Vining, L.C. (1967) *Can. J. Biochem.*, **45**, 809-23.
7. Helbling, A.M. and Viscontini, M. (1976) *Helv. Chim. Acta*, **59**, 2284-9.
8. McInnes, A.G., Smith, D.G., Walter, J.A. *et al.* (1979) *Can. J. Chem.*, **57**, 3200-4.
9. Wright, J.L.C., Vining, L.C., McInnes, A.G. *et al.* (1977) *Can. J. Biochem.*, **55**, 678-85.
10. Leete, E., Kowanko, N., Newmark, R.A. *et al.* (1975) *Tetrahedron Lett.*, 4103-6.
11. Iwai, Y., Kumano, K. and Omura, S. (1978) *Chem. Pharm. Bull. (Japan)*, **26**, 736-9.
12. Tanabe, M. and Urano, S. (1983) *Tetrahedron*, **39**, 3569-74.
13. Birch, A.J. and Simpson, T.J. (1979) *J. Chem. Soc. Perkin I*, 816-22.
14. Byrne, K.M., Hilton, B.D., White, R.J. *et al.* (1985) *Biochemistry*, **24**, 478-86.

15. Jeffs, P.W. and McWilliams, D. (1981) *J. Amer. Chem. Soc.*, **103**, 6185–92.
16. Shimada, H., Noguchi, H., Iitaka, Y. and Sankawa, U. (1981) *Heterocycles*, **115**, 1141–6.
17. Vining, L.C. and Wright, J.L.C. (1977) In *Biosynthesis (Specialist Periodical Reports)* (ed. J.D. Bu'Lock), The Chemical Society, London, vol. 5, pp. 240–305.
18. Allen, C.M., jun. (1973) *J. Amer. Chem. Soc.*, **95**, 2386–7.
19. Marchelli, R., Dossena, A. and Casnati, G. (1975) *J. Chem. Soc. Chem. Comm.*, 779–80.
20. Baldas, J., Birch, A.J. and Russell, R.A. (1974) *J. Chem. Soc. Perkin I*, 50–2.
21. MacDonald, J.C. and Slater, G.P. (1975) *Can. J. Biochem.*, **53**, 475–8.
22. Kirby, G.W. and Narayanaswami, S. (1976) *J. Chem. Soc. Perkin I*, 1564–7.
23. MacDonald, J.C. (1965) *Biochem. J.*, **96**, 533–8.
24. Akers, H.A., Llinás, M. and Nielands, J.B. (1972) *Biochemistry*, **11**, 2283–91.
25. Hanke, T. and Diekmann, H. (1974) *Arch. Microbiol.*, **95**, 227–36.
26. Barrow, K.D., Colley, P.W. and Tribe, D.E. (1979) *J. Chem. Soc. Chem. Comm.*, 225–6.
27. Steyn, P.S. and Vleggaar, R. (1983) *J. Chem. Soc. Chem. Comm.*, 560–1.
28. Kirby, G.W., Patrick, G.L. and Robins, D.J. (1978) *J. Chem. Soc. Perkin I*, 1336–8.
29. Johns, N. and Kirby, G.W. (1985) *J. Chem. Soc. Perkin I*, 1487–90.
30. Kirby, G.W. and Varley, M.J. (1974) *J. Chem. Soc. Chem. Comm.*, 833–4.
31. Ferezou, J.-P., Quesneau-Thierry, A., Servy, C. *et al.* (1980) *J. Chem. Soc. Perkin I*, 1739–46.
32. Voigt, S., El Kousy, S., Schwelle, N. *et al.* (1978) *Phytochemistry*, **17**, 1705–9.
33. Hurley, L.H. (1980) *Acc. Chem. Res.*, **13**, 263–9.
34. Brahme, N.M., Gonzalez, J.E., Rolls, J.P. *et al.* (1984) *J. Amer. Chem. Soc.*, **106**, 7873–8.
35. Brahme, N.M., Gonzalez, J.E., Mizsak, S. *et al.* (1984) *J. Amer. Chem. Soc.*, **106**, 7878–83.
36. Floss, H.G. (1976) *Tetrahedron*, **32**, 873–912.
37. Lee, S.-L., Floss, H.G. and Heinstein, P. (1976) *Arch. Biochem. Biophys.*, **177**, 84–94.
38. Plieninger, H., Meyer, E., Maier, W. and Gröger, D. (1978) *Annalen*, 813–7.
39. Floss, H.G., Tcheng-Lin, M., Chang, C. *et al.* (1974) *J. Amer. Chem. Soc.*, **96**, 1898–909.
40. Abe, M. (1970) paper presented at the first International Symposium on Genetics of Industrial Micro-organisms, Prague, (see ref. [36]).
41. Gröger, D., Mothes, K., Floss, H.G. and Weygand, F. (1963) *Z. Naturforsch.*, **18b**, 1123–4.
42. Gröger, D., Johne, S. and Härtling, S. (1974) *Biochem. Physiol. Pflanzen.*, **166**, 33–43.
43. Floss, H.G., Tcheng-Lin, M., Kobel, H. and Stadler, P. (1974) *Experientia*, **30**, 1369–70.
44. Quigley, F.R. and Floss, H.G. (1981) *J. Org. Chem.*, **46**, 464–6.
45. McGrath, R.M., Stein, P.S., Ferreira, N.P. and Neethling, D.C. (1976) *Bio-org. Chem.*, **5**, 11–23.
46. Chalmers, A.A., Gorst-Allman, C.P. and Steyn, P.S. (1982) *J. Chem. Soc. Chem. Comm.*, 1367–8.

47. Yamasaki, K., Kaneda, M., Watanabe, K. *et al.* (1983) *J. Antibiot.*, **36**, 552-8.
48. Woodard, R.W., Mascaro, Jr., L., Hörhammer, R. *et al.* (1980) *J. Amer. Chem. Soc.*, **102**, 6314-8.
49. Gould, S.J. and Cane, D.E. (1982) *J. Amer. Chem. Soc.*, **104**, 343-6.
50. Erickson, W.R. and Gould, S.J. (1987) *J. Amer. Chem. Soc.*, **109**, 620-1.
51. Chang, C.J., Floss, H.G., Hook, D.J. *et al.* (1981) *J. Antibiot.*, **34**, 555-66.
52. van Pée, K.-H., Salcher, O., Fischer, P. *et al.* (1983) *J. Antibiot.*, **36**, 1735-42.
53. Ritter, C. and Luckner, M. (1971) *Eur. J. Biochem.*, **18**, 391-400.
54. Butenandt, A. and Beckmann, R. (1955) *Z. Physiol. Chem.*, **301**, 115-7.
55. Herbert, R.B. (1974) *Tetrahedron Lett.*, 4525-8.
56. Etherington, T., Herbert, R.B., Holliman, F.G. and Sheridan, J.B. (1979) *J. Chem. Soc. Perkin I*, 2416-9.
57. Bying, G.S. and Turner, J.M. (1977) *Biochem. J.*, **164**, 133-45.
58. Römer, A. and Herbert, R.B. (1982) *Z. Naturforsch.*, **37c**, 1070-4.
59. Herbert, R.B., Holliman, F.G. and Ibberson, P.N. (1972) *J. Chem. Soc. Chem. Comm.*, 355-6.
60. Flood, M.E., Herbert, R.B. and Holliman, F.G. (1972) *J. Chem. Soc. Perkin I*, 622-6.
61. Simonsen, J.N., Paramasigamani, K., Vining, L.C. *et al.* (1978) *Can. J. Microbiol.*, **24**, 136-42.
62. Kibby, J.J., McDonald, I.A. and Rickards, R.W. (1980) *J. Chem. Soc. Chem. Comm.*, 768-9.
63. Hatano, K., Akiyama, S., Asai, M. and Rickards, R.W. (1982) *J. Antibiot.*, **35**, 1415-17.
64. Ghisalba, O. and Nüesch, J. (1981) *J. Antibiot.*, **34**, 64-71.
65. Hornemann, U., Kehrer, J.P. and Eggert, J.H. (1974) *J. Chem. Soc. Chem. Comm.*, 1045-6.
66. Haber, A., Johnson, R.D. and Rinehart, K.L., jun. (1977) *J. Amer. Chem. Soc.*, **99**, 3541-4.
67. Rinehart, K.L., jun., Potgieter, M. and Wright, D.A. (1982) *J. Amer. Chem. Soc.*, **104**, 2649-52.
68. Rinehart, K.L., jun., Potgieter, M. and Delaware, D.L. (1981) *J. Amer. Chem. Soc.*, **103**, 2099-101.
69. Milavetz, B., Kakinuma, K., Rinehart, K.L. *et al.* (1973) *J. Amer. Chem. Soc.*, **95**, 5793-5.
70. Ghisalba, O., Traxler, P. and Nüesch, J. (1978) *J. Antibiot.*, **31**, 1124-31.
71. Wu, T.S., Duncan, J., Tsao, S.W. *et al.* (1987) *J. Nat. Prod.*, **50**, 108-18.
72. Hatano, K., Mizuta, E., Akiyama, S. *et al.* (1985) *Agric. Biol. Chem.*, **49**, 327-33.
73. Zmijewski, M.J., jun., Wong, R., Paschal, J.W. and Dorman, D.E. (1983) *Tetrahedron*, **39**, 1255-63.
74. Wyss, R., Tamm, Ch. and Vederas, J.C. (1980) *Helv. Chim. Acta*, **63**, 1538-41.
75. Probst, A. and Tamm, Ch. (1981) *Helv. Chim. Acta*, **64**, 2065-77.
76. Mohr, P. and Tamm, Ch. (1981) *Tetrahedron*, **37**, Supplement No. 1, 201-12.
77. Nadzan, A.M. and Rinehart, K.L., jun. (1976) *J. Amer. Chem. Soc.*, **98**, 5012-4.
78. Palaniswamy, V.A. and Gould, S.J. (1986) *J. Amer. Chem. Soc.*, **108**, 5651-2.
79. Mikami, Y., Takahashi, K., Yazawa, K. *et al.* (1985) *J. Biol. Chem.*, **260**, 344-8.

80. Hurley, L.H. and Rokem, J.S. (1983) *J. Antibiot.*, **36**, 383–90.
81. Parry, R.J. and Buu, H.P. (1982) *Tetrahedron Lett.*, **23**, 1435–8.
82. Baldwin, J.E., Bansal, H.S., Chondrogianni, J. *et al.* (1985) *Tetrahedron*, **41**, 1931–8.
83. Puar, M.S., Munayyer, H., Hegde, V. *et al.* (1985) *J. Antibiot.*, **38**, 530–2.
84. Achenbach, H. and König, F. (1972) *Ber.*, **105**, 784–93.
85. Cable, K.M., Herbert, R.B. and Mann, J. (1987) *Tetrahedron Lett.*, **28**, 3159–62.
86. Cable, K.M., Herbert, R.B. and Mann, J. (1987) *J. Chem. Soc. Perkin I*, 1593–8.
87. Herbert, R.B. and Mann, J. (1984) *Tetrahedron Lett.*, **25**, 4263–6.
88. Garson, M.J. (1986) *J. Chem. Soc. Chem. Comm.*, 35–6.
89. Hagadone, M.R., Scheur, P.J. and Holm, A. (1984) *J. Amer. Chem. Soc.*, **106**, 2447–8.
90. Hillis, L.R. and Gould, S.J. (1985) *J. Amer. Chem. Soc.*, **107**, 4593–4.
91. Okuyama, A., Ohuchi, S., Tanaka, T. *et al.* (1986) *Biochem. Int.*, **12**, 485–91.
92. Queener, S.W. and Neuss, N. (1982) In *Chemistry and Biology of β-Lactam Antibiotics* (ed. R.B. Morin and M. Gorman), Academic Press, New York, Vol. 3, pp. 1–81.
93. J.E. Baldwin and E. Abraham (1988) *Nat. Prod. Rep.*, **5**, 129–45.
94. Neuss, N., Nash, C.H., Lemke, P.A. and Grutzner, J.B. (1971) *Proc. Roy. Soc.*, **B179**, 335–44.
95. Banko, G., Demain, A.L. and Wolfe, S. (1987) *J. Amer. Chem. Soc.*, **109**, 2858–60.
96. Baldwin, J.E., Johnson, B.L., Usher, J.J. *et al.* (1980) *J. Chem. Soc. Chem. Comm.*, 1271–3.
97. Jensen, S.E., Westlake, D.W.S. and Wolfe, S. (1982) *J. Antibiot.*, **35**, 483–90.
98. Baxter, R.L., McGregor, C.J. and Thomson, G.A. (1985) *J. Chem. Soc. Perkin I*, 369–72.
99. Carr, L.G., Skatrud, P.L., Scheetz II, M.E. *et al.* (1986) *Gene*, **48**, 257–66.
100. Baldwin, J.E., Killin, S.J., Pratt, A.J. *et al.* (1987) *J. Antibiot.*, **40**, 652–9.
101. Baldwin, J.E., Adlington, R.M., Robinson, N.G. and Ting, H.-H. (1986) *J. Chem. Soc. Chem. Comm.*, 409–11.
102. Baldwin, J.E., Adlington, R.M., Domayne-Hayman, B.P. *et al.* (1986) *J. Chem. Soc. Chem. Comm.*, 110–3.
103. Kluender, H., Huang, F.-C., Fritzberg, A. *et al.* (1974) *J. Amer. Chem. Soc.*, **96**, 4054–5.
104. Aberhart, D.J., Chu, J.Y.-R., Neuss, N. *et al.* (1974) *J. Chem. Soc. Chem. Comm.*, 564–5.
105. Pang, C.-P., White, R.L., Abraham, E.P. *et al.* (1984) *J. Biol. Chem.*, **222**, 777–88.
106. Townsend, C.A. and Barrabee, E.B. (1984) *J. Chem. Soc. Chem. Comm.*, 1586–8.
107. Jensen, S.E., Westlake, D.W.S. and Wolfe, S. (1985) *J. Antibiot.*, **38**, 263–5.
108. Baldwin, J.E., Adlington, R.M., Crabbe, M.J. *et al.* (1987) *Tetrahedron*, **43**, 3009–14.
109. Baldwin, J.E., Adlington, R.M., Kang, T.W. *et al.* (1987) *J. Chem. Soc. Chem. Comm.*, 104–6.

110. Baldwin, J.E., Adlington, R.M., Moroney, S.E. *et al.* (1984) *J. Chem. Soc. Chem. Comm.*, 984-6.

111. Townsend, C.A., Ho, M.-f. and Mao, S.-s. (1986) *J. Chem. Soc. Chem. Comm.*, 638-9.

112. Townsend, C.A. and Ho, M.-f. (1985) *J. Amer. Chem. Soc.*, **107**, 1066-8.

113. Gutman, A.L., Ribon, V., and Boltanski, A. (1985) *J. Chem. Soc. Chem. Comm.*, 1627-9.

114. Townsend, C.A. and Mao, S.-s. (1987) *J. Chem. Soc. Chem. Comm.*, 86-9.

115. Townsend, C.A. and Salituro, G.M. (1984) *J. Chem. Soc. Chem. Comm.*, 1631-2.

116. Houck, D.R., Kobayashi, K., Williamson, J.M. and Floss, H.G. (1986) *J. Amer. Chem. Soc.*, **108**, 5365-6.

117. Müller, B., Hädener, A. and Tamm, Ch. (1987) *Helv. Chim. Acta*, **70**, 412-22.

118. Wasserman, H.H., Shaw, C.K., Sykes, R.J. and Cushley, R.J. (1974) *Tetrahedron Lett.*, 2787-90.

119. Gerber, N.N., McInnes, A.G., Smith, D.G. *et al.* (1978) *Can. J. Chem.*, **56**, 1155-63.

120. Parry, R.J. and Mueller, J.V. (1984) *J. Amer. Chem. Soc.*, **106**, 5764-5.

121. Yamato, M., Takeuchi, T., Umezawa, H. *et al.* (1986) *J. Antibiot.*, **39**, 1263-9.

122. Martinkus, K.J., Tann, C.-H. and Gould, S.J. (1983) *Tetrahedron*, **39**, 3493-505.

123. Thiruvengadam, T.K., Gould, S.J., Aberhart, D.J. and Lin, H.-J. (1983) *J. Amer. Chem. Soc.*, **105**, 5470-6.

124. Ju, S., Palaniswamy, V.A., Yorgey, P. and Gould, S.J. (1986) *J. Amer. Chem. Soc.*, **108**, 6429-30.

125. Beale, J.M., Lee, J.P., Nakagawa, A. *et al.* (1986) *J. Amer. Chem. Soc.*, **108**, 331-2.

126. Kohl, W., Irschik, H., Reichenbach, H. and Höfle, G. (1984) *Annalen*, 1088-93.

127. Trowitzsch-Kienast, W., Wray, V., Gerth, K. *et al.* (1986) *Annalen*, 93-6.

128. Kohl, W., Witte, B., Kunze, B. *et al.* (1985) *Annalen*, 2088-97.

129. Kobayashi, H., Iwasaki, S., Yamada, E. and Okuda, S. (1986) *J. Chem. Soc. Chem. Comm.*, 1702-3.

130. Schipper, D., van der Baan, J.L., Harms, N. and Bickelhaupt, F. (1982) *Tetrahedron Lett.*, **23**, 1293-6.

131. LeFebre, J.W. and Kingston, D.G.I. (1984) *J. Org. Chem.*, **49**, 2588-93; see *ibid.*, 1985, **50**, 4666 for structure correction.

132. Purvis, M.B., Kingston, D.G.I., Fujii, N. and Floss, H.G. (1987) *J. Chem. Soc. Chem. Comm.*, 302-3.

133. Reed, J.W. and Kingston, D.G.I. (1986) *J. Nat. Prod.*, **49**, 626-30.

134. Leeper, F.J., Padmanabhan, P., Kirby, G.W. and Sheldrake, G.N. (1987) *J. Chem Soc. Chem. Comm.*, 505-6.

# Index